Advanced Biosignal Processing

A. Naït-Ali (Ed.)

Advanced Biosignal Processing

Springer

Assoc. Prof. Amine Naït-Ali
Université Paris 12
Labo. Images, Signaux et
Systèmes Intelligents (LISSI) EA 3956
61 av. du Général de Gaille
94010 Créteil
France
naitali@univ-paris12.fr

ISBN 978-3-540-89505-3 e-ISBN 978-3-540-89506-0

DOI 10.1007/978-3-540-89506-0

Library of Congress Control Number: 2008910199

Cover design: WMXDesign GmbH, Heidelberg

Printed on acid-free paper

9 8 7 6 5 4 3 2 1

springer.com

Preface

Generally speaking, Biosignals refer to signals recorded from the human body. They can be either electrical (e.g. Electrocardiogram (ECG), Electroencephalogram (EEG), Electromyogram (EMG), etc.) or non-electrical (e.g. breathing, movements, etc.). The acquisition and processing of such signals play an important role in clinical routines. They are usually considered as major indicators which provide clinicians and physicians with useful information during diagnostic and monitoring processes. In some applications, the purpose is not necessarily medical. It may also be industrial. For instance, a real-time EEG system analysis can be used to control and analyze the vigilance of a car driver. In this case, the purpose of such a system basically consists of preventing crash risks. Furthermore, in certain other applications, a set of biosignals (e.g. ECG, respiratory signal, EEG, etc.) can be used to control or analyze human emotions. This is the case of the famous polygraph system, also known as the "lie detector", the efficiency of which remains open to debate!

Thus when one is dealing with biosignals, special attention must be given to their acquisition, their analysis and their processing capabilities which constitute the final stage preceding the clinical diagnosis. Naturally, the diagnosis is based on the information provided by the processing system. In such cases, huge responsibility is placed on this system and in some countries, legislation relating to clinical practices is particularly important!

Therefore, specific attention should be paid to the way these signals have to be processed. As you are aware, clinicians dealing with processed signals care little about the algorithm that has been implemented in their system to extract the required information. For them, the final results are all that counts! We share this opinion!

Another remark: it should be noted that complex and sophisticated algorithms do not systematically lead to better results! It is clear that this doesn't mean that one should use simple algorithms instead. In fact, in such cases everything is relative! Hence, the following question arises: what is meant exactly by the phrase 'good results'? What does a sophisticated algorithm mean? We can guess that more than one definition can be employed. A given result is evaluated in terms of the purpose of the application and the criterion used for the evaluation. Generally speaking, signal processing and, in particular digital signal processing, has today become a huge universe, incorporating a wide range of techniques and approaches. Making oneself aware of the whole corpus of published techniques is far from straightforward. In fact, many fields deal with digital signal processing including mathematics, electrical engineering, computer engineering and so on. Consequently, the same problem will never be solved from the same point of view. In other words, in some scientist communities, one may develop a signal processing tool before verifying whether it can fit a given application. Conversely, other scientists study and analyze a problem then try to develop a technique that fits the concerned application. So which strategy should one use? I am quite sure that the reader will favour one strategy above another depending on his background. There is potential for huge debate on this subject! From my point of view, both have their mertis and are sometimes complementary. Combining approaches, ideas and experiences, especially when working on a specific problem may lead to interesting results.

In this book, the purpose is not to offer the reader an account of all the techniques which can be used to process any given biosignal. This would be very ambitious and unrealistic. The reason for this being that a given biosignal can be processed differently depending on the target defined by the application. Moreover, we will not be focusing on classical techniques that might also be efficient. Hence, the reader shouldn't confuse simplicity with the efficiency. As emphasized by the book's title, "Advanced Biosignal Processing", which could also, implicitly be read as: "Advanced Digital Signal Processing Techniques applied on Biosignals"; the purpose is to provide, in some way, an in-depth consideration of the particular orientations regarding the way one can process a specific biosignal using various recent signal processing tools.

Many scientists have contributed to this book. They represent several laboratories based around the world. As a result, the reader can gain access to a wide panel of strategies for dealing with certain biosignals. I say "certain biosignals", because our choice is, in fact, restricted mainly to "bio-electrical signals" and especially to those most frequently used in clinical routines such as ECG, EEG, EPs and EMG. For other types of biosignals, one may refer directly to other works published specifically on these.

The idea of this book is to explore the maximum level of advanced signal processing techniques rather than scanning the maximum number of biosignals. The intention is to assist the reader in making decisions regarding his own approach in his project of interest; perhaps by mixing two or more techniques, or improving some techniques and, why ever not, proposing entirely new ones? As you are aware, there is no such thing as perfection in research. So the reader can always think about making something "even better". For instance, this might take the form: "how can I get the same result but faster?" (think about the complexity of your algorithms); or "how can I get better results without increasing the complexity" and so on.

On the other hand, researchers might face problems when evaluating a new approach. This problem basically concerns the database upon which one has to evaluate his algorithms. In some cases, the evaluation is achieved on a confidential local database provided by a given medical institution. The constraint in such situations is that these data cannot be shared.

In some other cases, some international databases available on the Internet can be used for this purpose (e.g. Physionet). Among these databases, special attention has been given to MeDEISA "Medical Database for the Evaluation of Image and Signal Processing Algorithms", at *www.medeisa.net*. The specificity of MeDEISA which has been associated with this book is that data can be posted from scientists owning some particular data, recorded under certain specific conditions. These data are downloadable in MATLAB format and can be subject to various processing. Therefore, since each signal is identifiable by its reference, we believe that this will be, in the future a good way to evaluate and compare objectively any published signal processing technique.

This book is intended for final year undergraduate students, postgraduate students, engineers and researchers in biomedical engineering and applied digital signal processing. It has been divided into four specific sections. Each section concerns one of the biosignals pointed out above, namely the ECG, EEG, EMG and the EPs. The "Epilogue" deals with some general purpose techniques and multimodal processing. Consequently, numerous advanced signal processing methods are studied and analyzed such as:

- *Source separation,*
- *Statistical models,*
- *Metaheuristics,*
- *Time frequency,*
- *Adaptive tracking,*
- *Wavelets neural networks and wavelet networks,*
- *Modeling and detection,*
- *Wavelet and Chirplet transforms,*
- *Non-linear and EMD approaches,*
- *Compression.*

Of course, to deal with these subjects, we assume that the reader is familiar with basic digital signal processing methods.

These techniques are presented through 17 chapters, structured as follows:

- **Chapter 1** "*Biosignals: properties and acquisition*": This can be regarded as an introductory chapter in which some generic acquisition schemes are presented. For obvious reasons, some well known general biosignal properties are also evoked.
- **Chapter 2** "*Extraction of ECG characteristics using source separation techniques: exploiting statistical independence and beyond*": This chapter deals with the Blind Source Separation (BSS) approach. Special attention is given to fetal ECG extraction.
- **Chapter 3** "*ECG processing for exercise test*": In this chapter concepts of modeling and estimation techniques are presented for the purpose of extracting functional clinical information from ECG recordings, during an exercise test.
- **Chapter 4** "*Statistical models based ECG classification*": The authors describe, over the course of this chapter, how one can use hidden Markov models and hidden Markov trees for the purpose of ECG beat modeling and classification.

- **Chapter 5** "*Heart Rate Variability time-frequency analysis for newborn seizure detection*": In this chapter, time-frequency analysis techniques are discussed and applied to the ECG signal for the purpose of automatic seizure detection. The authors explain how the presented technique can be combined with the EEG-based methodologies.
- **Chapter 6** "*Adaptive tracking of EEG frequency components*": The authors address the problem of tracking oscillatory components in EEG signals. For this purpose, they explain how one can use an adaptive filter bank as an efficient signal processing tool.
- **Chapter 7** "*From EEG signals to brain connectivity: methods and applications in epilepsy*": In this chapter, 3 different approaches, namely, linear and nonlinear regression, phase synchronization, and generalized synchronization will be reviewed for the purpose of EEG analysis.
- **Chapter 8** "*Neural Network approaches for EEG classification*": This chapter provides a state-of-the-art review of the prominent neural network based approaches that can be employed for EEG classification.
- **Chapter 9** "*Analysis of event-related potentials using wavelet networks*": In this chapter wavelet networks are employed to describe automatically ERPs using a small number of parameters.
- **Chapter 10** "*Detection of evoked potentials*": This chapter is based on decision theory. Applied to visual evoked potentials, it will be shown how the stimulation and the detection can be combined suitably.
- **Chapter 11** "*Visual Evoked Potential Analysis Using Adaptive Chirplet Transform*". After explaining the transition from the wavelet to the chirplet, this new transform is applied and evaluated on VEPs.
- **Chapter 12** "*Uterine EMG analysis: time-frequency based techniques for preterm birth detection*". In this chapter, global signal processing wavelet-based and neural network-based systems will be described for the purpose of: detection, classification, identification and diagnostic of the uterine EMG.
- **Chapter 13** "*Pattern classification techniques for EMG signal decomposition*". The electromyographic (EMG) signal decomposition process is addressed by developing different approaches to pattern classification. For this purpose, single classifier and multiclassifier approaches are presented.
- **Chapter 14** "*Parametric modeling of biosignals using metaheuristics*". Two main metaheuristic techniques will be presented, namely Genetic Algorithms and the Swarm Particle Optimization algorithm. They will be used to model some biosignals, namely Brainstem Auditory Evoked Potentials, Event Related Potentials and ECG beats.
- **Chapter 15** "*Nonlinear analysis of physiological time series*": The aim of this chapter is to provide a review of the main approaches to nonlinear analysis (fractal analysis, chaos theory, complexity measures) in physiological research, from system modeling to methodological analysis and clinical applications.
- **Chapter 16** "*Biomedical data processing using HHT: a review*": In this chapter, biomedical data processing is reviewed using Hilbert-Huang Transform, also called Empirical Mode Decomposition (EMD).

- **Chapter 17** "*Introduction to multimodal compression of biomedical data*": The aim of this chapter is to provide the reader with a new vision of compressing both medical "*images/videos*" and "bio*signals*" jointly. This type of compression is called "multimodal compression".

Through these chapters, I hope that the reader will find this book useful and constructive and that the evoked approaches will contribute efficiently, by providing innovative ideas to be applied in this, so fascinating a field, by which I mean, of course, Biosignal Processing!

Finally, I would like to thank all the authors for their active and efficient contribution.

Créteil, France *A. Naït-Ali*

Contents

Contributors

Kamiar Aminian Laboratory of Movement Analysis and Measurement, Ecole Polytechnique Federale de Lausanne (EPFL), Switzerland, kamiar.aminian@epfl.ch

Rodrigo Varejão Andreão CEFETES, Coord. Eletrotécnica, Av. Vitória, 1729, Jucutuquara, Vitória – ES, Brazil, rodrigova@cefetes.br

Karim Ansari-Asl INSERM, U642, Rennes, F-35000, France; Université de Rennes 1, LTSI, Campus de Beaulieu, Université de Rennes 1, 263 Avenue du General Leclerc – CS 74205 – 35042 Rennes Cedex, France, karim.ansari@scu.ac.in

Malarvili Balakrishnan Perinatal Research Centre, University of Queensland Centre for Clinical Research, The University of Queensland, Australia, m.balakrishnan@uq.edu.au

Grégory Blain Laboratory I3S, CNRS/University of Nice-Sophia Antipolis, France; Faculty of sports sciences-Nice, Laboratory of physiological adaptations, motor performance and health, gregory.blain@unice.fr

Boualem Boashash Perinatal Research Centre, University of Queensland Centre for Clinical Research, The University of Queensland, Australia & College of Engineering, University of Sharjah, Sharjah, UAE, b.boashash@uq.edu.au

Jean-Marc Boucher Telecom Bretagne, UMR CNRS 3192 Lab_Sticc, CS 83818 29238 Brest, France, jm.boucher@telecom-bretagne.eu

Jérôme Boudy Institut Telecom; Telecom & Management Sud Paris; Dép. d'Electronique & Physique, 9 r. Charles Fourier, 91011 Evry, France, jerome.boudy@int-evry.fr

Amitava Chatterjee Electrical Engineering Department, Jadavpur University, Kolkata, West Bengal, India. PIN – 700 032, cha_ami@yahoo.co.in

Marwa Chendeb Electronic Department, Faculty of Engineering, Lebanese University, Elkoubbeh, Tripoli, Lebanon; ICD, FRE CNRS 2848, University of Technology of Troyes, 12 rue Marie Curie, BP 2060, 10010 Troyes, France, m_chendeb@hotmail.com

Paul B. Coldiz Perinatal Research Centre, University of Queensland Centre for Clinical Research, The University of Queensland, Australia, p.colditz@uq.edu.au

Jie Cui School of Health Information Sciences, University of Texas Health Science Center at Houston, 7000 Fannin Street, Suite 600, Houston, TX 77054, U.S.A., richard.cui@utoronto.ca

Mohamad Diab Université de Technologie de Compiègne – CNRS UMR 6600 Biomécanique et Bioingénierie, BP 20529, 60205 Compiègne Cedex, France; Islamic university of Lebanon, biomedical department, B.P. 30014, Khaldeh, Lebanon, mhmd_dib10@hotmail.com

Hartmut Dickhaus Medical Informatics, University of Heidelberg, Germany, Hartmut.Dickhaus@med.uni-heidelberg.de

Bernadette Dorizzi Institut Telecom; Telecom & Management Sud Paris; Dép. d'Electronique & Physique, 9 r. Charles Fourier, 91011 Evry, France, Bernadette.Dorizzi@it-sudparis.eu

Xavier Drouot AP-HP, Groupe Henri-Mondor Albert-Chènevièr, Service de Physiologie, Créteil, F-94010 France, xavier.drouot@hmn.aphp.fr

Cédric Duchêne Swiss Federal Institute of Technology Lausanne (EPFL), Institute of Electrical Engineering (IEL), EPFL-STI-IEL-LTS1, Station 11, 1015, Lausanne, Switzerland, cedric.duchene@epfl.ch

Jacques Duchêne ICD, FRE CNRS 2848, University of Technology of Troyes, 12 rue Marie Curie, BP 2060, 10010 Troyes, France, jacques.duchene@utt.fr

Salim Graja Telecom Bretagne, UMR CNRS 3192 Lab_Sticc, CS 83818 29238 Brest, France, salim.graja@telecom-bretagne.eu

Hartmut Heinrich Child & Adolescent Psychiatry, University of Erlangen, Germany; Heckscher Klinikum, Munich, Germany, hheinri@arcor.de

Norden E. Huang Research Center for Adaptive Data Analysis, National Central University, Chungli 32001; Research Center for Applied Sciences, Academia Sinica, Nankang, Taipei 11529, Taiwan, norden@ncu.edu.tw

Peter Husar Technische Universität Ilmenau, Germany, Institute of Biomedical Engineering and Informatics, peter.husar@tu-ilmenau.de

Patrick Karasinski Université Paris 12, Laboratoire, Image, Signaux et Systèmes Intelligents, LiSSi, EA. 3956. 61, avenue du Général de Gaulle, 94010, Créteil, France, karasinski@univ-paris12.fr

Mohamad Khalil Lebanese University, Faculty of Engineering, Section 1, El Koubbe, Tripoli, Lebanon, m.khalil@ul.edu.lb

Catherine Marque Université de Technologie de Compiègne – CNRS UMR 6600 Biomécanique et Bioingénierie, BP 20529, 60205 Compiègne Cedex, France, catherine.marque@utc.fr

Mostefa Mesbah Perinatal Research Centre, University of Queensland Centre for Clinical Research, The University of Queensland, Australia, m.mesbah@uq.edu.au

Olivier Meste Laboratoire d'Informatique, Signaux et Systèmes de Sophia, Antipolis, Université de Nice – Sophia Antipolis, CNRS, Les Algorithmes – Euclide-B, 2000 route des Lucioles, BP 121, 06903 Sophia, Antipolis Cedex, France, meste@i3s.unice.fr

Micah M. Murray Electroencephalography Brain Mapping Core, Center for Biomedical Imaging of Lausanne and Geneva; Functional Electrical Neuroimaging Laboratory, Neuropsychology and Neurorehabilitation Service & Radiology Service, Centre Hospitalier Universitaire Vaudois and University of Lausanne, Rue du Bugnon 46, Radiology, BH08.078, CH-1011 Lausanne, Switzerland, micah.murray@chuv.ch

Amine Naït-Ali Université Paris 12, Laboratoire, Image, Signaux et Systèmes Intelligents, LiSSi, EA. 3956. 61, avenue du Général de Gaulle, 94010, Créteil, France, naitali@univ-paris12.fr

Amir Nakib Université Paris 12, Laboratoire Image, Signaux et Systèmes Intelligents, LiSSi, EA. 3956 61 avenue du Général de Gaulle, 94010, Créteil, France, nakib@univ-paris12.fr

Anisoara Paraschiv-Ionescu Laboratory of Movement Analysis and Measurement, Ecole Polytechnique Federale de Lausanne (EPFL), Switzerland, anisoara.ionescu@epfl.ch

Yann Prudat Swiss Federal Institute of Technology Lausanne (EPFL), Institute of Electrical Engineering (IEL), EPFL-STI-IEL-LTS1, Station 11, 1015 Lausanne, Switzerland, yann.prudat@epfl.ch

Sarbast Rasheed Department of Systems Design Engineering, University of Waterloo, Waterloo, Ontario N2L 3G1, Canada, s.rasheed@ieee.org

Hervé Rix Laboratoire d'Informatique, Signaux et Systèmes de Sophia, Antipolis, Université de Nice – Sophia Antipolis, CNRS, Les Algorithmes – Euclide-B, 2000 route des Lucioles, BP 121, 06903 Sophia, Antipolis Cedex, France, Herve.Rix@i3s.unice.fr

Lotfi Senhadji INSERM, U642, Rennes, F-35000, France; Université de Rennes 1, LTSI, Campus de Beaulieu, Université de Rennes 1, 263 Avenue du General Leclerc – CS 74205 – 35042 Rennes Cedex, France, lotfi.senhadji@univ-rennes1.fr

Patrick Siarry Université Paris 12, Laboratoire, Image, Signaux et Systèmes Intelligents, LiSSi, EA. 3956.61, avenue du Général de Gaulle, 94010, Créteil, France, siarry@univ-paris12.fr

Daniel Stashuk Department of Systems Design Engineering, University of Waterloo, Waterloo, Ontario N2L 3G1, Canada, stashuk@pami.uwaterloo.ca

Laurent Uldry Swiss Federal Institute of Technology Lausanne (EPFL), Institute of Electrical Engineering (IEL), EPFL-STI-IEL-LTS1, Station 11, 1015 Lausanne, Switzerland, laurent.uldry@epfl.ch

Jean-Marc Vesin Swiss Federal Institute of Technology Lausanne (EPFL), Institute of Electrical Engineering (IEL), EPFL-STI-IEL-LTS1, Station 11, 1015 Lausanne, Switzerland, jean-marc.vesin@epfl.ch

Virginie Van Wassenhove California Institute of Technology, Division of Biology, 1200 East California BlvdM/C 139-74, Pasadena, CA 91125 USA; Comissariat à l'Energie Atomique, NeuroSpin, Cognitive Neuroimaing Unit, Gif-sur-Yvette, 91191 France, Virginie.van.Wassenhove@gmail.com

Fabrice Wendling INSERM, U642, Rennes, F-35000, France; Université de Rennes 1, LTSI, Campus de Beaulieu, Université de Rennes 1, 263 Avenue du General Leclerc - CS 74205 - 35042 Rennes Cedex, France, fabrice.wendling@univ-rennes1.fr

Willy Wong Department of Electrical and Computer Engineering, University of Toronto, 10 Kings College Road, Toronto, ON M5S 3G4, Canada, willy@eecg.utoronto.ca

Ming-Chya Wu Research Center for Adaptive Data Analysis and Department of Physics, National Central University, Chungli 32001, Taiwan; Institute of Physics, Academia Sinica, Nankang, Taipei 11529, Taiwan, mcwu@ncu.edu.tw

Vicente Zarzoso Laboratoire d'Informatique, Signaux et Systèmes de Sophia Antipolis, Université de Nice – Sophia Antipolis, CNRS,
Les Algorithmes – Euclide-B, 2000 route des Lucioles, BP 121, 06903 Sophia Antipolis Cedex, France, zarzoso@i3s.unice.fr

Emre Zeybek Université Paris 12, Laboratoire, Image, Signaux et Systèmes Intelligents, LiSSi, EA. 3956. 61, avenue du Général de Gaulle, 94010, Créteil, France, emre.zeybek@univ-paris12.fr

Chapter 1
Biosignals: Acquisition and General Properties

Amine Naït-Ali and Patrick Karasinski

Abstract The aim of this chapter is to provide the reader with some basic and general information related to the most common biosignals, in particular biopotentials, encountered in clinical routines. For this purpose, the chapter will be divided into two main sections. In Sect. 1.1, we will consider the basis of bipotential recording (i.e., electrodes, artifact rejection and safety). In the second section, some general properties of a set of biosignals, will be introduced briefly. This will concern essentially, ECG, EEG, EPs and EMG. This restriction is required to ensure an appropriate coherency over the subsequent chapters which will deal primarily with these signals.

1.1 Biopotential Recording

As mentioned previously in the introduction to this book, biosignals are intensively employed in various biomedical engineering applications. From unicellular action potential to polysomnogram, they concern both research and clinical routines. Since this book deals specifically with biopotentials (i.e. bioelectrical signals), a special focus on their acquisition is provided in this section.

As is the case in any common instrumentation system, biopotential recording schemes include an observed process, a sensor and an amplifier. In our case, the observed process is recorded from a human body which requires particular precautions to be taken into account.

Consequently, the following three most important aspects will be underlined in this section:

1. The sensor: electrode description and its modeling will be given in Sect. 1.1.1.
2. The power supply artifact: this point will be discussed in Sect. 1.1.2, in which we provide a description of some common schemes,
3. Safety: constraints and solutions are presented in Sect. 1.1.3.

A. Naït-Ali (✉)
Université Paris 12, Laboratoire, Image, Signaux et Systèmes Intelligents, LiSSi, EA. 3956. 61, avenue du Général de Gaulle, 94010, Créteil, France
e-mail: naitali@univ-paris12.fr

A. Naït-Ali (ed.), *Advanced Biosignal Processing*,
DOI 10.1007/978-3-540-89506-0_1, © Springer-Verlag Berlin Heidelberg 2009

1.1.1 Biopotentials Recording Electrodes

Generally speaking, to ensure an appropriate interface between living tissue and a conductor, specific sensors are required to transform ionic concentrations to electronic conductions. This sensor is, actually, an electrode in which a chemical reaction produces this transformation.

The biopotentials are produced from cell activity that changes the ionic concentration in intra and extra cellular environment. In electrical devices, electron activity produces tensions and currents. Both of them are electrical phenomenon but charge carriers are different. A current in an electronic circuit results from an electron movement and from the displacement ions in living tissue. As mentioned above, electrodes ensure the transformation from ionic conduction to electronic conduction through a chemical reaction. Biopotential electrodes are regarded as second kind electrodes (i.e. they are composed of metal, saturated salt from the same metal and an electrolyte made with common ion). For example, the most common electrode uses silver, silver chloride (Ag/AgCl) and a potassium or sodium chloride electrolyte. These electrodes present good stability because their potentials depend on common ion concentration and temperature. Moreover, electrodes are sensitive to movement. In a steady state, electrochemical reactions create a double layer of charges that might be a source of artifacts during any possible movement (e.g. patient movement). To reduce the effect of these artifacts, one has to ensure that the electrodes are properly fixed.

1.1.1.1 Equivalent Circuit

The electrode interface can be modeled by an equivalent electrical circuit, shown in (Fig. 1.1a). Here, E1 is the electrode potential. The charge double layer acts as a capacitor denoted by Ce, whereas Re stands for the conduction. It is worth noting that a more complex model has been proposed in the literature to take the skin structure into account [9]. The corresponding electrical circuit is given in Fig. 1.1b. The paste-epidermis junction produces another DC voltage source, denoted E2. Epidermis can be modeled by an RC network.

Generally speaking, any equivalent circuit can be reduced to a DC voltage, placed in serial with an equivalent impedance.

1.1.1.2 Other Electrodes

Even if Ag/AgCl electrodes are commonly used in clinical routines, one can evoke other existing types, namely:

- glass micropipet electrodes for unicellular recording,
- needle electrodes,
- micro array electrodes,
- dry electrodes.

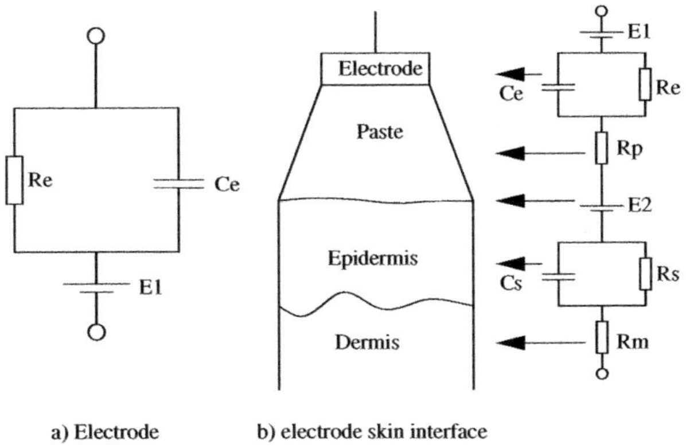

Fig. 1.1 (**a**) Electrode equivalent circuit, (**b**) electrode skin interface equivalent circuit

Of course, this is a non exhaustive list. For more details, the reader can refer, for instance, to [9].

Special attention is given to, dry electrodes which are capacitive effect based. These electrodes are well suited for both long-duration and ambulatory recording. Associated with wireless transmission systems, they could represent the biopotential recording standard of the future.

1.1.2 Power Supply Artifact Rejection

The power supply artifact is not specific to biopotential measurements. This problem concerns, rather, any instrumentation systems. Even if one can protect a strain gauge or any industrial sensor by a shielded box; building a Faraday cage around a patient, in an operating room is impossible.

If the power supply artifact rejection is something new for you, you can carry out a very simple experiment using an oscilloscope, as follows: connect a coaxial cable at the input and take the hot point between two fingers. You will see on the screen a 50 or 60 Hz sinusoidal noisy signal. The magnitude depends on your electrical installation. It can reach a few volts. Now, touch the grounding point with the same or the other hand; the signal will decrease. Finally, put down the coaxial cable and increase the oscilloscope gain to its maximum value; you will notice that the signal is still present. This "strange" signal is the power supply artifact.

From this experiment, one can wonder about this signal's sources which make it still present with two, one or even without any measurement point. Actually, the physical phenomenon at the signal origin is a capacitive coupling created by displacement currents, described in the following section.

1.1.2.1 Displacement Currents

Displacement currents are described in electromagnetism theory, in particular, in Maxwell's fourth equations:

$$c^2 \nabla \times \boldsymbol{B} = \frac{\boldsymbol{j}}{\varepsilon_0} + \frac{\partial \boldsymbol{E}}{\partial t}$$

This equation establishes the equivalence between the conduction current density and the temporal electrical field variation. Both of them can produce a magnetic field. The temporal electrical field variation term is useful for dielectric medium. It explains the reason for which an alternative current seems to go through a capacitor despite the presence of an insulator [5]. By analogy to the conduction current, this term is called displacement current.

The most important drawback when recording bio-potentials is that every conductor separated by an insulator forms a capacitor, including the human body. Actually, a human is more of a conductor than an insulator. For this reason capacitors are formed by power supply wires, including ground wires. In this context, the oscilloscope experiment can be modeled by the scheme shown in Fig. 1.2.

The capacitors represent a capacitive coupling between power supply wires and the patient. The Cp value is approximately in the range 2 to 3 pF and 200 to 300 pF for Cn according to the following references: [3, 10, 11]. Thus, the patient is located at the center of the capacitor voltage divider which explains the presence of "our low amplitude sinusoidal signal" on the oscilloscope whatever the measurement performed on the human body surface might be. Consequently, amplifying biopotentials using a standard amplifier is inappropriate. An instrumentation amplifier offers an initial solution since it amplifies the electrical tension between two inputs In+ and In- (Fig. 1.3).

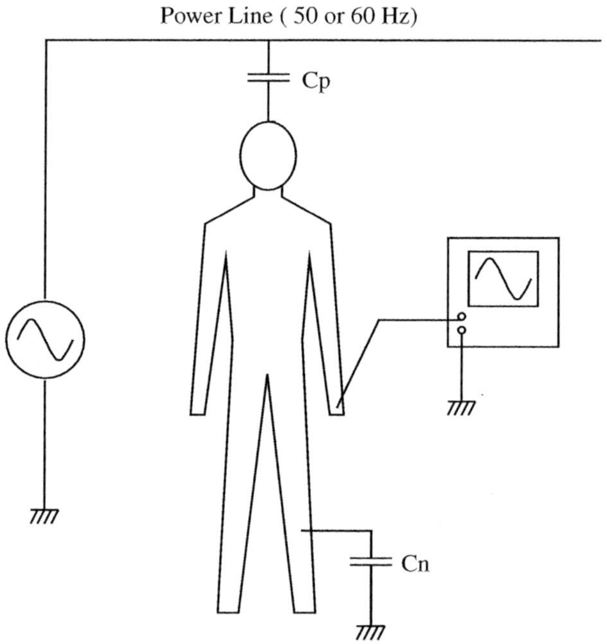

Fig. 1.2 Oscilloscope experiment

Fig. 1.3 First solution: Instrumentation amplifier

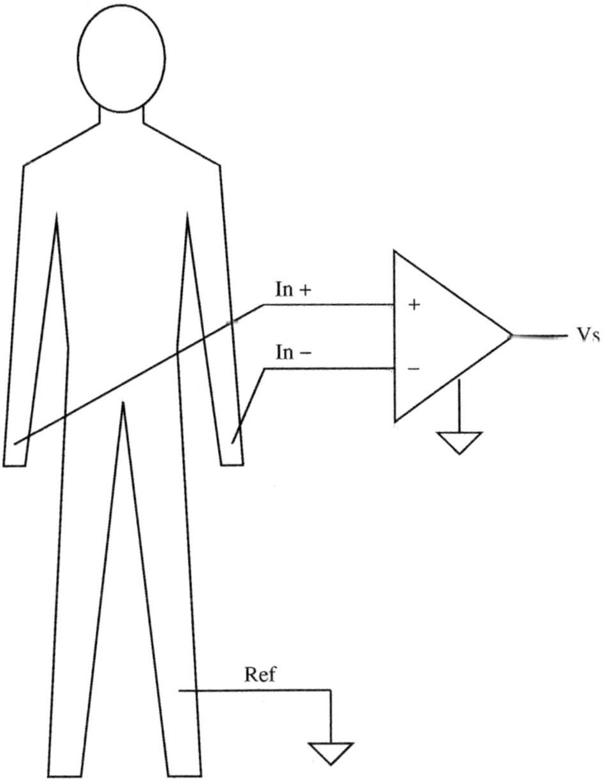

1.1.2.2 Common Mode Rejection

In theory, an amplifier provides at its output a voltage Vs, expressed by: Vs = Gd (V_{In+} - V_{In-}) where Gd is the differential gain. In this case, the power supply artifact will be canceled through subtraction. When two inputs are used, one should take into account the common mode voltage.

On the other hand, things are different in practice. The output voltage Vs will be expressed by: Vs = Gd (V_{In+} - V_{In-}) + Gc (V_{In+} + V_{In-})/2, where Gc is the common mode gain. In this case, the power supply artifact generated by displacement currents is still present. In addition, the power supply artifact is never the same at In+ and In-. Hence, it appears as well in the differential mode.

1.1.2.3 Magnetic Induction

Excepting the displacement current, another physical phenomenon such as the magnetic induction can produce power supply artifacts. Patient leads as well as the amplifier form a loop in which temporal magnetic field variations induce a voltage (Faraday law). A magnetic field is produced by transformers and any surrounding electrical motors. In order to reduce magnetic field effects, one avoid being in close proximity to magnetic sources. Moreover, one has to minimize the surface loop. In such situations, shielding is useless.

The solution illustrated in Fig. 1.3 is not appropriate in terms of safety. In such situations, the patient is directly connected to the ground.

1.1.2.4 The Right Leg Driver

Power supply artifact effects have been modeled through the equation described by Huhta and Webster [2]. This famous equation contains five terms, namely, the magnetic induction, the displacement current in leads, the displacement currents in the human body, the common mode gain and finally the effect of electrode impedances, unsteadiness.

Reducing artifact magnitude under a given threshold can be achieved with a specific reference device. The idea consists in suppressing the common mode voltage Vc using a -Vc voltage as reference. In others words, the patient is situated in a feedback loop whose aim is the common mode suppression [10]. A basic scheme is illustrated in Fig. 1.4.

The amplifier A1 provides a differential voltage Vs and a common voltage Vc between R1 and R2. A2 amplifies Vc with a negative gain. Hence, –GVc is applied to the human body through R3 which acts as a current limiter. The third amplifier, denoted A3, allows an operating mode without using any offset voltage. Moreover, one has to point out that all DC voltages provided by electrode skin interfaces (Fig. 1.1b) are not particularly the same and can produce undesirable offset voltage. In this scheme, R4 and C2 determine the low cutoff frequency.

Nowadays, this circuit has become somewhat classic. It is commonly used as an example in numerous amplifier design datasheets and application notes.

Generally speaking, the right leg driver circuit can be used for all biopotential recordings (e.g., EEG, EMG, EOG, etc.). It is also called the active reference circuit.

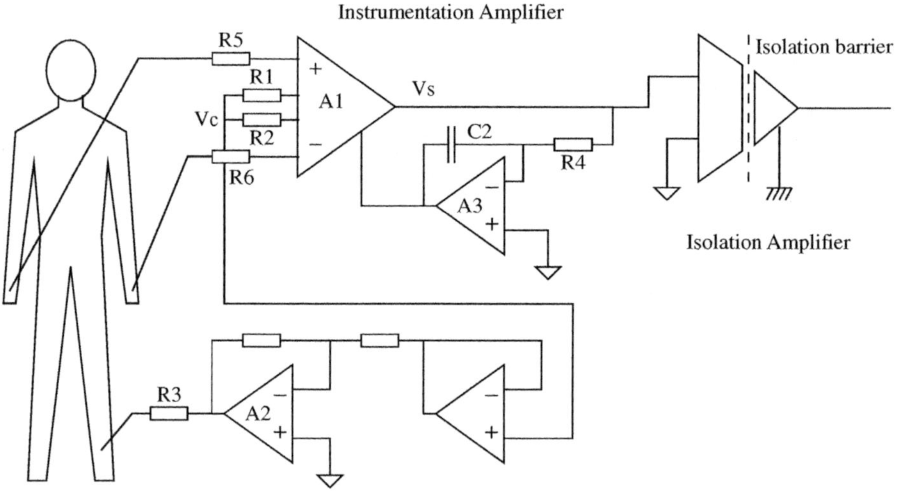

Fig. 1.4 Driven right leg circuit

On the other hand, it is also important to underline the fact that the configuration of the two electrodes is also used in some specific applications such as in telemetry or ambulatory recording. Consequently, it is still subject to modeling and design improvement. For instance, the reader can refer to: [8, 11, 7, 1].

1.1.3 Safety

Safety objectives deal with unexpected incidents. Misused and defective devices are two risky aspects of biopotentials recording. Maybe one cannot cite the entire list of possible scenarios, but it seems obvious that one has to avoid any sensitive, or even lethal electrical stimulation.

1.1.3.1 Current Limitation

All the leads connected to the patient are potentially dangerous. For example, in the right led driver (Fig. 1.4) there are three contacts with the patient. The resistor R3 limits the current supplied by the voltage source A2. The inputs In+ and In- shouldn't supply any current except in case of A1 dysfunction, In+ or In- can became a DC voltage source. Therefore, R4 and R3 operate as a current limiter as well and are useless for amplification.

1.1.3.2 Galvanic Isolation

Grounding is another important default source. If several devices are connected to the patient, the ground loop can generate sensitive stimulation. Obviously, a more dangerous case occurs when the patient is accidentally in contact with the power supply line. An electrocution is unavoidable if the current finds, through the patient, a pathway to the ground!

Safety grounding requires an efficient galvanic isolation (i.e. elimination of all electrical links between electronic devices inside the patient side and the power supply system). Electronic manufacturers propose isolation amplifiers that provide isolated DC supply, isolated ground and an analog channel through the isolation barrier. Galvanic isolation justifies the two different grounds symbols used in Fig. 1.4. Isolated ground (or floating ground) inside the patient is totally disconnected from the power supply ground. There is no pathway whatsoever for the power line current.

1.1.4 To Conclude this Section...

In this field, technological progress has tended to take the form of system miniaturizations; low consumption battery powered systems; hybrid designs (by including digital systems), secured transmission and so on.

What about tomorrow? Certainly, this trend will persist and one can imagine a single chip device that integrates electrodes, amplifiers, codecs, digital data processing codes, as well as a wireless transmission system.

In the next section, the reader will find some basic biopotential properties that naturally cannot be exhaustive due to the numerous different cases and various applications.

1.2 General Properties of Some Common Biosignals

As explained earlier, the purpose of this section is to present some basic generalities related to the most common biosignals used in clinical routines (i.e., ECG, EEG, EP and EMG).

1.2.1 The Electrocardiogram (ECG)

The ECG is an electrical signal generated by the heart's muscular activity. It is usually recorded by a set of surface electrodes placed on the thorax. The number of channels depends on the application. For instance, it could be 1, 2, 3, 6, 12 or even more, in some cases such as in mapping protocols (e.g. 64 or 256 channels).

Generally speaking, the ECG provides a useful tool for monitoring a patient, basically when the purpose consists in detecting irregular heart rhythms or preventing myocardial infarctions.

A typical ECG beat mainly has 5 different waves (P, Q, R, S and T), as shown in Fig. 1.5. These waves are defined as follows:

- **P wave:** this corresponds to atrial depolarisation. Its amplitude is usually lower than 300 μV and its duration is less than 0.120 s. Furthermore, its frequency may vary between 10 and 15 Hz,
- **QRS complex:** This is produced after the depolarisation process in the right and left ventricles. Its duration usually varies from 0.070 s to 0.110 s and its amplitude

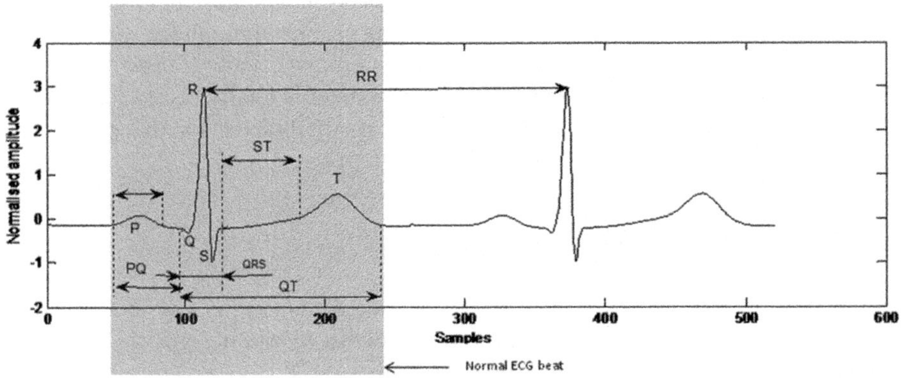

Fig. 1.5 Normal heart beats showing some essential indicators, generally measured by clinicians for the purpose of diagnosis

is around 3 mV. It should also be pointed out that the QRS complex is often used as a reference for automatic heart beat detection algorithms,

- **T wave**: this low frequency wave corresponds to the ventricular polarisation;
- **ST segment**: this corresponds to the time period during which the ventricles remain in a depolarised state,
- **RR interval**: this interval may be used as an indicator for some arrhythmias,
- **PQ and QT intervals**: they are also used as essential indicators for diagnostic purposes.

As it is well known, heart rhythm varies according to a person's health (fatigue, effort, emotion, stress, disease etc.). For instance, in the case of cardiac arrhythmias, one can emphasize the following types: ventricular, atrial, junctional, atrio-ventricular and so on [6]. Special cases, including advanced processing techniques will be presented in Chaps. 2, 3, 4 and 5.

1.2.2 The Electroencephalogram (EEG)

The EEG is a physiological signal related to the brain's electrical activity. Its variation depends on numerous parameters and situations such as whether the patient is healthy pathological, awake, asleep, calm and so on. This signal is recorded using electrodes placed on the scalp. The number of electrodes depends on the application. Generally speaking, the EEG may be used to detect potential brain dysfunctions, such as those causing sleep disorders. It may also be used to detect epilepsies known as "paroxysmal" identified by peaks of electrical discharges in the brain.

A considerable amount of the EEG energy signal is located in low frequencies (i.e., between 0 and 30 Hz). This energy is mainly due to five rhythms, namely, δ, θ, α, β and γ. They are briefly described as follows:

1. ***δ rhythm***: consists of frequencies below 4 Hz; it characterizes cerebral anomalies or can be considered as a normal rhythm when recorded in younger patients,
2. ***θ rhythm***: having a frequency around 5 Hz, it often appears amongst children or young adults,
3. ***α rhythm***: generated usually when the patient closes his eyes, its frequency is located around 10 Hz,
4. ***β rhythm***: frequencies around 20 Hz may appear during a period of concentration or during a phase of high mental activity,
5. ***γ rhythm***: its frequency is usually above 30 Hz; it may appear during intense mental activity, including perception.

Above 100 Hz, one can note that the EEG energy spectrum varies according to a $1/f$ function, where f stands for the frequency. When recording the EEG signal peaks and waves may appear at random epochs (e.g. with cases of epilepsy). Moreover, it is important to note that other biosignals may interfere with the EEG signal during

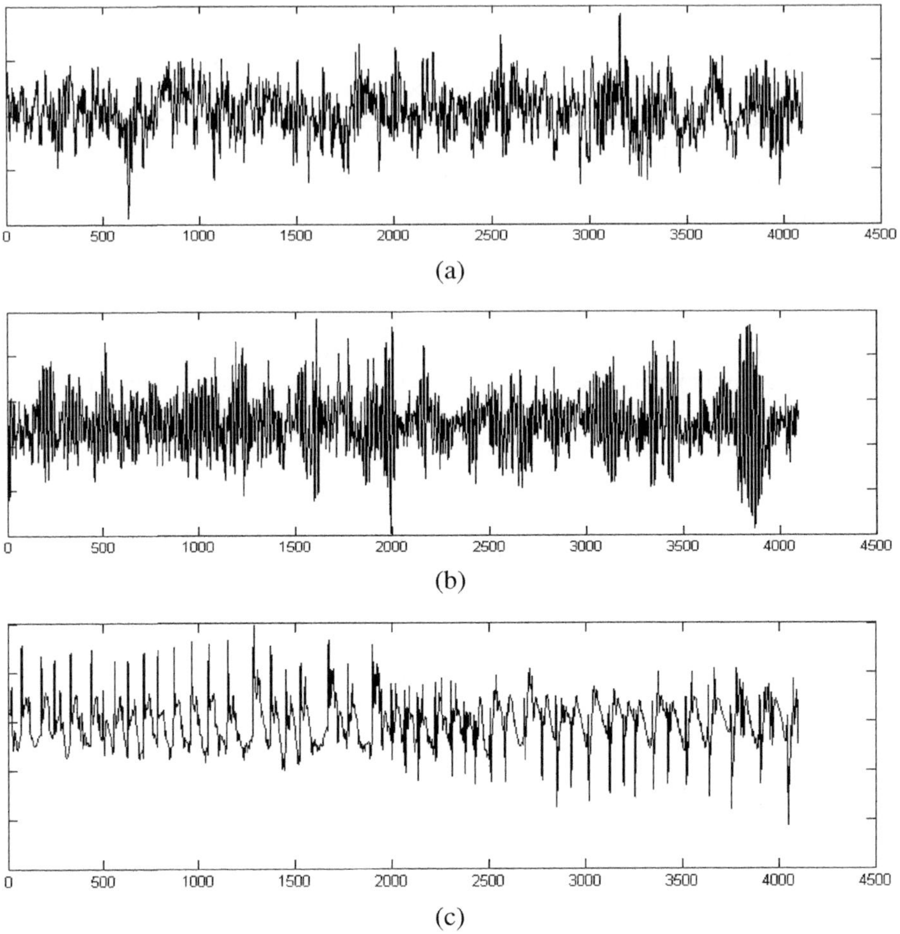

Fig. 1.6 Recorded EEG signals: (**a**) From a healthy patient (eyes open) (**b**) from a healthy patient (eyes closed) – (**c**) from an epileptic patient

the acquisition phase (e.g. ECG or EMG). The amplitude of the EEG signals varies from a few micro volts up to about 100 μV. As mentioned above, the number of electrodes required for the acquisition depends on the application. For instance, in some applications, a standard such as 10–20 system may be used.

For the purpose of illustration, some EEG signals recorded from a healthy patient (eyes open, eyes closed) as well as for an epileptic patient are presented in Fig. 1.6. Additionally, the reader can refer to Chaps. 6, 7 and 8 for advanced processed techniques.

1.2.3 Evoked Potentials (EP)

When a sensory system is stimulated, the corresponding produced response is called "Evoked Potential" (EP). Nervous fibres generate synchronized low-amplitude action potentials, also called spikes. The sum of these action potentials provides an EP that should be extracted from the EEG, considered here as noise. Generally, EPs are used to diagnose various anomalies such as auditory or visual pathways (see also Chaps. 9, 10, 11 and 14).

There are three major categories of evoked potentials:

1. **Somatosensory evoked potentials (SEP)**: these are obtained through muscle stimulations,
2. **Visual evoked potentials (VEP)**: for which a source of light is used as a stimulus,
3. **Auditory Evoked Potentials (AEP)**: they are generated by stimulating the auditory system with acoustical stimuli. In Fig. 1.7, we represent a simulated signal showing a Brainstem Auditory Evoked Potentials (BAEP), thalamic sub-cortical potentials and late potentials (cortical origin).

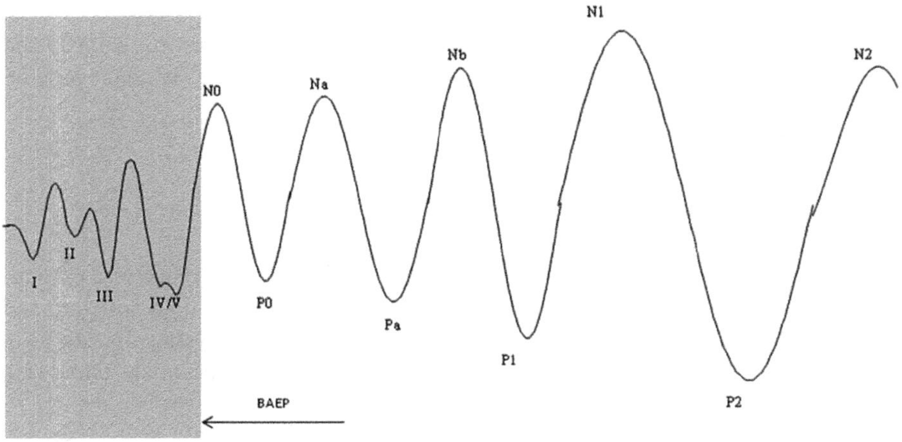

Fig. 1.7 Simulated Auditory Evoked Potentials. BAEP (waves: I- II, II, III, IV and V); Thalamic and sub-cortical potentials (waves: N_O, O_O, N_a, P_a, *and* N_b); Late potentials (waves: *P1, N1, P2 and N2*)

1.2.4 The Electromyogram (EMG)

EMG is a recording of potential variations due to voluntary or involuntary muscle activities. The artefact's amplitude (about 5 μV) resulting from muscular contraction is superior to that of the EEG and the time period varies between 10 and 20 ms. This signal can be used to detect some specific abnormalities related to the electrical activity of a muscle. For instance, this can be related to certain diseases including:

- muscular dystrophy,
- amyotrophic lateral sclerosis,
- peripheral neuropathies,
- disc herniation.

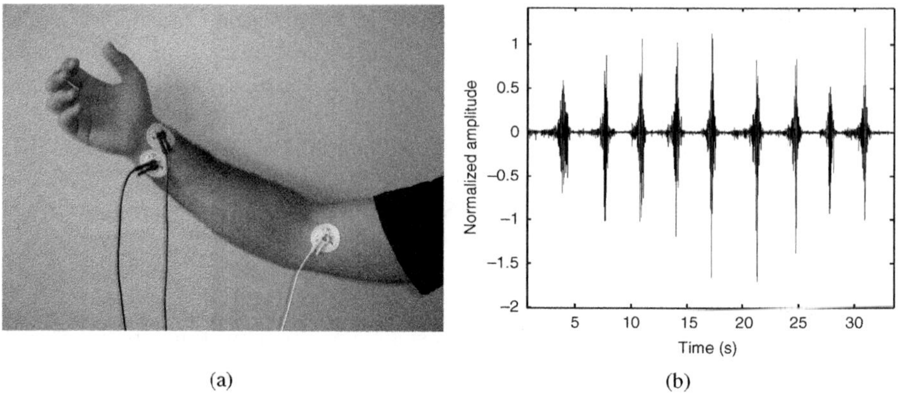

Fig. 1.8 Arm EMG acquisition, (**a**) electrods position, (**b**) corresponding recorded signal for a periodic "open-close" hand movement

- From Fig. 1.8(a,b), we show respectively an example of an arm muscle EMG and its corresponding recorded signal due to a periodic "open/close" hand movement. (See also Chaps. 12 and 13).

1.3 Some Comments. . .

Generally speaking, an efficient biomedical engineering system requires a particular optimization of its various components which might include both software and hardware. For this purpose, implementing advanced signal processing algorithms in such a system becomes interesting, mainly if the following aspects are taken into account:

1. ***Acquisition system***: its performance, its power consumption and its size, are regarded as essential technical features,
2. ***Processing algorithms***: their efficiency is clearly important, but what about their complexity? Can the processing be achieved in real-time?
3. ***Processing system***: which platform is best suited for a given algorithm? One with a mono-processor or a multi-processor? Should the algorithm be implemented in a mobile system? In such cases, what about power consumption?
4. ***Transmission system***: does the application require a real-time transmission of data? Which protocol should be used? Does one have enough bandwidth? Should we compress data [4]? Is there any protection against channel errors?
5. ***Data security***: since we deal with medical signals, how should one protect the data? Is any watermarking or encryption required? What about the local legislation?

In addition, another non-technical aspect should be taken into account. It essentially concerns the "development cost". This important financial consideration should never be overlooked!

1.4 Conclusion

We have tried throughout this first chapter to provide the reader with the basics of biopotential acquisition systems as well as some common general properties of ECG, EEG, EPs and EMG signals. As one will notice, no specific cases have been evoked. This can be explained by the fact that some of them will be subject to advanced analysis and study in subsequent chapters.

Finally, we advise the reader to pay special attention to the references proposed by the authors at the end of each chapter.

References

1. Dobrev D, Neycheva T and Mudrov N (2005) Simple two-electrode biosignal amplifier Med. Biol. Comput. 43:725–730
2. Huhta J C and Webster J G (1973) 60 Hz Interference in Electrocardiography. IEEE Trans. Biomed. Eng. 20:91–101
3. Metting-Van-Rijn A C, Peper A A et al. (1990) High-quality recording of bioelectric events. Part 1 Interference reduction theory and practice. Med. Biol. Comput. 28:389–397
4. Naït-Ali A and Cavaro-Menard C (2008) Compression of biomedical images and signals. ISTE-WILEY
5. Feynman R P, Leigthon R B et al. (1964) The Feynman lectures on physics. Addison-Wesley, Boston, MA
6. Sörnmo L and Laguna P (2005) Bioelectrical signal processing in cardiac and neurological applications, Elsevier Academic Press, New York
7. Spinelli E M and Mayosky M A (2005) Two-electrode biopotential measurements: power line interference analysis. IEEE Trans. Biomed. Eng. 52:1436–1442
8. Thakor N and Webster J G (1980) Ground free ECG recording with two electrodes. IEEE Trans. Biomed. Eng. 20:699–704
9. Webster J G (1998) Medical instrumentation Application and Design, Third Ed.
10. Winter B B and Webster J G (1983) Driven-right-Leg Circuit Design. IEEE Trans. Biomed. Eng. 30:62–66
11. Wood D E, Ewins D J et al. (1995) Comparative analysis of power line interference between two or three electrode biopotentials amplifiers. Med. Biol. Comput. 43:63–68

Chapter 2
Extraction of ECG Characteristics Using Source Separation Techniques: Exploiting Statistical Independence and Beyond

Vicente Zarzoso

Abstract The extraction of signals of interest from electrocardiogram (ECG) recordings corrupted by noise and artifacts accepts a blind source separation (BSS) model. The BSS approach aims to estimate a set of underlying source signals of physiological activity from the sole observation of unknown mixtures of the sources. The statistical independence between the source signals is a physiologically plausible assumption that can be exploited to achieve the separation. The mathematical foundations, advantages and limitations of the most common BSS techniques based on source independence, namely, principal component analysis (PCA) and independent component analysis (ICA), are summarized. More recent techniques taking advantage of prior knowledge about the signal of interest or the mixing structure are also briefly surveyed. The performance of some of these methods is illustrated on real ECG data. Although our focus is on fetal ECG extraction from maternal skin potential recordings and atrial activity extraction in surface ECG recordings of atrial fibrillation, the BSS methodology can readily be extended to a variety of problems in biomedical signal processing and other domains.

2.1 Introduction

Extracting signals of interest from the observation of corrupted measurements is a fundamental signal processing problem arising in numerous applications including, but not limited to, biomedical engineering. In biomedical applications, an accurate signal estimation and interference cancellation step is often necessary to ensure the success and increase the performance of subsequent higher-level processing stages

---

V. Zarzoso (✉)
Laboratoire d'Informatique, Signaux et Systémes de Sophia Antipolis,
Université de Nice – Sophia Antipolis, CNRS, Les Algorithmes – Euclide-B, 2000 route des Lucioles, BP 121, 06903 Sophia Antipolis Cedex, France
e-mail: zarzoso@i3s.unice.fr

The core of the present chapter was delivered as a lecture at the 6th International Summer School on Biomedical Signal Processing of the IEEE Engineering in Medicine and Biology Society, Certosa di Pontignano, Siena, Italy, July 10–17, 2007.

A. Naït-Ali (ed.), *Advanced Biosignal Processing*,
DOI 10.1007/978-3-540-89506-0_2,

such as wave detection, signal compression and computer-aided diagnosis. In electrocardiography, a classical instance of this problem is encountered when the fetal heartbeat signal is to be observed from cutaneous potential recordings of a pregnant woman. An example is shown in Fig. 2.1. The first five channels have been obtained from abdominal electrodes, so that fetal cardiac activity is visible. The 4th abdominal lead presents an important baseline wander, probably due to the mother's respiration. The last three channels belong to the thoracic region and contain mainly maternal cardiac components. As can be observed on the first lead output, the more powerful maternal heartbeat hinders the observation of the low-amplitude fetal signal. To evaluate the fetus' well-being, currently employed techniques basically rely on Doppler ultrasound heart-rate monitoring [45]. Obtaining a clean fetal electrocardiogram (FECG) non-invasively can provide a more accurate estimate of the heart rate variability, but also constitutes a safe cost-effective method to perform a more detailed morphological analysis of the fetal heartbeat. This goal requires the suppression of the maternal ECG (MECG) and other artifacts present in the mother's skin electrode recording.

A second interesting problem is the analysis of atrial fibrillation (AF), the most prevalent cardiac arrhythmia encountered by physicians. This trouble, often associated with minor symptoms, can also entail potentially more serious complications such as thromboembolic events. A typical standard 12-lead electrocardiogram (ECG) of an AF episode is shown in Fig. 2.2. The P-wave corresponding to well-organized atrial activation in normal sinus rhythm is replaced by chaotic-like oscillations known as f-waves, which contain useful information about the condition. In Fig. 2.2, the f-waves are easily observed during the T-Q intervals of lead V1, for example, but are masked by the QRS complex precisely when the atrial activity (AA) signal could provide crucial information about physiological phenomena such as the ectopic activation of the atrio-ventricular node [9]. Diagnostic and prognostic information about AF can be derived from the time-frequency analysis of the AA signal. The dominant frequency of the AA signal, typically located between 5 and 9 Hz, is closely related to the atrial cycle length and the refractory period of atrial myocardium cells, and thus to the stage of evolution and degree of organization of the disease [7, 8]. In particular, a decreasing trend in the main frequency is associated with a higher probability of spontaneous termination (cardioversion) of the fibrillatory episode. The non-invasive analysis of the AA signal calls for the suppression of the ventricular activity (VA) as well as other artifacts and noise contributing to the surface ECG.

Depending on the signal model and assumptions, the signal extraction problem can be approached from different perspectives. Classical approaches include frequency filtering, average beat subtraction [61] and Wiener's optimal filtering [68]. The first approach requires the desired and the interfering signals to lie in distinct frequency bands. The second assumes a regular morphology for the interfering signal waveform. The third methodology relies on reference measurements correlated with the interference but uncorrelated to the desired signal. However, in many practical scenarios like the above two examples, the signal of interest and the interference often overlap in the frequency spectrum, the morphology of the interfering wave-

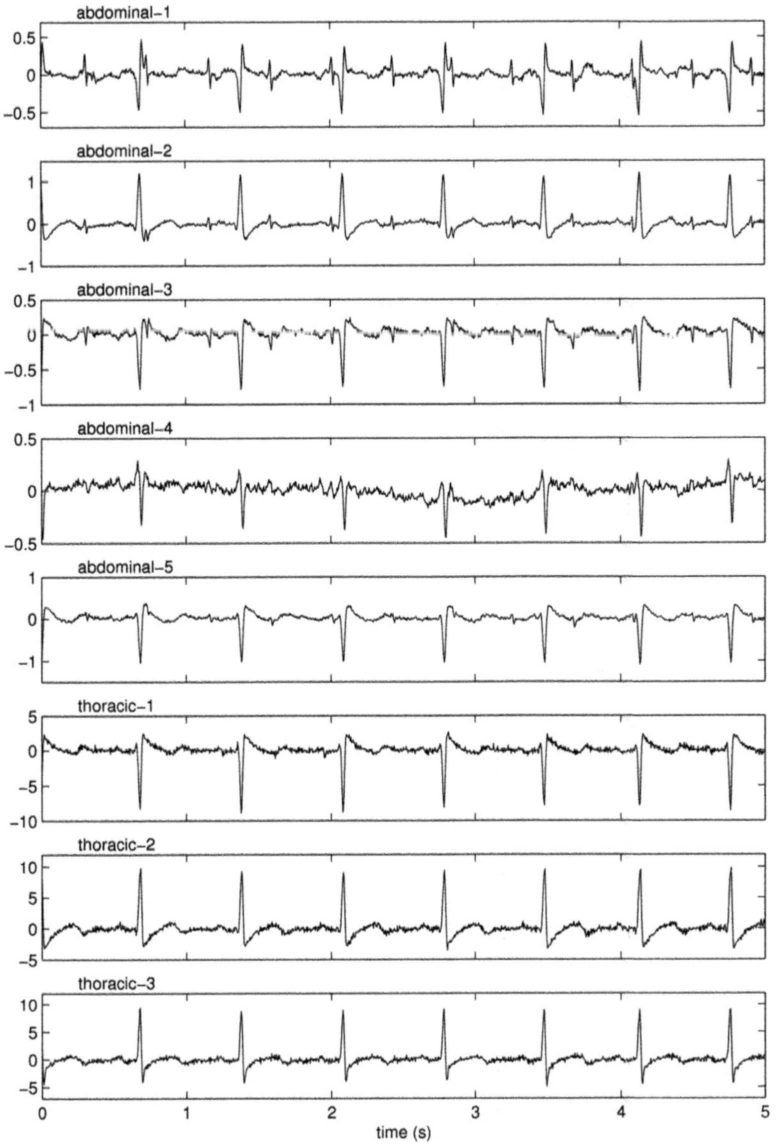

Fig. 2.1 Cutaneous electrode potential recording from a pregnant woman. Channels 1–5: abdominal electrodes. Channels 6–8: thoracic electrodes. A 5-s segment is shown, the available data consisting of 10 s. Only the relative amplitudes are important. Sampling frequency: 500 Hz3[1]

[1]ECG recording courtesy of L. de Lathauwer and B. de Moor, from K. U. Leuven, Belgium.

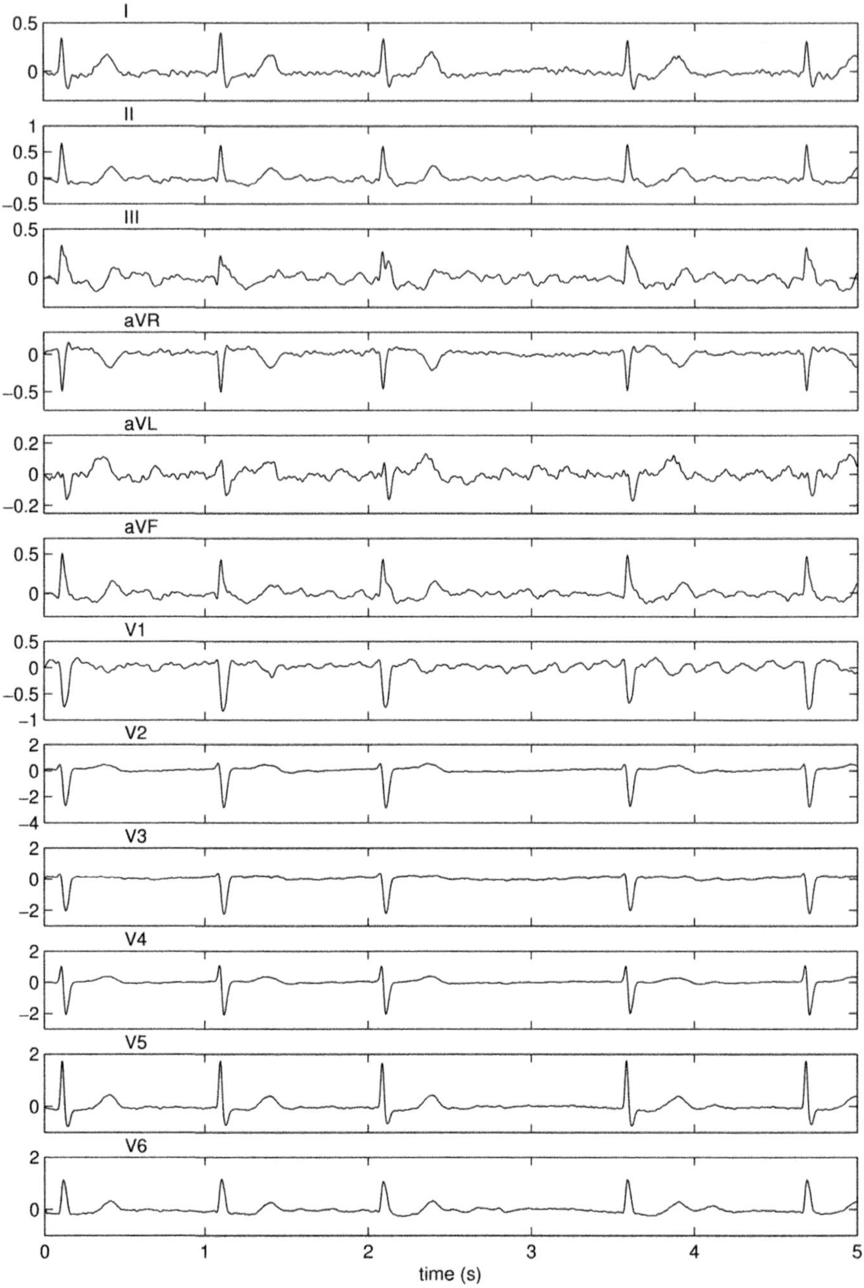

Fig. 2.2 Standard 12-lead ECG recording from an AF patient. Only the first 5 s of a 12-s recording are displayed for clarity. Only the relative amplitudes are important. Sampling frequency: 1000 Hz[2]

[2]Data recorded at the Hemodynamic Department, Clinical University Hospital, Valencia, Spain.

form may be irregular, and obtaining pure reference signals uncorrelated with the signal of interest may be a difficult task.

The *blind source separation (BSS)* approach, introduced now over two decades ago [32], provides a more general versatile framework. In its instantaneous linear mixture formulation, BSS assumes that the desired and interfering signals, so-called sources, may have arbitrary morphology with possibly overlapping spectra and may all appear mixed together in each of the observations. The estimation of appropriate extracting filters, and thus the estimation of the source waveforms from the observed mixtures, is achieved by recovering a known or assumed property of the sources. Some of the most common properties include statistical independence, non-stationarity, cyclo-stationarity, finite alphabet (enabled in digital communications) and, more recently, sparseness. The assumption of statistical independence is very plausible in many biomedical applications and will be our main focus.

The present chapter summarizes the basic concepts behind independence-exploiting BSS, gives an overview of some specific techniques, and discusses their application to ECG signal extraction. The chapter is structured as follows. Section 2.2 surveys traditional approaches to signal extraction. Section 2.3 is devoted to the BSS approach based on statistical independence. Two main families of techniques can be distinguished, depending on the degree of statistical independence being exploited: second order or higher order. The former give rise to techniques based on *principal component analysis (PCA)* and its variants. The latter comprises the *independent component analysis (ICA)* approach. The advantages and limitations of the different methods are illustrated on the problems of FECG extraction during pregnancy and AA extraction in AF episodes using the real non-invasive recordings of Figs. 2.1 and 2.2, respectively. It is important to remark that BSS is a rather general methodology not restricted to signal estimation in the ECG, but also applicable to other biomedical problems (e.g., in electroencephalography, electromyography, etc.) as well as non-biomedical areas (e.g., communications, audio, seismic exploration, data compression, data classification, finance, etc.). The performance of BSS techniques can be improved by suitable modifications capitalizing on the prior knowledge of the problem under examination. The chapter concludes with some of these recent lines of research aiming to improve the performance of BSS in specific biomedical applications.

The first part of the chapter (Sects. 2.3, 2.4, 2.5 and 2.6) is mainly addressed to readers who have little or no familiarity with the topic of source separation, but could also be useful to more experienced practitioners as a brief reference and an introduction to ECG applications of BSS. The second part (Sect. 2.7) is devoted to more recent developments that should be of interest to readers acquainted with the fundamentals of BSS.

In the sequel, scalar, vector and matrix quantities are denoted in lightface lowercase, boldface lowercase and boldface uppercase letters, respectively. Symbol $E\{.\}$ is the mathematical expectation, and $(.)^T$ represents the transpose operator. For the sake of simplicity, the following exposition will be constrained to real-valued signals and mixtures, more typically encountered in ECG signal processing. Nevertheless, many results can easily be extended to the complex-valued case.

2.2 Approaches to Signal Extraction in the ECG

A variety of different approaches have been proposed to cancel artifacts and enhance signals of interest in the ECG. In AA extraction, the analysis of ECG segments outside the QRST interval is probably the simplest possible option [54], but is not suitable when a continuous monitoring is required or in patients with high ventricular rates. This option is readily discarded in FECG extraction, where the different heart-rates of mother and child cause the respective QRST complexes to overlap in time. An alternative is frequency filtering. However, very often the spectra of the desired signal (FECG, AA) and the interference (MECG, VA) overlap, rendering this approach ineffective.

More successful techniques focus on the explicit cancellation of the most significant features of the interfering cardiac waveform, that is, the maternal heartbeat in FECG extraction or the patient's QRST complex in AA extraction. The average beat subtraction (ABS) method [61, 8, 35] computes a template of the interfering complex by synchronized averaging and then subtracts it, after appropriate scaling, from the lead output. The technique relies on the assumptions that the interference and the signal of interest are uncoupled, and that the former presents a repetitive regular waveform. The ABS approach requires beat detection and classification before averaging, is thus sensitive to the morphology of the ventricular beats, and is unable to suppress noise and artifacts uncoupled with the interfering signal (e.g., noise from electronic equipment).

To mitigate the sensitivity to local QRST morphological variations caused by minor changes in the electrical axis of the heart (due, e.g., to respiration), the spatiotemporal QRST cancellation (STC) technique [62] takes into account the average beats from adjacent leads via weighted least-square fitting before subtraction. Like ABS, STC requires a sufficient number of beats with similar morphology in order to obtain a significant QRST average and ensure the proper cancellation of AA in the averaged beat. Alternative methods extract the VA using artificial neural networks [67], or are based on the decomposition of the ECG using discrete packet wavelet transforms [56].

All the above techniques are unable to efficiently exploit the diversity provided by the spatially-separated electrodes. Indeed, the standard ECG employed in clinical practice is composed of 12 leads, while more sophisticated recording equipment used for body surface potential mapping (BSPM) may include up to hundreds of leads. Each lead captures a different mixture of bioelectrical phenomena of interest, artifacts, interference and noise. This spatial diversity can be efficiently exploited by processing, in a suitable manner, all leads simultaneously [80]. The spatial information that can be derived by exploiting this kind of diversity may provide new insights into how the physiological mechanisms of interest (fetal heartbeat, atrial activation) reflect on the surface ECG, and may help assess the prognostic features of external recordings, currently not fully understood.

A classical approach based on this observation is Widrow's multi-reference adaptive noise canceller, based on Wiener's optimal spatial filtering theory [68]. A primary input to the canceller is given by a lead containing a mixture of the desired signal and the interference. Some reference leads correlated to the interference but uncorrelated with the desired signal are processed by finite impulse response filters, whose outputs are added together to form an estimate of the interference in the primary input. The filter weights are adaptive updated to minimize the mean square difference (or a stochastic approximation thereof) between the primary input and the estimated interference. The availability of reference leads linked to the interference but free from the desired signal is a crucial assumption for the success of this method, and introduces strong constraints on the electrode location [74, 80].

The first plot of Fig. 2.3 displays the FECG signal reconstructed on the 4th abdominal lead of Fig. 2.1 by the Wiener-based multi-reference noise canceling technique of Widrow et al. [68] (see also [80]). The optimal Wiener-Hopf solution has been computed using the three thoracic leads as references. The strong baseline wander corrupting the primary lead cannot be removed, as it hardly contributes to the reference leads. Similar results are obtained on the AF recording of Fig. 2.2. Using all leads except V1 as references, the AA signal estimated by the Wiener filtering approach on lead V1 is shown in the first plot of Fig. 2.4 and the corresponding spectrum appears in Fig. 2.5. The estimated signal does not completely explain all the AA present in that lead, especially around the QRST complexes. Nevertheless, the spectral characteristics of the extracted signal (Fig. 2.5) present the expected typical shape of AA in AF, with an estimated dominant frequency around 5 Hz and an important concentration around the main peak. The spectral concentration is computed as the percentage of signal power contained in the $[0.82f_p, 1.17f_p]$ frequency band, where f_p denotes the dominant frequency of the signal spectrum [20].

2.3 Blind Source Separation

The BSS approach provides a more general framework for signal extraction whereby each of the recorded signals may contain a contribution from the desired signal and the interference. Moreover, the sources may contain overlapping time-frequency spectra with possibly non-repetitive irregular waveforms. In its basic formulation, BSS assumes that each of the observations is an unknown instantaneous linear mixture of the sources, and aims at inverting the mixture in order to recover the sources. The FECG extraction was originally formulated as a BSS problem in [27, 28]. The AA extraction problem was first approached from the BSS perspective in [51, 52].

2.3.1 Signal Model and Assumptions

Mathematically, BSS assumes the following relationship between sources and observations:

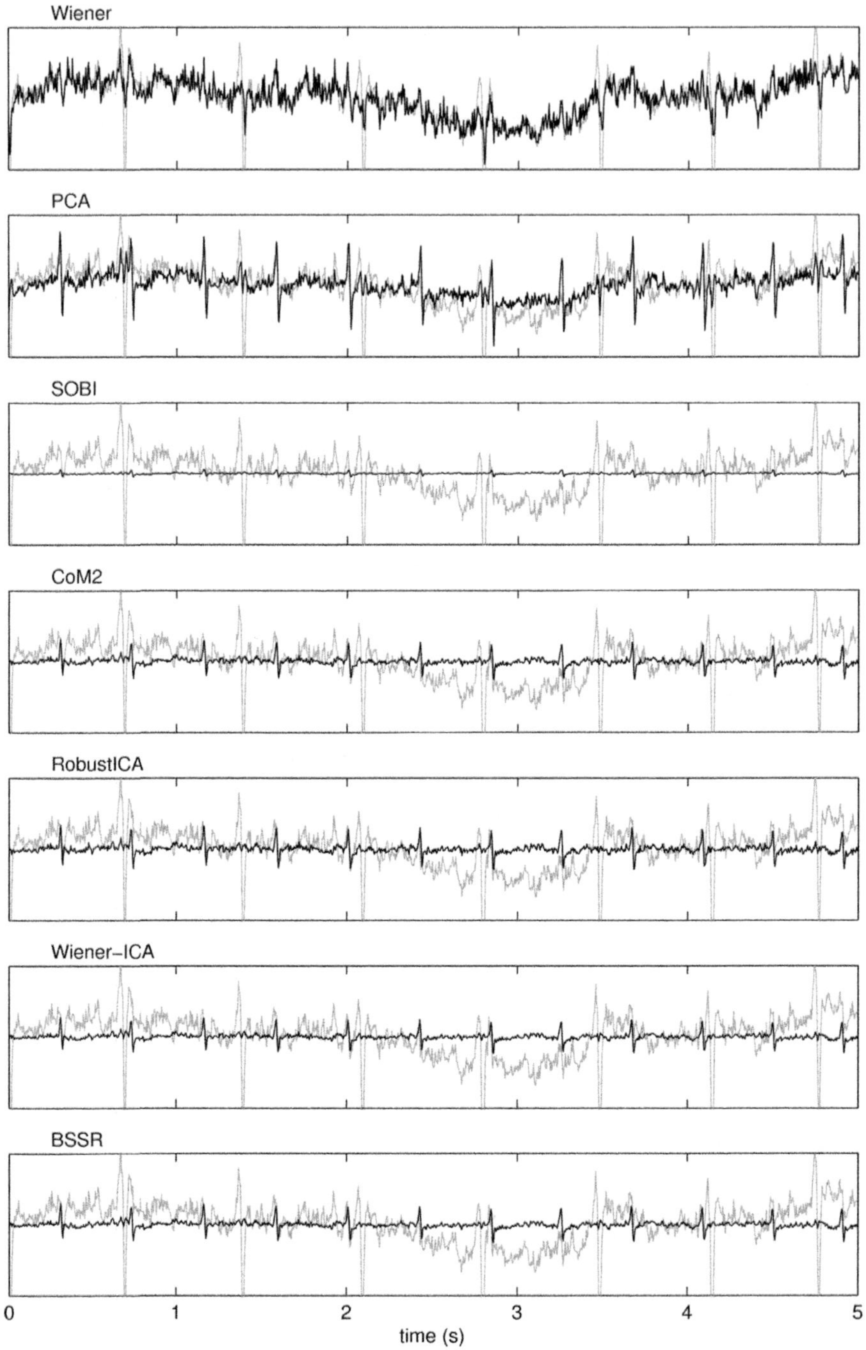

Fig. 2.3 FECG contributions to the 4th abdominal lead of Fig. 2.1 estimated by some of the signal extraction methods surveyed in this chapter. For comparison, the signal recorded on the 4th abdominal lead is plotted in a lighter colour

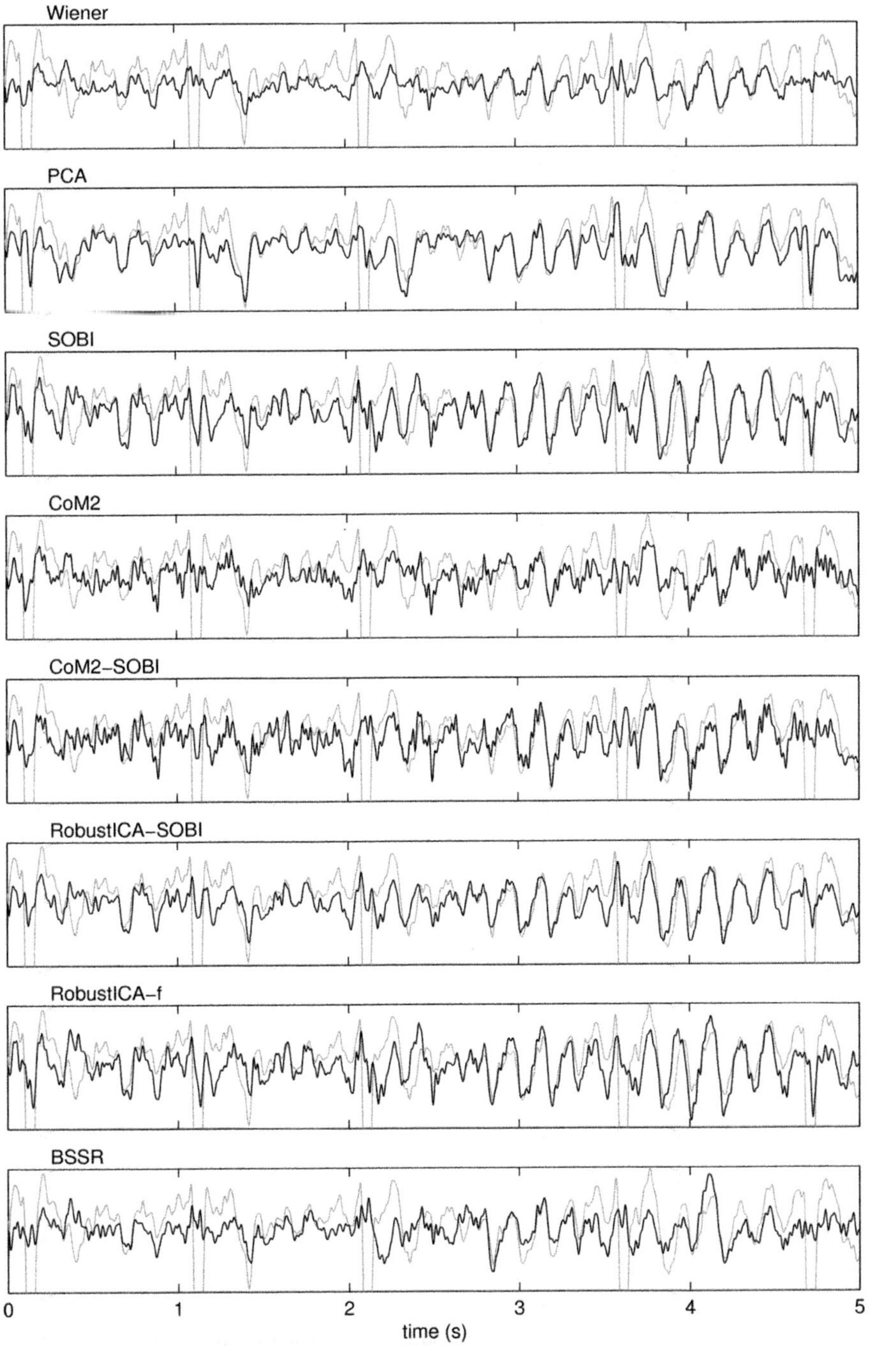

Fig. 2.4 AA contributions to lead V1 estimated by some of the signal extraction methods surveyed in this chapter. For comparison, the signal recorded on lead V1 is plotted in a lighter colour

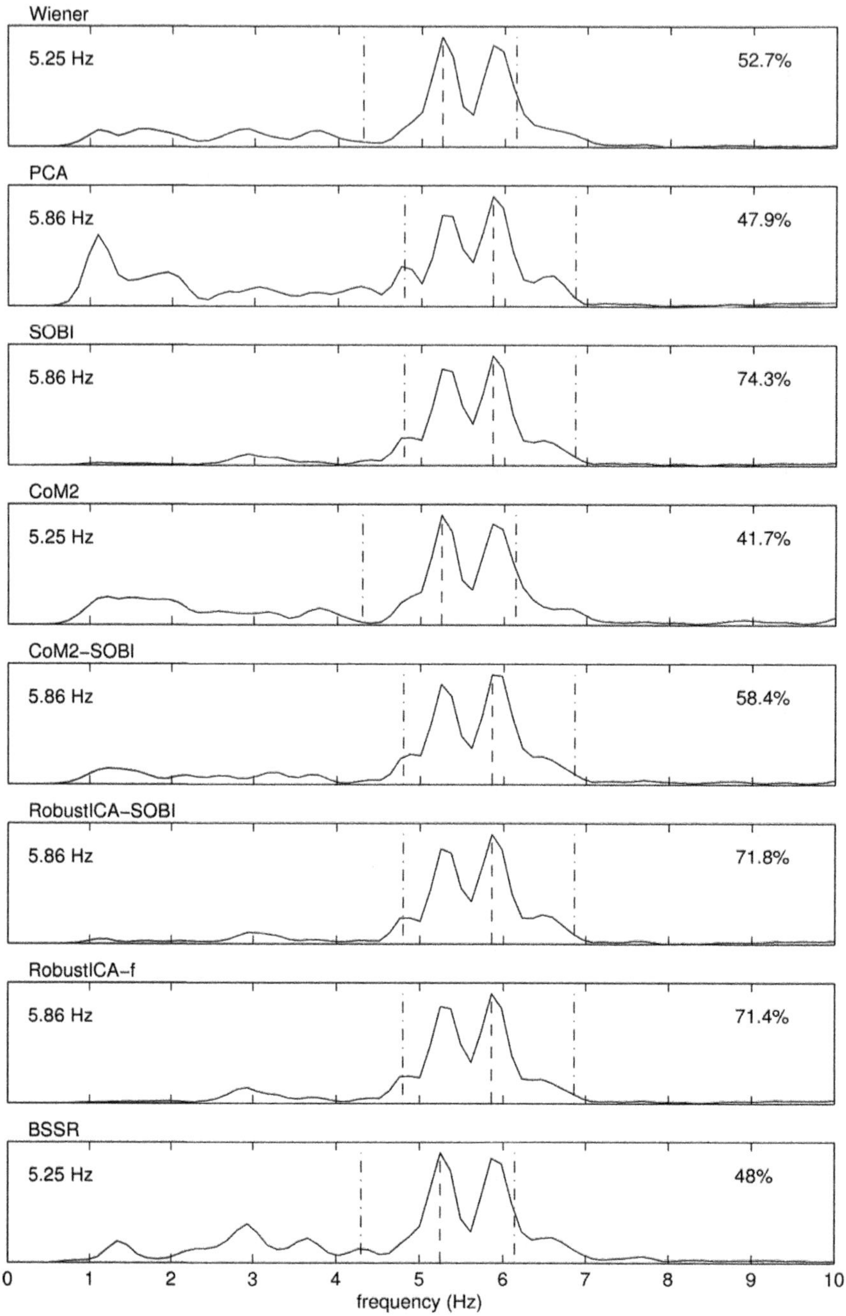

Fig. 2.5 Spectral characteristics of the reconstructed AA contributions to lead V1 shown in Fig. 2.4. Values on the left and vertical dashed lines represent the main frequency peak locations. Values on the right denote the spectral concentration. Vertical dash-dotted lines mark the bounds of the frequency band used to compute the spectral concentration. Both frequency spectrum and spectral concentration are estimated as in [20]

$$\mathbf{x}(t) = \mathbf{H}\mathbf{s}(t) = \sum_{i=1}^{n} \mathbf{h}_i s_i(t) \tag{2.1}$$

where vector $\mathbf{x}(t) = [x_1(t), x_2(t), \ldots, x_m(t)]^{\mathrm{T}}$ contains the observed signal mixtures, vector $\mathbf{s}(t) = [s_1(t), s_2(t), \ldots, s_n(t)]^{\mathrm{T}}$ the unknown zero-mean source signals, and $\mathbf{H}$ is the unknown $(m \times n)$ mixing matrix whose coefficient $h_{ij} = [\mathbf{H}]_{ij}$ represents the contribution or projection of source j onto observation i. The mixing matrix columns, $\{\mathbf{h}_i, i = 1, 2, \ldots, n\}$, are also known as source directions or transfer vectors. Their entries reflect the spatial pattern or topography of the relative power contribution described by the associated sources onto the spatially-separated sensors, and correspond to potential field spatial distributions in the case of bioelectrical signals.

In the FECG extraction problem, $\mathbf{s}(t)$ comprises the sources of fetal and maternal cardiac activity, noise and interference [27, 28]. In the AA extraction problem, the physiological sources are mainly composed of atrial and ventricular activity as well as noise [51, 52]. In both scenarios, the mixing matrix coefficients associated with the cardiac sources are defined by the propagation of physiological signals from the heart to the electrodes through the body tissues. Due to the distance between heart and electrodes, the speed of propagation of electrical signals across the body and the bandwidth of the phenomena of interest, the transfer between sources and sensors can reasonably be considered as linear and instantaneous. These facts support the suitability of Eq. (2.1) to describe the signal generation model not only in the two biomedical problems in hand, but in numerous problems in many other areas. Given T samples or realizations of the observed vector $\mathbf{x}(t)$, the objective of BSS is to reconstruct the corresponding T realizations of $\mathbf{s}(t)$. The separation may be performed with or without a previous explicit identification of the mixing matrix $\mathbf{H}$. Once the source separation and mixing matrix identification has been carried out, the contributions of the desired signals to the recordings can be reconstructed by isolating the sources of interest and their transfer vectors in expression (2.1).

The second part of model (2.1) provides an interesting geometrical insight, as it signifies that each source amplitude s_i is projected on its transfer vector $\mathbf{h}_i$ before adding up to the observations. If two of these columns are parallel, the corresponding sources will vary along the same direction and cannot be distinguished from each other. Hence, a necessary condition for model (2.1) to be invertible is that the source directions be linearly independent or, equivalently, the mixing matrix be full column rank. This requires, in particular, that $m \geq n$, i.e., the number of observations be equal to, or larger than, the number of sources. Mixtures of this type are called *overdetermined.* In the following, we will assume *square* overdetermined mixtures, in which $m = n$. If the sensor positions are known or can be accurately modeled, the source spatial locations can be determined from the estimated transfer vectors. The problem of source localization, however, is beyond our scope.

2.3.2 *Why Blind?*

The term *blind* means that the source signals and the mixture coefficients are unknown, and little prior knowledge about the sources and the mixture is assumed.

In conventional array processing techniques such as MUSIC [59] and its variants, the mixing matrix is modeled as a function of the angles whereby the source propagation wavefronts arrive at the sensor array. This requires accurate knowledge of sensor positions. If the actual positions differ from the modeled positions, calibration errors occur. By reducing the model assumptions, blind processing is more robust to calibration errors. The blind methodology proves particularly interesting in biomedical problems, where parametric approaches may be cumbersome. Indeed, the use of parameters would require time-consuming calibration protocols and may easily be subjected to a large patient-to-patient variability. Parameters may also be expected to evolve with time in the same patient. In addition, it is important that methods be capable of providing satisfactory performance in a number of potential patho-physiological conditions. Hence, the blind approach arises as a judicious option in this uncertain environment [27, 28]. In the context of ECG processing, another important advantage of the fully-blind approach is that it does not require a previous heartbeat detection and classification stage and, as a result, is essentially unaffected by wave morphology variations. Interesting examples illustrating this robustness to ectopic beats in pregnancy and AF ECG episodes are reported in [28, 21], respectively.

2.3.3 Achieving the Separation

Source separation is carried out by estimating a separating matrix $\mathbf{W}$ such that the separator output $\mathbf{y}(t) = \mathbf{W}^{\mathrm{T}}\mathbf{x}(t)$ contains an estimate of the sources. Each column of $\mathbf{W}$ represents a *spatial filter* for the extraction of a single source, $y_i(t) = \mathbf{w}_i^{\mathrm{T}}\mathbf{x}(t)$. It is clear that, without further information, the estimation of the sources and the mixing matrix from model (2.1) is an ill-posed problem. This is because exchanging an arbitrary invertible matrix between the source vector and the mixing matrix does not modify the observations at all, i.e., the pair $(\mathbf{HA}^{-1}, \mathbf{As}(t))$ produces exactly the same sensor output as $(\mathbf{H}, \mathbf{s}(t))$, for any invertible matrix $\mathbf{A}$. To render the problem solvable, the source signals must possess certain measurable property that can be exploited to achieve the separation, such as mutual statistical independence, non-Gaussian distribution, distinct frequency spectra, or known discrete support (as in digital communications). The estimation of the extracting filters, and thus of the source waveforms, is achieved by recovering at the separator output one of these known or assumed property of the sources. Not all properties constitute valid separation criteria; the concept of *contrast function* defines the conditions to be fulfilled [24] (see also Sect. 2.6).

A very plausible property is statistical independence. Indeed, the fetal and maternal heartbeats can be considered as independent [27, 28], and so can the atrial and ventricular activities in AF [51, 52]. Non-biological sources such as thermal noise and mains interference are also typically independent among themselves as well as from the biological activity sources. Hence, statistical independence is a sensible hypothesis in these scenarios. Depending on the order of independence exploited, two main types of techniques can be distinguished, namely, those based on second-order statistics and those relying on higher-order statistics, as detailed in the next sections.

Under the independence assumption (see Sect. 2.6 for details), the sources and their directions can be recovered up to two indeterminacies. Firstly, a scale factor can be exchanged between a source and its direction (scale ambiguity) without altering the observations nor the source independence. Hence, the sources can be assumed, without loss of generality, to have unit variance, $E\{s_i^2(t)\} = 1$, $i = 1, 2, \ldots, n$. Since the information of interest is often contained in the source waveform rather than in its power, this scale normalization is admissible. Secondly, if the sources and their directions are rearranged accordingly, the observations do not change, nor does the source independence. Without further information about the sources, their ordering is immaterial. As a result, one can hope to find, at most, a source estimate of the form $\mathbf{y}(t) = \mathbf{PDs}(t)$, where $\mathbf{P}$ is a permutation and $\mathbf{D}$ an invertible diagonal matrix. This scale and permutation indeterminacy can be reduced even further under additional hypotheses about the sources and/or the mixture.

2.4 Second-Order Approaches: Principal Component Analysis

2.4.1 Principle

PCA, a method widely used in biomedical signal analysis [22], is probably the simplest approach to BSS under the independence assumption. The PCA of observed vector $\mathbf{x}(t)$ can be briefly expressed as:

1. Find vector $\mathbf{w}_1$ maximizing $E\{y_1^2(t)\}$, with $y_1(t) = \mathbf{w}_1^T\mathbf{x}(t)$, subject to $||\mathbf{w}_1||^2 = 1$.
2. For $k = 2, 3, \ldots, n$:
   Find vector $\mathbf{w}_i$ maximizing $E\{y_i^2(t)\}$, with $y_i(t) = \mathbf{w}_i^T\mathbf{x}(t)$, subject to $||\mathbf{w}_i||^2 = 1$ and $\mathbf{w}_i^T\mathbf{w}_j = 0, j = 1, \ldots, i-1$.

As noted in [66, 27], each $\mathbf{w}_k$ represents a spatial filter orthogonal to $\{\mathbf{w}_1, \mathbf{w}_2, \ldots, \mathbf{w}_{k-1}\}$ whose output has maximal power. It is well known that the solution to the above problem is given by the eigenvalue decomposition (EVD) of the observation covariance matrix $\mathbf{R_x} = E\{\mathbf{x}(t)\mathbf{x}(t)^T\}$. The EVD yields $\mathbf{R_x} = \mathbf{U}\Sigma\mathbf{U}^T$, where the $(n \times n)$ matrix $\mathbf{U}$ is orthonormal, $\mathbf{U}^T\mathbf{U} = \mathbf{U}\mathbf{U}^T = \mathbf{I}_n$, and contains the eigenvectors in its columns; matrix $\Sigma = \mathrm{diag}(\sigma_1^2, \sigma_2^2, \ldots, \sigma_n^2)$ is composed of the eigenvalues. The PCA solution, which corresponds to the source estimate in the BSS context, is given by $\mathbf{y}_{PCA}(t) = \Sigma^{-1/2}\mathbf{U}^T\mathbf{x}(t)$. This source estimate is obtained by the separating matrix $\mathbf{W}_{PCA} = \mathbf{U}\Sigma^{-1/2}$. The term $\Sigma^{-1/2}$ guarantees that the source unit-variance assumption is respected, but does not alter the orthogonality of the estimated source directions. PCA relies on *second-order statistics (SOS)* only.

2.4.2 Covariance Matrix Diagonalization

To understand the use of PCA for BSS, we need to examine the above solution in the light of model (2.1) under the independence assumption. At second order, independence reduces to uncorrelation, mathematically expressed as $\mathrm{E}\{s_i(t)s_j(t)\} = 0, \forall i \neq j$, which together with the unit-variance assumption result in an identity source covariance matrix, $\mathbf{R}_s = \mathrm{E}\{\mathbf{s}(t)\mathbf{s}(t)^{\mathrm{T}}\} = \mathbf{I}_n$. Accordingly, the observation covariance matrix can be expressed as $\mathbf{R}_x = \mathbf{H}\mathbf{R}_s\mathbf{H}^{\mathrm{T}} = \mathbf{H}\mathbf{H}^{\mathrm{T}}$, where the first equality stems from model (2.1) and the second from the source covariance matrix structure. Comparing this expression with the EVD of $\mathbf{R}_x$ results in the mixing matrix estimate $\hat{\mathbf{H}} = \mathbf{W}_{\mathrm{PCA}}^{-\mathrm{T}} = \mathbf{U}\Sigma^{1/2}$. On the other hand, the application of $\mathbf{W}_{\mathrm{PCA}}^{\mathrm{T}}$ on to the observations diagonalizes their covariance matrix, so that the covariance matrix of the estimated sources, $\mathbf{y}_{\mathrm{PCA}}(t)$, is also the identity. Hence, PCA recovers the source covariance matrix structure and yields source estimates that are uncorrelated, or independent up to the second order, just like the actual sources. Consequently, PCA can be seen as exploiting the independence assumption at order two.

Many algorithms are available to obtain the EVD of a symmetric matrix $\mathbf{A}$ [31]. The Jacobi technique for symmetric matrix diagonalization applies planar Givens rotations to each component pair (i, j) to cancel entries (a_{ij}, a_{ji}) until convergence. Alternatively, PCA can be carried out in a numerically more reliable manner by performing the singular value decomposition (SVD) [31, 66, 12] of the data matrix $\mathbf{X}_T = [\mathbf{x}(0), \mathbf{x}(1), \mathbf{x}(2), \ldots, \mathbf{x}(T-1)]$.

2.4.3 Limitations of Second-Order Statistics

Despite its simplicity, PCA fails to perform the separation in the general case. Firstly, the mixing matrix comprises n^2 unknowns, whereas uncorrelation can only impose $n(n-1)/2$ constraints. Secondly, due to the structure of $\mathbf{U}$ and Σ, the columns of $\hat{\mathbf{H}}$ are mutually orthogonal: $\hat{\mathbf{h}}_i^{\mathrm{T}}\hat{\mathbf{h}}_j = \sigma_i \mathbf{u}_i^{\mathrm{T}}\mathbf{u}_j\sigma_j = 0$. Thirdly, if two eigenvalues are equal, the corresponding eigenvectors can at most be estimated up to a rotation. From the above observations, it follows that the success of the method requires a mixing matrix composed of orthogonal columns with different norms. In general, the PCA source estimate is linked to the actual sources through an orthonormal matrix $\mathbf{Q}$:

$$\mathbf{z}(t) = \mathbf{y}_{\mathrm{PCA}}(t) = \mathbf{Q}\mathbf{s}(t) \tag{2.2}$$

a relation easily found by realizing that $\mathbf{R}_z = \mathbf{R}_s = \mathbf{I}_n$. Vector $\mathbf{z}(t)$ contains the so-called *prewhitened* observations. As a result, the mixing matrix can actually be decomposed as $\mathbf{H} = \mathbf{W}_{\mathrm{PCA}}^{-\mathrm{T}}\mathbf{Q}$, matrix $\mathbf{Q}$ being characterized by the $n(n-1)/2$ parameters (rotation angles) that PCA is unable to identify. To complete the separation, this matrix needs to be estimated in a second step involving an additional property such as time coherence or higher-order independence. Consequently, PCA is unable to perform the separation but simplifies the problem to less that half the number of unknowns. This process, also known as *prewhitening*, introduces a bound on the achievable separation performance [13].

In the biomedical problems under study, the structure of the mixing matrix estimated by PCA constrains the location of electrodes, as they must be placed to guarantee the orthogonality of the transfer vectors. The use of thoracic as well as abdominal electrodes aims at the orthogonality between the subspaces spanned by the FECG and MECG source directions [66]. Nevertheless, this requirement can be relaxed. It is proved in [66] that the direction of the most powerful source estimated by PCA is generally very accurate, even if the source directions are not orthogonal. Moreover, the stronger the first source, the better it will be suppressed in the estimation of the second source. Thoracic leads record strong MECG signals and will thus improve their cancellation from the FECG. The argument is similar to Widrow's approach, where reference leads must be clean from the desired signal in order to enhance the interference suppression and prevent the signal of interest being cancelled from the filter input. In the AA extraction problem, however, the transfer vector orthogonality is more difficult to achieve, due to the spatial proximity of the atrial and ventricular sources.

The application of the SVD-based PCA on the maternal recording of Fig. 2.1 yields the FECG contribution on the 4th abdominal lead shown in Fig. 2.3; the estimated sources are reported in [75]. Although the recovered FECG is clearer than in Wiener's approach, the fetal peak amplitudes are overestimated and the baseline wander is not suppressed. In Fig. 2.4, the AA signal estimated by PCA from the recording of Fig. 2.2 is a more faithful approximation of the AA observed in the T-Q segments of lead V1 than Wiener's. However, an important residual ventricular activity remains around the heartbeats, leading to an increase in low-frequency content and a consequent decrease in spectral concentration relative to the previous approach (Fig. 2.5).

A related SVD-based method for FECG extraction can be found in [2]. References [40, 19] (see also [80], Sect. 4) use a single lead by relying on the repetitive character of the interference, assumed to have a waveform morphology with little variability (the MECG, the VA). These two approaches require the prior detection of the main features (R peaks) of interfering waveform in the recording.

2.5 Second-Order Approaches Exploiting Spectral Diversity

As we have just seen, forcing uncorrelation at the separator output does not constitute a valid source separation criterion, since PCA is unable to identify the unknown unitary mixing matrix $\mathbf{Q}$ in the general case. To estimate this matrix and complete the separation, one may resort to the time coherence of the source signals, that is, to their correlation at different time lags. Let us define the correlation function of the ith source as $r_i(\tau) = \mathrm{E}\{s_i(t)s_i(t-\tau)\}$. Under the independence assumption, the source correlation matrix at any time lag τ is diagonal: $\mathbf{R}_\mathbf{s}(\tau) = \mathrm{E}\{\mathbf{s}(t)\mathbf{s}(t-\tau)^\mathrm{T}\} = \mathrm{diag}(r_1(\tau), r_2(\tau), \ldots, r_n(\tau))$. From Eq. (2.2), it follows that

$$\mathbf{R}_\mathbf{z}(\tau) = \mathrm{E}\left\{\mathbf{z}(\tau)\mathbf{z}(t-\tau)^\mathrm{T}\right\} = \mathbf{Q}\mathbf{R}_\mathbf{s}(\tau)\mathbf{Q}^\mathrm{T}, \forall\tau. \tag{2.3}$$

Hence, the application of matrix $\mathbf{Q}^\mathrm{T}$ on the whitened observations diagonalizes their correlation matrices at any lag. As a consequence, matrix $\mathbf{Q}$ can be uniquely identified from the EVD of $\mathbf{R}_\mathbf{z}(\tau_0)$ if the source correlation functions at lag τ_0 are all different. This is the basis of the *algorithm for multiple signal extraction (AMUSE)* by Tong et al. [64]. In practice, it is not easy to find a lag fulfilling this condition, and the method may fail due to the presence of close eigenvalues (eigenspectrum degeneracy).

To surmount this difficulty, the *second-order blind identification (SOBI)* method by Belouchrani et al. [5] performs the simultaneous approximate diagonalization of a matrix set $\{\mathbf{R}_\mathbf{z}(\tau_k)\}_{k=1}^{K}$:

$$\begin{aligned}\hat{\mathbf{Q}}_{\mathrm{SOBI}} &= \arg\min_{\mathbf{V}} \sum_{k=1}^{K} \mathrm{off}\left(\mathbf{V}^{\mathrm{T}}\mathbf{R}_\mathbf{z}(\tau_k)\mathbf{V}\right) \\ &= \arg\max_{\mathbf{V}} \sum_{k=1}^{K} \|\mathrm{diag}(\mathbf{V}^{\mathrm{T}}\mathbf{R}_\mathbf{z}(\tau_k)\mathbf{V})\|^2 \text{with } \mathbf{V}^{\mathrm{T}}\mathbf{V} = \mathbf{I}\end{aligned} \tag{2.4}$$

where, for an arbitrary $(n \times n)$ matrix $\mathbf{A}$, $\mathrm{off}(\mathbf{A}) = \sum_{1\le i\neq j\le n} |a_{ij}|^2$ and $\mathrm{diag}(\mathbf{A}) = [a_{11}, a_{22}, \ldots, a_{nn}]^{\mathrm{T}}$. According to equality (2.3), the last sum in expression (2.4) becomes $\sum_{i=1}^{n}\sum_{k=1}^{K} r_i^2(\tau_k)$ when matrix $\mathbf{V}$ equals $\mathbf{Q}$. As a result, SOBI is naturally suited to separating sources with long autocorrelation functions or, equivalently, narrowband spectra. This joint diagonalization can be seen as an extension of the Jacobi-based diagonalization of a single symmetric matrix, and can also be carried out iteratively by means of Givens rotations at an affordable computational cost. The condition for a successful source separation is now relaxed: for each source pair, it suffices to include a time lag for which their correlation function is different. Asymptotically, as the number of lags increases, the condition becomes that the sources have different frequency spectra.

The application of SOBI on the recording of Fig. 2.1 produces the estimated FECG on the 4th abdominal lead shown in Fig. 2.3. The 169 prime numbers between 1 and 1000 – spanning a total duration of 2 s – have been used as time lags. Specifically designed to exploit the time coherence of narrowband signals, SOBI neatly separates the baseline interference from the mixture. However, it clearly underestimates the FECG contribution to that lead, although the hardly perceptible fetal R-peaks appear at the right positions. Due to its ability to deal with narrowband sources, the method is more successful in extracting the AA from the recording of Fig. 2.2. Although the amplitude of the recovered AA seems overestimated at some points (Fig. 2.4), the residual low-frequency content is considerably reduced, resulting in a high spectral concentration (Fig. 2.5). In this example, the correlation matrices at 17 equally-spaced time lags spanning the maximum expected period of the AA (around 340 ms) were jointly diagonalized, as proposed by Castells et al. [20].

Recently, a modification of the AMUSE method taking into account the quasi-periodic structure of the ECG has been proposed by Sameni et al. [55] relying on the concept of *periodic component analysis (πCA)* by Saul and Allen [58]. The basic ingredient of this modification is the definition of a suitable piecewise linear function indicating the position (phase) of each sample in a beat relative to the R peak. A correlation matrix of the whitened data aligned according to the phase function is computed at a one-beat lag, and diagonalized instead of $\mathbf{R}_\mathbf{z}(\tau_0)$. In this manner, the procedure tends to emphasize the signal components presenting the quasi-periodicity described by the phase function. The method requires the prior R-peak detection of the desired or the interfering signal, but can be used to enhance or suppress ECG-correlated signals.

2.6 Higher-Order Approaches: Independent Component Analysis

2.6.1 Contrast Functions, Independence and Non-Gaussianity

As recalled in the preceding section, BSS can be performed if the sources present sufficient spectral diversity. Alternatively, time coherence may be ignored and the property of independence may be exploited up to orders higher than two, leading to the concept of *independent component analysis (ICA)*. The first thorough specific mathematical framework for BSS via ICA was established by Comon [24].[3] A key concept in this formulation was the definition of *contrast function*, a functional $\Psi(\mathbf{y})$ in the distribution of the separator output measuring an assumed property of the sources. By virtue of three essential features characterizing a contrast (invariance, domination and discrimination), its optimization is achieved if and only if the sources are separated up to acceptable indeterminacies. Hence, a contrast implicitly quantifies the degree of separation, and the sources are recovered by recovering their property at the separator output through contrast optimization. As mentioned in Sect. 2.3.3, under the independence assumption the sources can only be separated up to scale and permutation indeterminacies.

Many contrasts stem from information-theoretical concepts and present insightful elegant connections [15, 16]. The starting point is the *maximum likelihood (ML)* principle, which searches for the mixing matrix maximizing the probability of the observed data, given the source distribution. In [15, 16], the ML is shown to be equivalent to minimizing the Kullback-Leibler divergence (distance) between the distribution of the sources and that of the separator output. The popular Bell and Sejnowski's Infomax method [4] can also be interpreted, asymptotically, from the ML standpoint [14]. The main limitation of the ML approach is that it requires the knowledge (or assumption) of the source probability density function (pdf), although it is quite robust to the source distribution mismatch [15].

An alternative criterion is mutual independence, giving rise to the concept of ICA. The purpose of ICA is finding a linear transformation maximizing the statistical independence between the components of the resulting random vector. A random vector $\mathbf{s}$ is made up of mutually independent components if and only if its

[3] See also [23] for an earlier reference in French.

joint pdf can be decomposed as the product of its marginal pdfs: $p_{\mathbf{s}}(\mathbf{s}) = \prod_{i=1}^{n} p_{s_i}(s_i)$. The *mutual information (MI)* is defined as the Kullback-Leibler divergence between the separator output pdf and the product of its marginal pdfs:

$$\Psi(\mathbf{y}) = \int_{Y} p_{\mathbf{y}}(\mathbf{u}) \log \frac{p_{\mathbf{y}}(\mathbf{u})}{\prod_{i=1}^{n} p_{y_i}(u_i)} d\mathbf{u}. \tag{2.5}$$

The MI can be seen as a measure of the distance to independence, as it is always positive and becomes zero if and only if the components of $\mathbf{y}$ are independent. Under the source independence assumption, it is not surprising that the minimization of MI at the separator output constitutes a valid contrast for BSS [24]. Consequently, BSS can be achieved by restoring the property of statistical independence at the separator output, and ICA arises as the natural tool for BSS under the source independence assumption. This criterion is found to be related to the ML principle, up to a mismatch on the source marginal pdfs [15], but the former has the advantage of sparing the knowledge of the source distributions. After prewhitening (PCA), this contrast is to be minimized subject to unitary transformations only, and reduces to minimizing the sum of *marginal entropies (MEs)* of the separator output components. Among the distributions with unbounded support, the Gaussian distribution maximizes Shannon's entropy. As a result, the minimum ME criterion is equivalent to maximizing non-Gaussianity at the separator output. This result is consistent with intuition and the Central Limit Theorem: as mixing random variables tends to increase Gaussianity, one should proceed in the opposite direction, decreasing Gaussianity, to achieve their separation. In short, the MI criterion (2.5) is shown to be linked to entropy and negentropy, both concepts in turn related with non-Gaussianity [24, 15, 16].

Despite their optimality, information-theoretical contrasts such as MI or ME involve pdfs, difficult to deal with in practice. To improve the numerical tractability, pdfs can be approximated by their truncated Edgeworth or Gram-Charlier expansions around a Gaussian distribution. These approximations lead to practical algorithms involving *higher-order statistics (HOS)*, such as the *cumulants*, easier to compute and to deal with. A variety of these cumulant-based approximations are addressed in [16]. The minimization of ME is shown to be equivalent, under unitary transformations, to the maximization of the square marginal cumulants of the separator output [24, 16], a criterion that, despite arising from a truncated pdf approximation, is also a contrast:

$$\Psi_{\mathrm{MI}}^{\mathrm{ort}}(\mathbf{y}) = \sum_{i=1}^{n} \kappa_{y_i}^2 \tag{2.6}$$

where, for a zero-mean variable y, the fourth-order cumulant (*kurtosis*) is defined as

$$\kappa_y = \mathrm{E}\{\,|y|^4\,\} - 2\mathrm{E}^2\{\,|y|^2\,\} - \left|\mathrm{E}\{y^2\}\right|^2 \tag{2.7}$$

and simplifies to $\kappa_y = \mathrm{E}\{y^4\} - 3\mathrm{E}^2\{y^2\}$ in the real case. The higher-order marginal cumulants of a Gaussian variable being null, this criterion is naturally connected to the maximization of non-Gaussianity of the separator output components. The *contrast maximization (CoM2)* method of [24] maximizes Eq. (2.6) iteratively by applying a planar Givens rotation to every signal pair until convergence, as in the Jacobi technique for matrix diagonalization [31], or in the joint approximate diagonalization technique of [17, 5]. In the real-valued case, the optimal rotation angle maximizing (2.6) is obtained by finding the roots of a 4th-degree polynomial. Although it becomes more involved, the method is also valid for complex-valued mixtures. Algebraically, this procedure can be considered as an extension of PCA in that it aims at diagonalizing the 4th-order cumulant tensor of the observations.

A contrast similar to (2.6) can be reached from an algebraic approach whereby the cumulants are arranged in multi-way arrays (tensors) and considered as linear operators acting on matrices. It is shown that matrix $\mathbf{Q}$ can be estimated from the eigendecomposition, or diagonalization, of any such cumulant matrices. To improve the robustness to eigenspectrum degeneration, a set of cumulant matrices can be exploited simultaneously, much in the spirit of SOBI [5] (see also Sect. 2.5) giving rise to the *joint approximate diagonalization of eigenmatrices (JADE)* method, characterized by an objective function very similar to (2.4). JADE presents the same asymptotic performance as CoM2, but requires the computation of the $O(n^4)$ entries of the fourth-order cumulant tensor, its complexity thus becoming prohibitive as the number of sources increases. Nonetheless, JADE is a very popular technique often used as a performance reference in the comparison of BSS-based methods.

In the maternal recording of Fig. 2.1, the CoM2 algorithm estimates a clear FECG contribution to the 4th abdominal lead, as shown in Fig. 2.3; the separated sources can be found in [75]. The ICA approach outperforms the three previous methods (Wiener, PCA and SOBI) in this example, since the FECG constitutes a strongly non-Gaussian signal which is easily estimated with the aid of HOS. In the AA extraction problem, the situation is different. The AA has a near-Gaussian probability distribution, which hampers its HOS-based extraction from Gaussian noise and interference. As a result, the AA estimated by CoM2 appears rather noisy (Fig. 2.4), with a significant low-frequency interference (Fig. 2.5).

2.6.2 Source Separation or Source Extraction?

The above family of methods are designed to estimate all sources jointly or simultaneously. An alternative approach is to extract one source after another, a process known as *sequential extraction* or *deflation* [29]. Regarding the use of SOS, the extension of the SOBI approach (Sect. 2.5) to source extraction is carried out in [41]; proof of these results can be found in [69]. The proposed alternating least squares algorithm, however, is not guaranteed to converge to the desired extracting solutions.

As far as HOS are concerned, it was proved by Delfosse and Loubaton [29] that the maximization of criterion

$$|\Psi_{\mathrm{KM}}(y)|, \quad \text{with} \quad \Psi_{\mathrm{KM}}(y) = \frac{\kappa_y}{\sigma_y^4} \tag{2.8}$$

is a valid contrast for the extraction of any source with non-zero kurtosis from model (2.1) in the real-valued case. The extractor output is given by $y(t) = \mathbf{q}^{\mathrm{T}}\mathbf{z}(t)$ and the unitary extracting vector $\mathbf{q}$ by the corresponding column of matrix $\mathbf{Q}$ linking the sources with the whitened observations in Eq. (2.2). Symbol σ_y^4 denotes the square variance of the extractor output. A similar result had been obtained a few years earlier by Donoho [30] and Shalvi and Weinstein [60] in the context of blind deconvolution and blind equalization of digital communication channels, a related but somewhat different problem. In [29], matrix $\mathbf{Q}$ is suitably parameterized as a function of angular parameters, and function (2.8) iteratively maximized with respect to these angles. This parameterization allows the reduction of the dimensionality of the observation space after each extraction, so that the size of the remaining orthogonal matrix decreases as the sources are estimated. The orthogonality between successive extracting vectors prevents the same source from being extracted twice.

The same contrast is proposed by Tugnait [65] for the convolutive mixture scenario, but without parameterization of matrix $\mathbf{Q}$. After convergence of the search algorithm, the contribution of the estimated source to the observations is computed via the minimum mean square error (MMSE) solution to the linear regression problem $\mathbf{x} = \hat{\mathbf{h}}\hat{s}$, given by

$$\hat{\mathbf{h}} = \arg\min_{\mathbf{h}} \mathrm{E}\left\{\|\mathbf{x} - \mathbf{h}\hat{s}\|^2\right\} = \mathrm{E}\{\mathbf{x}\hat{s}\}/\mathrm{E}\left\{\hat{s}^2\right\}. \tag{2.9}$$

The observations are then 'deflated' as $\mathbf{x} \leftarrow \mathbf{x} - \hat{\mathbf{h}}\hat{s}$ before re-initializing the algorithm in the search for the next source. This regression-based deflation is an alternative method to avoid extracting the same source more than once.

In its original definition, the popular FastICA algorithm [36–38] aimed at the maximization of contrast (2.8). Some simplifications finally lead to the following update rule for extracting vector $\mathbf{q}$:

$$\mathbf{q}' = \mathbf{q} - \frac{1}{3}\mathrm{E}\{(\mathbf{z}^{\mathrm{T}}\mathbf{q})^3\mathbf{z}\}. \tag{2.10}$$

Vector $\mathbf{q}$' is then projected on the subspace orthogonal to the previously estimated extracting vectors, and normalized to keep the denominator of (2.8) constant. This approach to sequential extraction is called *deflationary orthogonalization* [36, 38]. Equation (2.10) represents the so-called FastICA method with cubic non-linearity and is shown to have, as the sample size tends to infinity, global cubic convergence. Nevertheless, update rule (2.10) turns out to be a gradient-descent algorithm with constant step size [77]. To improve robustness to outliers, the method can incorporate other non-linear functions (hyperbolic tangent, sigmoid, etc.) at the expense of slowing down convergence. Under analogous simplifications than its real-valued counterpart, the extension of FastICA with cubic non-linearity to the complex-valued scenario [6] neglects the non-circular part in the general definition of kurtosis (2.7) and is thus restricted to circular source distributions.

2.6.3 *Optimal Step-Size Iterative Search*

Despite the simplifying assumptions (e.g,. prewhitening, real-valued sources and mixtures, circular complex sources etc.) made in previous works, criterion (2.8) is actually quite general. Indeed, it is a valid contrast for the extraction of a non-zero kurtosis source from mixture (2.1) whatever the type (real or complex-valued) of sources and mixtures, and regardless of whether prewhitening has been carried out. More interestingly, this contrast can be maximized by an effective, computationally efficient search algorithm. Assuming an extractor output $y = \mathbf{w}^{\mathrm{T}}\mathbf{x}$, a quite natural update rule for the extracting vector $\mathbf{w}$ along an appropriate search direction $\mathbf{g}$ (e.g., the gradient) is given by $\mathbf{w}' = \mathbf{w} + \mu\mathbf{g}$, where the real-valued μ is the step size or adaption coefficient. In conventional search algorithms, μ is set to a constant or possibly time-varying value trying to balance a difficult trade-off between convergence speed and final accuracy. Rather, we are interested in the value of μ that *globally* maximizes the normalized kurtosis contrast in the search direction:

$$\mu_{\mathrm{opt}} = \arg\max_{\mu} |\Psi_{\mathrm{KM}}(y + \mu g)| \tag{2.11}$$

where $g = \mathbf{g}^{\mathrm{T}}\mathbf{x}$. It was first remarked by Comon [25, 26] that $\partial\Psi_{\mathrm{KM}}(y + \mu g)/\partial\mu$ is a rational function in μ with a fourth-degree polynomial as numerator. Hence, μ_{opt} can be computed algebraically by finding the roots of this polynomial in the step size. Its coefficients, obtained in [70, 77], are functions of the observed data fourth-order statistics and the extracting vector and search direction at the current iteration. The optimal step size is the root maximizing expression (2.11). The resulting method is referred to as RobustICA.

In the numerical results of [70, 77], RobustICA demonstrates a very fast convergence measured in terms of source extraction quality against number of operations. The optimal-step size update rule provides the method with some robustness to saddle points and spurious local extrema in the contrast function, which tend to appear when short data blocks are processed [63]. The generality of contrast (2.8) guarantees that RobustICA is able to separate real and complex (possibly non-circular) sources without any modification. In particular, the method can easily process signals in the frequency domain, an interesting possibility that can be exploited in AA extraction [72] (see also Sect. 2.7.1). In addition, the method does not require prewhitening, thus avoiding the associated performance limitations. Deflation must then be carried out through linear regression, as in Eq. (2.9). Prewhitening can also be used, in conjunction with regression, or with deflationary orthogonalization as in FastICA.

2.7 Exploiting Prior Information

The above BSS techniques make little assumptions about the sources or the mixture. As recalled in Sect. 2.3.2, the strength of the blind approach – its robustness to modeling errors – lies precisely in its freedom from model assumptions. Despite the success witnessed by independence-exploiting BSS techniques over the last decades, their performance may be unsatisfactory in certain applications. For instance, the AA signal is often near-Gaussian, so that its separation from Gaussian noise and interference is compromised when relying on HOS only, as illustrated by the results of the CoM2 algorithm in Sect. 2.6.1 (Figs. 2.4, 2.5). As noted in [33, 34], statistical independence alone is sometimes unable to produce physiologically meaningful results in biomedical signal processing applications. In these conditions, source extraction performance can be improved by taking into account additional assumptions about the signals of interest or the mixing structure other than independence. Furthermore, the exploitation of prior knowledge may enable the resulting algorithms to focus on the extraction of the signal(s) of interest only, thus avoiding the unnecessary complexity of a full separation and the permutation ambiguity of conventional ICA.

Different kinds of prior knowledge have recently been considered by researchers in the field. These include partial information about the source statistics [18, 20, 21, 46–49], the availability of reference signals correlated with the source of interest [3, 39, 42, 43, 57], or constraints on the structure of the transfer vectors or spatial topographies [10, 11]. A Bayesian formulation is theoretically optimal but usually impractical, as it involves the specification of probability distributions for the parameters associated with the prior information. Determining such distributions may be difficult or simply unfeasible in certain scenarios. Moreover, the convergence of Bayesian model estimation methods (e.g., the expectation-maximization algorithm) is often very slow. This has motivated the search for suboptimal but more practical alternatives for exploiting prior knowledge in BSS. Some of these more recent approaches are briefly surveyed next.

2.7.1 Source Statistical Characterization

In many biomedical applications, some statistical features of the source(s) of interest may be known in advance. The AA waveform in atrial flutter episodes typically shows a sawtooth shape that can be characterized as a sub-Gaussian distribution, and becomes near-Gaussian in more disorganized states of AF observed as the disease evolves. In the separation of the FECG from maternal skin recordings, the sources of interest, the fetal heartbeat signals, are usually impulsive and thus present heavy tails in their pdfs; they can be considered as super-Gaussian random variables. This prior information can be capitalized on in several fashions to enhance signal extraction performance.

2.7.1.1 Combining Non-Gaussianity and Spectral Features

A hybrid approach is proposed by Castells et al. [20] to improve the performance of conventional ICA in AA extraction. The idea is to exploit a well-known feature of the AA signal: its time coherence, reflected on a quasi-periodic autocorrelation function and a narrowband frequency spectrum (Fig. 2.5). HOS-based ICA (Sect. 2.6.1) is first applied on the full ECG recording of an AF episode in order to estimate the strongly non-Gaussian VA components. The other sources estimated by ICA usually contain mixtures of near-Gaussian AA and noise, with low kurtosis values. The SOBI technique [5] (see also Sect. 2.5) is then used to extract the AA from the remaining mixtures by exploiting its time structure. The AA estimation results obtained by this hybrid method on the recording of Fig. 2.2 are summarized in the 5th plot of Figs. 2.4, 2.5. After applying the CoM2 algorithm as initial ICA stage, four near-Gaussian sources (with kurtosis below 1.5) are passed on to the SOBI step, which extracts an AA with higher spectral concentration than using CoM2 alone (4th plot).

A similar idea is developed in [18], where the initial stage consists of an ML estimate of the unknown (2×2) rotation after prewhitening in the two-signal case. The ventricular and atrial signal distributions are approximated as a Laplacian and a uniform pdf, respectively, and the resulting log-likelihood function is maximized by a gradient-based search. Although the extension of this ML solution to a full 12-lead ECG recording is unclear, good performance seems to be achieved even if the SOBI step is omitted [21].

2.7.1.2 Extraction of Sources with Known Kurtosis Sign

The above implementations perform a full source separation. When only a few sources are of interest, separating the whole mixture incurs an unnecessary computational cost and, in the case of sequential extraction, an increased source estimation inaccuracy due to error accumulation through successive deflation stages. A more judicious alternative is extracting the desired type of sources exclusively.

Functional $\Psi_p(\mathbf{y}) = \sum_{i=1}^{n} \varepsilon_i \kappa_{y_i}$, where $\varepsilon_i = \text{sign}(\kappa_{s_i})$ and $p \leq n$ denotes the number of sources with positive kurtosis, is a valid orthogonal contrast for BSS under the source independence assumption [79]. Moreover, the criterion is able to arrange the estimated sources in two groups according to their kurtosis sign, thus partially resolving the permutation ambiguity of ICA. This contrast is linked to the ML criterion and can easily be optimized through a Jacobi-like procedure involving cost-efficient closed-form solutions [73, 76, 78]. Although originally designed for joint separation, the contrast can easily be adapted to perform sequential separation or single-source extraction by keeping fixed one of the signals and sweeping over the rest in the pairwise algorithm. The criterion has been applied to AA extraction [46, 47] by assuming that the kurtosis of the desired signal is negative. The spectral concentration of the resulting signal is tested within the Jacobi sweep to ensure that the atrial signal estimation is improved at each pairwise iteration.

The deflation-based RobustICA method described in Sect. 2.6.3 aims at maximizing the absolute normalized kurtosis, and is thus also able to extract sources with positive or negative kurtosis (i.e., super-Gaussian or sub-Gaussian). RobustICA can easily be modified to target a source with specific kurtosis sign ϵ. After computing the roots of the step-size polynomial, one simply needs to replace (2.11) by

$$\mu_{\text{opt}} = \arg\max_{\mu} \varepsilon \Psi_{\text{KM}}(y + \mu g) \tag{2.12}$$

as best root selection criterion. If no source exists with the required kurtosis sign, the algorithm may converge to a non-extracting local extrema, but will tend to produce components with maximal or minimal kurtosis from the remaining signal subspace when $\varepsilon = 1$ or $\varepsilon = -1$, respectively. The algorithm can also be run by combining global line maximizations (2.12) and (2.11) for sources with known and unknown kurtosis sign, respectively, in any desired order. The freely available implementation of the RobustICA algorithm[4] incorporates this feature.

Using $\varepsilon = 1$ on the recording of Fig. 2.1, RobustICA finds two FECG signals among the first five extracted sources, yielding the FECG contributions to the 4th abdominal lead shown in the 5th plot of Fig. 2.3. The estimated signal is identical to CoM2's, but RobustICA only required half the iterations while sparing the separation of the whole mixture. Using $\varepsilon = -1$ to estimate a six-dimensional minimal-kurtosis source subspace from the recording of Fig. 2.2 and followed by SOBI, RobustICA yields the AA estimate shown in the 6th plot of Figs. 2.4, 2.5. Although it achieves a slightly lower spectral concentration than SOBI, the recovered waveform seems a more accurate fit to the actual AA time course observed in lead V1 [71].

A signal is referred to as *sparse* if it takes non-zero amplitude values with low probability. The Fourier transform can be seen as a sparsifying transformation for narrowband signals, as their frequency support is bounded. Moreover, it is not difficult to prove that the more sparse a signal is, the more super-Gaussian its amplitude probability distribution becomes. Hence, the AA signal, although near-Gaussian in its time-domain representation, is expected to become highly non-Gaussian in the frequency domain. Capitalizing on this observation, the RobustICA algorithm is run on the FFT of the ECG data in Fig. 2.2, using $\varepsilon = 1$, and the estimated independent component is then transformed back to the time domain through the inverse FFT. The 3rd extracted source corresponds to a clear AA signal, yielding the contribution shown in the 7th plot ('RobustICA-f') of Figs. 2.4, 2.5. The result is very similar to that of SOBI and RobustICA-SOBI, at a small fraction of the computational cost (just 16 iterations per source as opposed to 179 iterations per source for RobustICA-SOBI in this particular example) [72]. Further improvements could be achieved by taking into account the frequency band on which AA typically occurs.

[4] http://www.i3s.unice.fr/~zarzoso/robustica.html

Related techniques exploiting the spectral characteristics of the AA signal to perform its extraction in the frequency domain are reported in [48, 50].

2.7.2 *Reference Signal*

Current fetal monitoring devices include Doppler ultrasound measurements of the fetal heart rate. Clearly, the Doppler signal is correlated with the FECG and can then be used as a reference to refine the fetal cardiac signal extraction from maternal potential recordings [57]. Likewise, the T-Q segments in an AF recording present mostly AA and noise, but are free of ventricular interference. It seems thus sensible to employ such segments as reference signals to aid in AA extraction in heartbeat intervals [10, 11]. External stimuli in event-related experiments also make good references. In general, any signal sufficiently correlated with the source of interest can be considered and exploited as a reference.

The use of reference signals for BSS is somewhat reminiscent of Wiener filtering and the related Widrow's noise cancellation approach [68, 80]. Indeed, in the absence of noise, the Wiener spatial filter

$$\mathbf{w}_0 = \arg\min_{\mathbf{w}} \mathrm{E}\left\{(y-r)^2\right\} = \mathbf{R}_{\mathbf{x}}^{-1}\mathrm{E}\{r\mathbf{x}\}, \text{ with } y = \mathbf{w}^{\mathrm{T}}\mathbf{x} \qquad (2.13)$$

performs exact source extraction, i.e., $y = \mathbf{w}_0^{\mathrm{T}}\mathbf{x} \equiv s_i$, when the reference r is correlated with the source of interest s_i but uncorrelated with the other sources, even without prewhitening; cf. [3]. Nevertheless, the Wiener extractor is bound to fail in the presence of spurious correlations of the reference signal with other sources, which often occurs in practice [43]. To overcome this drawback, Barros et al. [3] propose to initialize the ICA iterative search with the Wiener solution (2.13) and keep the ICA update $\mathbf{w}_k$ close enough to the initialization by replacing it, if necessary, with $\mathbf{w}_0$ plus a small random term so that $\|\mathbf{w}_k - \mathbf{w}_0\| < \zeta$, for a positive constant ζ. Interestingly, a similar ICA-aided algorithm is later put forward to improve the Wiener (MMSE) receiver in CDMA telecommunication systems, a very different application [53].

A reference signal is generated by centering 100-ms positive amplitude square waves around the R peaks detected on the 4th plot of Fig. 2.3, corresponding to the FECG signal estimated by the fully-blind CoM2 method on the 4th abdominal lead of Fig. 2.1. Its amplitude is set to zero elsewhere and its power is normalized to unity. The Wiener-ICA technique described above is applied on the whitened data (the PCA sources) using the FastICA update (2.10) initialized with (2.13). In a few iterations, the method recovers an FECG signal virtually identical to CoM2's, as shown in Fig. 2.3.

The reference signal of the previous example is unrealistic, as it has been derived from a clean FECG estimate. In practice, the Doppler ultrasound signal can employed as reference for FECG extraction [57]. A higher-order correlation between the extractor output y and the reference r, subject to a constraint on the extractor vector norm, is put forward as an objective function:

$$L(\mathbf{w}) = \frac{1}{c}\mathrm{E}\{y^c r^c\} - \frac{\lambda}{2}\left(\|\mathbf{w}\|^2 - 1\right) \tag{2.14}$$

where c is a positive integer. This problem can be solved in closed form: at order 1, it reduces to Wiener's solution (2.13); at order 2, the optimal spatial filter is the dominant eigenvector of the reference-weighted covariance matrix $\mathrm{E}\{\mathbf{x}\mathbf{x}^{\mathrm{T}} r^2\}$. At orders greater than two, however, no algebraic solution exists. The maximization of this Lagrangian can then be achieved by an iterative update of the form

$$\begin{aligned} \mathbf{w}_{k+1} &= \frac{\mathrm{E}\{y_k^{c-1} r^c \mathbf{x}\}}{\mathrm{E}\{y_k^c r^c\}} \\ y_{k+1} &= \mathbf{w}_{k+1}^{\mathrm{T}}\mathbf{x} \end{aligned} \tag{2.15}$$

This method is referred to as *BSS with reference (BSSR)*. Note that the expectation in the denominator of $\mathbf{w}_{k+1}$ can be spared if the extractor vector is divided by its norm after its update. Synchronized averaging of the signal estimated by this method can reveal the fetal P and T waves in addition to the R wave, thus providing additional diagnostic information about the fetal heart. However, the method is applied to the recordings after MECG suppression, performed by a least-squares procedure that assumes a maternal heartbeat signal subspace of dimension two only. This seems to contradict previous results in which the maternal subspace is usually found to be three-dimensional [12, 27, 28, 66]. The robustness of this approach against the quality of the reference signal is analyzed in [44]. The form of the above cost function lends itself to the iterative search technique used in RobustICA (Sect. 2.6.3), but with a step-size polynomial of degree (c–1).

Using the same reference signal as in the Wiener-ICA method, the closed-form solution to the 2nd-order BSSR criterion (2.14) applied on the whitened data yields the same FECG contribution to the 4th abdominal lead of Fig. 2.1 as in the previous method, as seen in the last plot of Fig. 2.3. Update rule (2.15) with an arbitrary extracting vector initialization converges in a few iterations to the same solution. From the AF recording of Fig. 2.2, a normalized reference signal is generated by setting its amplitude to a constant positive value around the manually selected T-Q segments of the V1 lead, and zero elsewhere. BSSR reconstructs the AA signal shown in the last plots of Figs. 2.4, 2.5. Although its time course does not seem very accurate, the dominant spectral shape is successfully recovered. In these examples, the reported results were the best among orders 1–5 of criterion (2.14). Although it demands the manual detection or segmentation of significant morphological features (R-wave, T-Q periods, etc.), the BSSR method has the potential of providing algebraically a good initial approximation of the desired signal. Depending of the quality of the reference, this initial estimate may be refined by later processing.

The prior information can be incorporated explicitly within the ICA update by means of appropriate constraints on the contrast function. This idea gives rise to a general framework called *constrained ICA (cICA)* [42, 43]. When prior knowledge is expressed in terms of reference signals, the approach is referred to as *ICA with reference (ICA-R)* and can be mathematically cast into the constrained optimization problem: maximize $\Psi(y)$ subject to $\varepsilon(y) \leq \xi$. In this expression, $\Psi(y)$ is a valid contrast (such as negentropy or absolute normalized kurtosis) for source extraction in model (2.1), $\varepsilon(y)$ represents a measure of similarity or closeness between the output and the reference (e.g., mean square error or correlation), and ξ is a suitable closeness threshold. This optimization problem is solved by a Newton-like algorithm on the associated Lagrangian function. The approach can be extended to several reference and output signals, and is successfully applied to the extraction of brain fMRI images [42, 43] and artifact rejection in electromagnetic brain signal recordings [39]. The algorithm is somewhat cumbersome in that it requires updating not only of the separating filter coefficients but also of other parameters included in the Lagrangian function; in turn, these updates are characterized by adaption coefficients that need to be appropriately selected. Although the method is locally stable, its global convergence depends on the closeness threshold, which must also be chosen with care: if too large, several possible solutions (local extrema) may appear and the source permutation problem may persist; if too small, the constraint may be unreachable and an unpredictable result obtained. In practice, ξ has to be modified adaptively. Simpler algorithms can be designed by introducing the reference signal as a constraint into the pairwise Jacobi-like iteration of the CoM2 method [24] (Sect. 2.6.1) and related contrasts.

Contrast functions based on reference signals have been developed in [1]. However, such references are defined as arbitrary unitary transformations acting on the original sources, and so they constitute a somewhat different concept than in the works reported in the above paragraphs.

2.7.3 Spatial Reference

The prior knowledge about our signal extraction problem can sometimes be captured by the structure or topography of the transfer vector associated with the source of interest, rather than by the time course of a reference signal. For instance, it is likely than the spatial pattern of the AA source during the T-Q segments be highly correlated with (if not the same as) that during a ventricular beat; that source is also expected to contribute to lead V1 more strongly than to other standard leads, due to the close proximity of that lead to the atria. For similar reasons, fetal cardiac signals are expected to contribute with higher variance to the abdominal electrodes. In the separation process, this information can be expressed mathematically as specific constraints on the direction of the corresponding mixing-matrix columns. One then speaks of *spatial references* or *reference topographies* [33, 34].

The degree of certainty on a given spatial reference can be reflected on the amount of deviation from the constraint allowed to the estimated transfer vector. Accordingly, three types of constraints are distinguished by Hesse and James [33, 34]: the estimated source direction is enforced to be equal to the spatial reference when *hard* constraints are employed; *soft* constraints allow for some discrepancy bounded by a closeness threshold; in *weak* constraints, the ICA extractor is simply initialized with the constraint, but otherwise left to run freely, much like in the Wiener-based initialization of Barros et al. [3] (see Sect. 2.7.2). The sources associated with the constrained transfer vectors are called *spatial components*, and are assumed to be independent of the other sources – the independent components – but not necessarily independent among themselves. A modification of the FastICA algorithm incorporating spatial constraints is developed in [33], and yields a satisfactory artifact extraction in electromagnetic brain signals [34]. More recently, some methods combining this idea with the narrowband character of the AA signal have been proposed for AA extraction in AF episodes [10, 11].

However, some theoretical aspects of this approach require further investigation. For instance, fixing the mixing matrix columns is likely to destroy the attractive equivariance property of BSS algorithms, whereby the source estimation performance becomes independent of the mixing matrix structure [13]. Whether the performance improvement brought about by the use of spatial references makes up for the loss of equivariance is yet unknown. Also, it is unclear whether appropriate sets of spatial and independent components can always be uniquely determined regardless of the dimension and relative orientation of their respective subspaces.

2.8 Conclusions and Outlook

Signal extraction and artifact rejection in surface ECG recordings can be modeled as a BSS problem of instantaneous linear mixtures. The pertinence of this approach is supported by considerations regarding the generation and propagation of electrophysiological signals across the body. Compared to alternative approaches such as multi-reference filtering, average beat subtraction or spatio-temporal cancellation, BSS does not assume any particular pattern for the contribution of the sources onto the electrodes, nor a specific morphology or repetitive pattern for the interfering waveforms. In problems such as FECG extraction from maternal skin electrodes and AA extraction in surface ECG recordings of AF, the independence between the sources of interest and the artifacts is a realistic assumption. The exploitation of independence at second order (PCA) requires a careful electrode placement to perform the separation unless additional properties are relied upon such as time coherence, non-stationarity or cyclo-stationarity. The concept of contrast function defines the conditions to be fulfilled for a source property to constitute a valid separation criterion. By imposing independence at orders higher than two, ICA is linked to contrasts capable of separating or extracting any kind of independent non-Gaussian sources. Although *blindness* is an attractive feature in the uncertainty of clinical environments, prior knowledge in the form of reference signals and spatial patterns can also be incorporated into the separation criteria to improve source separation performance.

Although the BSS approach has proven its potential in a variety of biomedical signal processing problems beyond ECG analysis, further research is necessary to answer some important open questions. A fundamental issue concerns the relationship between the signals estimated by source separation techniques and the actual internal sources of electrophysiological activity. In turn, shedding light on this link should help discern the clinical and physiological knowledge to be gained from the analysis of the estimated signals. In FECG extraction, fetal source typically contributes more strongly to abdominal electrodes, whereas in AA extraction, the atrial source is expected to appear predominantly in the V1 lead. The mathematical formulation of these fuzzy constraints and their incorporation into BSS criteria are other interesting problems to be tackled. A related issue is how to best exploit and combine various kinds of available prior information to improve separation performance while maintaining the robustness of the blind approach. In particular, the optimal use of the variety of information provided by simultaneous recordings in different modalities (e.g., ECG in combination with Doppler ultrasound) constitutes a major research challenge in the field of biomedical signal extraction.

References

1. Adib A, Moreau E, Aboutajdine, D (2004) Source separation contrasts using a reference signal. IEEE Signal Processing Letters 11(3):312–315
2. Al-Zaben A, Al-Smadi A (2006) Extraction of foetal ECG by combination of singular value decomposition and neuro-fuzzy inference system. Physics in Medicine and Biology 51(1):137–143
3. Barros AK, Vigário R, Jousmäki V et al. (2000) Extraction of event-related signals from multichannel bioelectrical measurements. IEEE Transactions on Biomedical Engineering 47(5):583–588
4. Bell AJ, Sejnowski TJ (1995) An information-maximization approach to blind separation and blind deconvolution. Neural Computation 7(6):1129–1159
5. Belouchrani A, Abed-Meraim K, Cardoso JF et al. (1997) A blind source separation technique using second-order statistics. IEEE Transactions on Signal Processing 45(2):434–444
6. Bingham E, Hyvärinen A (2000) A fast fixed-point algorithm for independent component analysis of complex valued signals. International Journal of Neural Systems 10(1):1–8
7. Bollmann A, Lombardi F (2006) Electrocardiology of AF. IEEE Engineering in Medicine and Biology Magazine 25(6):15–23
8. Bollmann A, Kanuru NK, McTeague KK et al. (1998) Frequency analysis of human AF using the surface electrocardiogram and its response to Ibutilide. American Journal of Cardiology 81(12):1439–1445
9. Bonizzi P, Meste O, Zarzoso V (2007) Atrio-ventricular junctionbehaviour during AF. In: Proc. 34th IEEE Annual Conference on Computers in Cardiology, Durham, North Carolina, USA, 561–564
10. Bonizzi P, Phlypo R, Zarzoso V et al. (2008a) The exploitation of spatial topographies for atrial signal extraction in AF ECGs. In: Proc. EMBC-2008, 30thAnnual International Conference of the IEEE Engineering in Medicine and Biology Society, Vancouver, Canada, 1867–1870
11. Bonizzi P, Phlypo R, Zarzoso V et al. (2008b) Atrial signal extraction in AF ECGs exploiting spatial constraints. In: Proc. EUSIPCO-2008, 16th European Signal Processing Conference, Lausanne, Switzerland
12. Callaerts D, De Moor B, Vandewalle J et al. (1990) Comparison of SVD methods to extract the foetal electrocardiogram from cutaneous electrode signals. Medical & Biological Engineering & Computing 28:217–224

13. Cardoso JF (1994) On the performance of orthogonal source separation algorithms. In: Proc. EUSIPCO-94, VII European Signal Processing Conference, Edinburgh, UK, 776–779
14. Cardoso JF (1997) Infomax and maximum likelihood in blind source separation. IEEE Signal Processing Letters 4(4):112–114
15. Cardoso JF (1998) Blind signal separation: statistical principles. Proceedings of the IEEE 86(10):2009–2025
16. Cardoso JF (1999) Higher-order contrasts for independent component analysis. Neural Computation 11:157–192
17. Cardoso JF, Souloumiac A (1993) Blind beamforming for non-Gaussian signals. IEE Proceedings-F 140(6):362–370
18. Castells F, Igual J, Rieta JJ et al. (2003) AF analysis based on ICA including statistical and temporal source information. In: Proc. ICASSP-2003, 28th IEEE International Conference on Acoustics, Speech and Signal Processing. Volume V., Hong Kong, China, 93–96
19. Castells F, Mora C, Rieta JJ et al. (2005a) Estimation of atrial fibrillatory wave from single-lead AF electrocardiograms using principal component analysis concepts. Medical & Biological Engineering & Computing 43(5):557–560
20. Castells F, Rieta JJ, Millet et al. (2005b) Spatiotemporal blind source separation approach to AA estimation in atrial tachyarrhythmias. IEEE Transactions on Biomedical Engineering 52(2):258–267
21. Castells F, Igual J, Millet JJ et al. (2005c) AA extraction from AF episodes based on maximum likelihood source separation. Signal Processing 85(3):523–535
22. Castells F, Laguna P, Sörnmo L et al. (2007) Principal component analysis in ECG signal processing. EURASIP Journal on Advances in Signal Processing, 21 pages
23. Comon P (1990) Analyse en composantes indpendantes et identification aveugle. Traitement du signal (Numéro spécial non linéaire et non gaussien) 7(3):435–450
24. Comon P (1994) Independent component analysis, a new concept? Signal Processing (Special Issue on Higher-Order Statistics) 36(3):287–314
25. Comon P (2002) Independent component analysis, contrasts, and convolutive mixtures. In: Proc. 2nd IMA Intl. Conference on Mathematics in Communications, Lancaster, UK, 10–17
26. Comon P (2004) Contrasts, independent component analysis, and blind deconvolution. International Journal of Adaptive Control and Signal Processing (Special Issue on Blind Signal Separation) 18(3):225–243
27. De Lathauwer L, Callaerts D, De Moor B et al. (1995) Fetal electrocardiogram extraction by source subspace separation. In: Proc. IEEE/ATHOS Signal Processing Conference on Higher-Order Statistics, Girona, Spain, 134–138
28. De Lathauwer L, De Moor B, Vandewalle J (2000) Fetal electrocardiogram extraction by blind source subspace separation. IEEE Transactions on Biomedical Engineering (Special Topic Section on Advances in Statistical Signal Processing for Biomedicine) 47(5):567–572
29. Delfosse N, Loubaton P (1995) Adaptive blind separation of independent sources: a deflation approach. Signal Processing 45(1):59–83
30. Donoho D (1980) On minimum entropy deconvolution. In: Proc. 2nd Applied Time Series Analysis Symposium, Tulsa, OK, USA, 565–608
31. Golub GH, Van Loan CF (1996) Matrix Computations. 3rd edn. The John Hopkins University Press, Baltimore, MD, USA (1996)
32. Hérault J, Jutten C, Ans B (1985) Détection de grandeurs primitives dans un message composite par une architecture neuromimétique en apprentissage non supervisé. In: Actes 10ème Colloque GRETSI, Nice, France, 1017–1022
33. Hesse CW, James CJ (2005) The FastICA algorithm with spatial constraints. IEEE Signal Processing Letters 12(11):792–795
34. Hesse CW, James CJ (2006) On semi-blind source separation using spatial constraints with applications in EEG analysis. IEEE Transactions on Biomedical Engineering 53(12): 2525–2534

35. Holm M, Pehrson S, Ingemansson M et al. (1998) Noninvasive assessment of the atrial cycle length during AF in man: introducing, validating and illustrating a new ECG method. Cardiovascular Research 38(1):69–81
36. Hyvärinen A (1999) Fast and robust fixed-point algorithms for independent component analysis. IEEE Transactions on Neural Networks 10(3):626–634
37. Hyvärinen A, Oja E (1997) A fast fixed-point algorithm for independent component analysis. Neural Computation 9(7):1483–1492
38. Hyvärinen A, Karhunen J, Oja E (2001) Independent Component Analysis. John Wiley & Sons, New York
39. James CJ, Gibson OJ (2003) Temporally constrained ICA: an application to artifact rejection in electromagnetic brain signal analysis. IEEE Transactions on Biomedical Engineering 50(9):1108–1116
40. Kanjilal P, Palit S, Saha G (1997) Fetal ECG extraction from single-channel maternal ECG using singular value decomposition. IEEE Transactions on Biomedical Engineering 44(1): 51–59
41. Li X, Zhang X (2007) Sequential blind extraction adopting second-order statistics. IEEE Signal Processing Letters, 14(1):58–61
42. Lu W, Rajapakse JC (2005) Approach and applications of constrained ICA. IEEE Transactions on Neural Networks 16(1):203–212
43. Lu W, Rajapakse JC (2006) ICA with reference. Neurocomputing 69:2244–2257
44. Netabayashi T, Kimura Y, Chida S et al. (2008) Robustness of the blind source separation with reference against uncertainties of the reference signals. In: Proc. EMBC-2008, 30th Annual International Conference of the IEEE Engineering in Medicine and Biology Society, Vancouver, BC, Canada, 1875–1878
45. Peters M, Crowe J, Piéri, JF et al. (2001) Monitoring the fetal heart non-invasively: a review of methods. Journal of Perinatal Medicine 29(5):408–416
46. Phlypo R, D'Asseler Y, Lemahieu I et al. (2007a) Extraction of the AA from the ECG based on independent component analysis with prior knowledge of the source kurtosis signs. In: Proc. EMBC-2007, 29th Annual International Conference of the IEEE Engineering in Medicine and Biology Society, Lyon, France, 6499–6502
47. Phlypo R, Zarzoso V, Comon P et al. (2007b) Extraction of AA from the ECG by spectrally constrained kurtosis sign based ICA. In: Proc. ICA-2007, 7th International Conference on Independent Component Analysis and Signal Separation, London, UK, 641–648
48. Phlypo R, Zarzoso V, Lemahieu I (2008a) Exploiting independence measures in dual spaces with application to atrial f-wave extraction in the ECG. In: Proc. MEDSIP-2008, 4th International Conference on Advances in Medical, Signal and Information Processing, Santa Margherita Ligure, Italy
49. Phlypo R, Zarzoso V, Comon P et al. (2008b): Cumulant matching for independent source extraction. In: Proc. EMBC-2008, 30th Annual International Conference of the IEEE Engineering in Medicine and Biology Society, Vancouver, BC, Canada, 3340–3343
50. Phlypo R, Zarzoso V, Lemahieu I (2008c) Eigenvector analysis for separation of a spectrally concentrated source from a mixture. In: Proc. EMBC-2008, 30th Annual International Conference of the IEEE Engineering in Medicine and Biology Society, Vancouver, BC, Canada, 1863–1866
51. Rieta JJ, Zarzoso V, Millet-Roig, J et al. (2000) AA extraction based on blind source separation as an alternative to QRST cancellation for AF analysis. In: Proc. Computers in Cardiology. Vol. 27, Boston, MA, USA, 69–72
52. Rieta JJ, Castells F, Sánchez C et al. (2004) AA extraction for AF analysis using blind source separation. IEEE Transactions on Biomedical Engineering 51(7):1176–1186
53. Ristaniemi T, Joutsensalo J (2002) Advanced ICA-based receivers for block fading DS-CDMA channels. Signal Processing 82(3):417–431
54. Rosenbaum DS, Cohen, RJ (1990) Frequency based measures of AF in man. In: Proc. 12th Annual International Conference of the IEEE Engineering in Medicine and Biology Society

55. Sameni R, Jutten C, Shamsollahi MB (2008) Multichannel electrocardiogram decomposition using periodic component analysis. IEEE Transactions on Biomedical Engineering, in press
56. Sánchez C, Millet J, aRieta JJ (2001) Packet wavelet decomposition: an approach for AA extraction. In: Computers in Cardiology. Volume 29, Rotterdam, The Netherlands, 33–36
57. Sato M, Kimura Y, Chida S et al. (2007) A novel extraction method of fetal electrocadiogram from the composite abdominal signal. IEEE Transactions on Biomedical Engineering 54(1):49–58
58. Saul LK, Allen JB (2000) Periodic component analysis: an eigenvalue method for representing periodic structure in speech. In: Advances in Neural Information Processing Systems 13, Denver, CO, USA, 807–813
59. Schmidt RO (1986) Multiple emitter location and signal parameter estimation. IEEE Transactions on Antennas and Propagation AP-34(3):276–280
60. Shalvi O, Weinstein E (1990) New criteria for blind deconvolution of nonminimum phase systems (channels). IEEE Transactions on Information Theory 36(2):312–321
61. Slocum J, Byrom E, McCarthy L et al. (1985) Computer detection of atrioventricular dissociation from surface electrocardiogram during wide QRS complex tachycardia. Circulation 72:1028–1036
62. Stridh M, Sörnmo L (2001) Spatiotemporal QRST cancellation techniques for analysis of AF. IEEE Transactions on Biomedical Engineering 48(1):105–111
63. Tichavský P, Koldovský Z, Oja E (2006) Performance analysis of the FastICA algorithm and Cramér-Rao bounds for linear independent component analysis. IEEE Transactions on Signal Processing 54(4):1189–1203
64. Tong L, Liu R, Soon VC et al. (1991) Indeterminacy and identifiability of blind identification. IEEE Transactions on Circuits and Systems 38(5):499–509
65. Tugnait JK (1997) Identification and deconvolution of multichannel non-Gaussian processes using higher order statistics and inverse filter criteria. IEEE Transactions on Signal Processing 45:658–672
66. Vanderschoot J, Callaerts D, Sansen W et al. (1987) Two methods for optimal MECG elimination and FECG detection from skin electrode signals. IEEE Transactions on Biomedical Engineering BME-34(3):233–243
67. Vásquez C, Hernández A, Mora F et al. (2001) AA enhancement by Wiener filtering using an artificial neural network. IEEE Transactions on Biomedical Engineering 48(8):940–944
68. Widrow B, Glover JR, McCool JM et al. (1975) Adaptive noise cancelling: principles and applications. Proceedings of the IEEE 63(12):1692–1716
69. Zarzoso V (2008) On an extended SOBI algorithm for bind source extraction. IEE Electronics Letters, to be submitted
70. Zarzoso V, Comon P (2007) Comparative speed analysis of FastICA. In: Proc. ICA-2007, 7th International Conference on Independent Component Analysis and Signal Separation, London, UK, 293–300
71. Zarzoso V, Comon P (2008a) Robust independent component analysis for blind source separation and extraction with application in electrocardiography. In: Proc. EMBC-2008, 30th Annual International Conference of the IEEE Engineering in Medicine and Biology Society, Vancouver, BC, Canada, 3344–3347
72. Zarzoso V, Comon P (2008b) Independent component analysis based on optimal step-size iterative search. IEEE Transactions on Signal Processing, to be submitted
73. Zarzoso V, Nandi AK (1999) Blind separation of independent sources for virtually any source probability density function. IEEE Transactions on Signal Processing 47(9):2419–2432
74. Zarzoso V, Nandi AK (2001) Noninvasive fetal electrocardiogram extraction: blind separation versus adaptive noise cancellation. IEEE Transactions on Biomedical Engineering 48(1): 12–18
75. Zarzoso V, Nandi AK, Bacharakis E (1997) Maternal and foetal ECG separation using blind source separation methods. IMA Journal of Mathematics Applied in Medicine & Biology 14(3):207–225

76. Zarzoso V, Nandi AK, Herrmann F et al. (2001) Combined estimation scheme for blind source separation with arbitrary source PDFs. Electronics Letters 37(2):132–133
77. Zarzoso V, Comon P, Kallel M (2006a) How fast is FastICA? In: Proc. EUSIPCO-2006, XIV European Signal Processing Conference, Florence, Italy
78. Zarzoso V, Murillo-Fuentes JJ, Boloix-Tortosa R et al. (2006b) Optimal pairwise fourth-order independent component analysis. IEEE Transactions on Signal Processing 54(8):3049–3063
79. Zarzoso V, Phlypo R, Comon P (2008a) A contrast for independent component analysis with priors on the source kurtosis signs. IEEE Signal Processing Letters, in press
80. Zarzoso V, Phlypo R, Meste O et al. (2008b) Signal extraction in multisensor biomedical recordings. In Verdonck, P., ed.: Advances in Biomedical Engineering. Volume 1. Elsevier, Amsterdam, The Netherlands, in press

Chapter 3
ECG Processing for Exercise Test

Olivier Meste, Hervé Rix and Grégory Blain

Abstract The specificity of processing the ECG signal recorded during an exercise test is analysed. After introducing the interest of such an experiment to catch physiological information, the acquisition protocol is first described. Then new results on heart rate variability estimation, using parametric and non parametric models are given, showing in the time-frequency plane the evolutions of cardiac and respiratory frequencies, together with the pedalling one. Methods for the estimation of PR intervals, when T and P waves are overlapped, are then described, which leads to the enhancement of hysteresis phenomenon for this signal during the phases of exercise and recovery. Finally, the modelling and estimation of shape changes along the test is developed with an application to P waves. The shape changes are modelled by simulation as changes in the relative propagations in the both auricles. In addition, alternatives to the classical signal averaging technique, including signal shape analysis, are discussed.

3.1 Introduction

While the ECG components are well described and understood at rest their global behavior remains unclear under intense exercise. One could wonder why this specific condition is of interest, since apparently the variability that characterises a healthy heart tends to disappear. The example of the random respiration input [1] is a good illustration of the application of an identification method to a physiological system. Following this idea, intense exercise conditions provide new system outputs that allow a better system characterization. Although pharmacological experiments have shown the role of the sympathetic and parasympathetic activities in the cardiac rhythm driving, advanced signal processing techniques now allow the study of

O. Meste (✉)
Laboratoire d'Informatique, Signaux et Systèmes de Sophia, Antipolis, Université de Nice – Sophia Antipolis, CNRS, Les Algorithmes – Euclide-B, 2000 route des Lucioles, BP 121, 06903 Sophia, Antipolis Cedex, France
e-mail: meste@i3s.unice.fr

A. Naït-Ali (ed.), *Advanced Biosignal Processing*,
DOI 10.1007/978-3-540-89506-0_3, © Springer-Verlag Berlin Heidelberg 2009

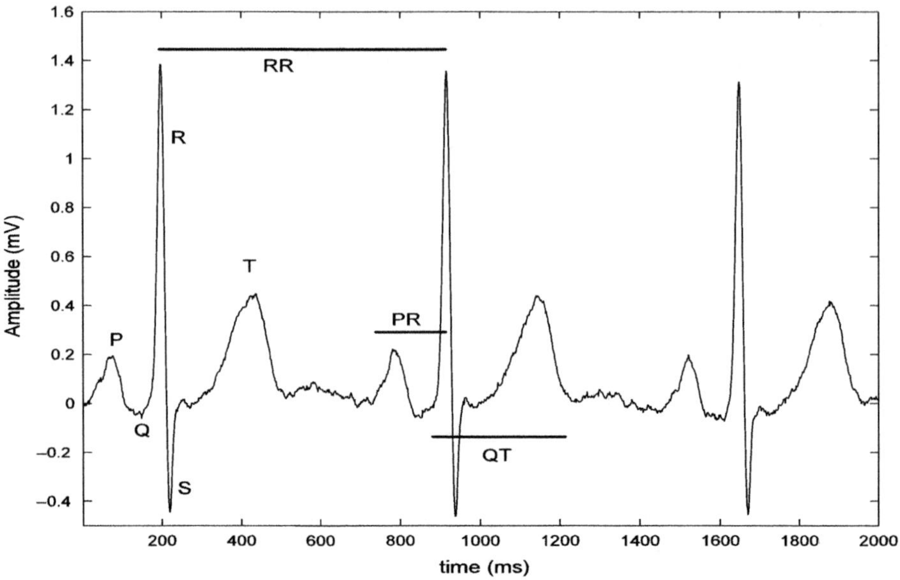

Fig. 3.1 An example of ECG recorded on the lead II. Waves and intervals of interest are clearly visible

finer regulations. The mechanical modulation of the cardiac automaticity is a good example of this fine regulation induced by the respiration and thanks to adapted processing this can be quantified at exercise [2].

The cardiac rhythm analysis is based on time-interval estimations. The temporal series constituted of the intervals separating consecutive R waves (see Fig. 3.1), called RR, is the most studied signal of this type. They are supposed to reveal the properties of the sinus node in response to neural or mechanical changes. Similarly, the PR interval series would contain information related to the atrioventricular node properties. The QT intervals also play an important role, because the short term and long term features correspond to the properties of the ventricles in the repolarization phase. These intervals of interest are quite easily estimated at rest when artifactual signals are neglectable. However, under exercise conditions classical estimators fail to provide reliable or unbiased results because waves overlap or wave shapes change. The overlap of the T and P waves for low RR values affects the PR estimation because of the bias introduced by this superimposition. Because of the repolarization adaptation of the myocardial cells to changes of the depolarization rate, the shape of the T waves varies with decreasing RR during exercise. This variation added to increasing noise level limits the choice of a proper time delay estimator.

Shape variations seem to exist solely as a perturbation in the estimation process. In fact, since the variation is a side effect of an underlying physiological process, it could provide information of interest. For instance, the shape analysis of P waves recorded during exercise test allows for valid PR estimates, despite the shape change. In addition, since the P wave is the sum of the two auricles activities, its shape variation will be related to electrophysiological changes of each individual auricle. The increase of the conduction velocity in nodal and non-nodal tissues due to sympathetic activation can fully explain these changes. This phenomenon is also present in exercise test ECG records.

The preceding examples show the interest in recording ECG during exercise tests. Although the objectives presented are mainly for electrophysiologic modelling and understanding, they provide reference levels or parameter values for diagnosis purposes. Because of the complexity of the recorded signal, specific tools are needed to get rid of artifactual signals impact or to explore additional information with respect to resting ECG conditions.

After a brief description of acquisition protocols for exercise test ECG recordings, general and ad-hoc signal processing tools will be presented.

3.2 ECG Acquisition During Exercise Test

During exercise, accurate estimation of the heart's electrical activity using electrocardiogram recording (ECG) is challenging, especially at high workloads or during prolonged exercise. Indeed, the ECG signal is distorted because of upper limb and trunk muscular activity, respiration, electrode artifacts due to skin impedance, perspiration and electrode movements.

Proper skin cleaning and use of specifically designed electrodes significantly reduce the electrode noise.

To reduce distortion due to muscular activity, accurate electrode placement is necessary. The most commonly used clinical ECG-system, the so-called 12-leads ECG system, consists of 6 frontal leads (I, II, III, and aV_R, aV_L, aV_F) and 6 precordial leads (V_1, V_2, V_3, V_4, V_5, V_6) [3]. During exercise, to minimize noise from muscular activation Mason and Likar [4] suggested to modify electrode placements of the standard I, II and III to the shoulders and to the hip instead of the arms and the legs. The right arm electrode is moved to a point in the subclavicular fossa medial to the border of the deltoid muscle and 2 cm below the lower border of the clavicle. The left arm electrode is located symmetrically on the left side. The left leg electrode is placed on top of the left iliac crest. The right leg electrode is placed in the region of the right iliac fossa. The precordial leads are located in the standard places of the 12-lead system. In practice, during cycling, our group obtained better ECG recordings when the left leg electrode was moved from the top of the iliac crest to the intersection of the anterior axillary line and the sixth intercostal space.

However, accurate electrode placement is insufficient to guarantee noise reduced ECG recordings. The nature of the exercise should also be chosen in order to limit movement-induced ECG distortion. Cycling tests are probably among the best exercise candidates because cycling limits the movements of the upper limbs and trunk compared to running, rowing or walking.

3.3 Interval Estimation and Analysis

In summary, the autonomic nervous system (ANS) drives the heart rate trough the sympathetic and parasympathetic branches. An increase of the sympathetic activity, as well as a parasympathetic tone withdrawal, increases the heart rate. Note that in that case we refer to a global trend and not to instantaneous or beat to beat quantities. This phenomenon occurs during continuously increasing effort such as during the stress test or during short-term effort. Periodicities in the heart rate variability (HRV) have been studied as a non invasive tool for the beat to beat quantification of the parasympathetic-sympathetic balance [5]. Spectral analysis of this variability shows components in the high-frequency band (0.15–0.4 Hz) that are essentially modulated by the parasympathetic branch while the low-frequency components (0.04–0.15 Hz) are affected by both [6]. Frequencies below 0.04 Hz exist and are mainly due to the regulation process. In addition, we show that a mechanical modulation exists [7] at very high exercise intensity that is beyond the range of the high-frequency band. A second mechanical modulation, correlated to the pedalling frequency, can be observed at higher frequency. Although evidences based on muscle pumping effect are questionable, its potentiality to produce misleading conclusions with respect to the ANS is actual.

Although the heart period is the most studied ECG interval, other intervals of interest are available and convey different kinds of information. Among them, the PR interval brings added value to the ANS characterization. Since this interval involves the atrioventricular node and thanks to the special innervation scheme of the latter [8] its analysis will provide deeper ANS understanding. The QT and RT intervals are also good markers because they are mainly affected by the repolarization phase of the ventricules. The analysis of these intervals as a function of the RR ones provides pathology characterizations and exhibits fast adaptation and long term memory behavior similar to the ventricule myocyte itself.

From this short introduction it is clear that the notion of interval is of interest for the characterization of the heart functional properties. As it will be shown in the sequel, the estimation of the intervals becomes difficult during intense exercise. The increase of the baseline wander level, the noise level, the shape changes and the overlapping waves phenomena hinder the information recovery from the intervals.

3.3.1 Heart Rate Variability Analysis

Strictly speaking, the heart period should refer to the PP intervals. Because of the weak variability of the PR intervals compared to the RR ones, the PP analysis is usually substituted by the RR analysis (see Fig. 3.1) assuming they convey the same information. In the following, the heart period (HP) $hp(k)$ is defined as:

$$hp(k) = t_k - t_{k-1}$$

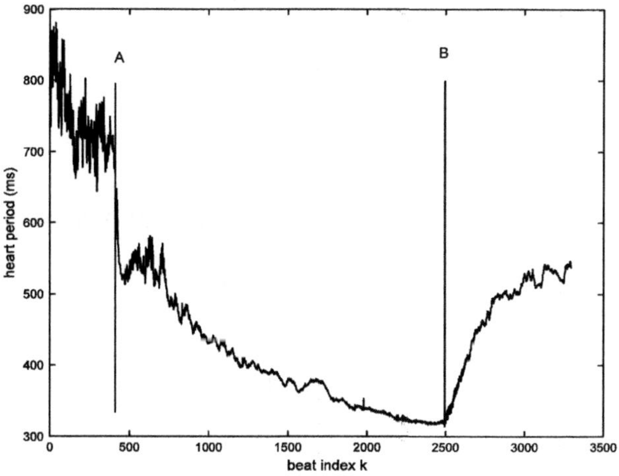

Fig. 3.2 RR intervals during exercise test. Resting, exercise and recovery periods correspond to intervals [0-A], [A-B], [B-end] respectively

where t_k is the occurrence time of the kth beat. In Fig. 3.2, a $hp(k)$ is given where the different stages, rest, exercise and recovery stages are clearly visible. From $hp(k)$, the trend $po(k)$ of the heart period is computed using an order 20 polynomial fitting. The variability $m(k)$ is the high-pass (normalized cutoff frequency 0.03) filtered residual. In Figs. 3.3 and 3.4, the trend and variability from HP in Fig. 3.2 are given, respectively.

It should be noted that $hp(k)$ is processed without resampling and corresponds to an unevenly sampled signal. Thus, in order to relate normalized frequency to Hertz the trend $po(k)$ will be considered as the time varying sampling period. The variability signal $m(k)$ is usually analyzed in the frequency domain. It is clear that the stationarity assumption is not valid under dynamic conditions such as this exercise

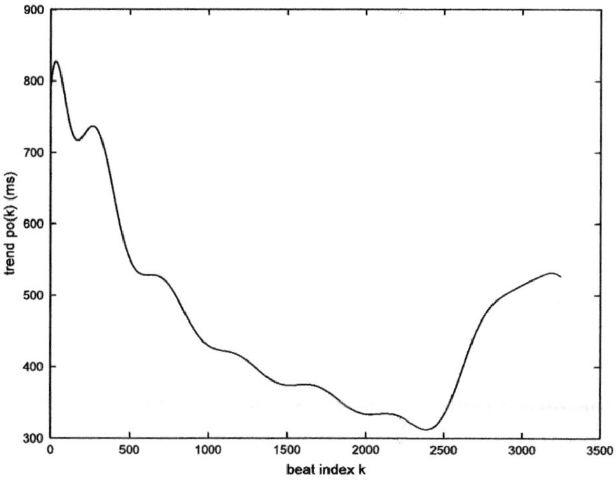

Fig. 3.3 The heart period trend or instantaneous mean heart period

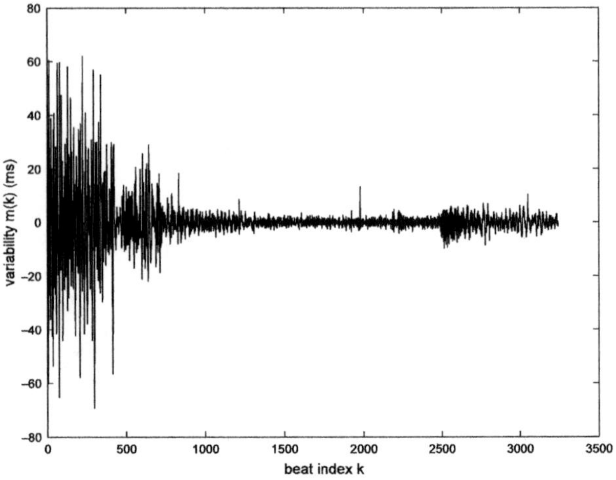

Fig. 3.4 The variability signal

test. To overcome this limitation the analysis has been addressed in a time-frequency manner using parametric [9] and non-parametric [10, 11] modelling. Apparently, one of the major components is related to the respiration, namely the respiratory sinus arrhythmia (RSA). It has been demonstrated that its influence is governed by the parasympathetic tone. Thus, the quantification of the components magnitude at the respiration frequency brings information with respect to the parasympathetic activity. The focus on the RSA component has lead to two methods that differ in their underlying assumptions.

The first approach [12] relies on a linear modelling of the time-varying respiration frequency. This assumption is fully exploited by using the smoothed instantaneous autocorrelation function that is approximated by a sum of complex damped sinusoids. In contrast to this approach where the information extraction is achieved from the transformed signal, a direct approach has been proposed in [13]. In that case, the variability *m(k)* is modeled as a time-varying AR model as:

$$m(k) = \sum_{i=1}^{p} a_i(k)m(k-i) + v(k), \quad p+1 \leq k \leq N \tag{3.1}$$

The time-varying parameters are not updated by the estimation process, such as a recursive least square, but are linearly modelled as:

$$a_i(k) = \sum_{l=0}^{q} a_{il}u_l(k) \tag{3.2}$$

where the functions $u_l(k)$ are chosen as orthogonal Fourier functions. Note that since the respiration frequency and the sampling frequency (*1/po(k)*) increase simultaneously during exercise the normalized (observed) frequency varies slowly. This property permits to choose a low order decomposition in (3.2). Thus, the model is fully linear with respect to the parameters a_{ij} that can be estimated with a least square estimator. From the set of a_{ij}, the time-varying AR coefficients $a_i(k)$ are computed thanks to (3.2). The idea behind this modelling is to estimate the frequency tracks of the variability components by using an AR model and to use these tracks to design a linear time variant filter.

In the stationary case, the relation linking the AR coefficients and the frequencies of the spectral lines is given by:

$$1 - \sum_{i-1}^{p} a_i z^{-i} = \prod_{i-1}^{p} (1 - z_i z^{-1}), \quad \text{where} \quad z_i = e^{j2\pi f_i} \tag{3.3}$$

From the poles z_i the frequencies f_i are computed providing the quantity of interest. In the non-stationary case, although the AR coefficients are slowly varying, the corresponding poles could vary in a very different manner. This is due to the relation linking the derivative of $z(k)$ with respect to the variable k, given by:

$$\dot{z}_i(k) = \sum_{n=1}^{p} \frac{\partial z_i}{\partial a_n}(k)\dot{a}_n(k) \tag{3.4}$$

where

$$\frac{\partial z_i}{\partial a_n}(k) = \frac{z_i^{p-n}(k)}{\prod_{l=1|l\neq i}^{p} (z_i(k) - z_l(k))} \tag{3.5}$$

It appears from (3.4) that it is a difficult task to maintain the continuity of the track of the frequencies f_i. An efficient solution that overcomes this difficulty is proposed [13] based on a factorization of (3.3) by order two polynomial functions.

Once the frequency tracks are computed, the one that lies in the respiration frequency band is retained for the amplitude estimation. In [14], this method has been applied to assess the ventilatory threshold during graded and maximal exercise where an additional signal from an ergospyrometer (breathing frequency) is not available. In [15], it has been shown that a priori information of respiratory frequency can be included in the instantaneous autocorrelation function based method [12]. In the latter case this information is extracted from the ECG itself [16] but could be computed from the ergospyrometer signal.

The amplitude estimation must account for the time-varying properties for the respiration signal or correspondingly for its frequency. To attain this goal, time-frequency representations are well adapted since the signal is nonstationary by nature (see also Chap. 5). Although a quadratic time-frequency representation is eligible for this kind of processing its quadratic property makes its inversion difficult. Linear transformations will be preferred such as the short-time Fourier transform defined as:

$$HP(k, f) = \sum_{u} m(u)h(u-k)e^{-j2\pi(l/K)u} \tag{3.6}$$

with $-K/2 \leq l \leq K/2-1$ integer and $f = l/K$. The function $h(u)$ is a weighting function and K an even number of frequencies. Once the frequency track $f_r(k)$ of the respiration has been calculated as above a binary time-frequency template $G(k,f)$ is designed such that

$$G(k, f) = \begin{cases} 1, & for\ |f| \in [f_r(k) - \delta; f_r(k) + \delta] \\ 0, & elsewhere \end{cases} \tag{3.7}$$

The selectivity of the time-varying filter is then adjusted by the correct selection of the δ value. This template can be used to directly compute the magnitude $M_r(k)$ of the frequency component $f_r(k)$ by using the relation:

$$M_r(k) = \sqrt{\frac{1}{K}\sum_f (G(k, f)HP(k, f))^2}, \tag{3.8}$$

or by computing the inversion of the short-time Fourier transform:

$$m_r(k) = \sqrt{\frac{1}{K}}\sum_u \sum_{l=-K/2}^{K/2-1} G\left(u, \frac{l}{K}\right) HP\left(u, \frac{l}{K}\right) h(k-u)e^{j2\pi(l/K)u} \tag{3.9}$$

providing the filtered version of $m(k)$ using the knowledge of $f_r(k)$. The envelope of this signal is obtained from the analytical version of $m_r(k)$ since it contains only one frequency component.

It should be recalled that the observed $m_r(k)$ is an indirect measurement of the RSA. The heart itself dictates the timing of the observations that produces a non linear transformation of the continuous modulation. Several models relate the continuous modulation to the heart timing or beat time occurrence. The Integral Pulse Frequency Modulation (IPFM) is the most studied model [17]. However, it fails to account for any type of heart rate modulation and can be replaced by the Pulse Frequency Modulation (PFM) that succeeded to analyze mechanically induced RSA [13, 2]. Although more complete and detailed description of the nonlinearities induced by the IPFM are available [18, 19], an approximation of the relation between the magnitude of the continuous modulating signal, i.e. the respiration, and the observed one is given in [13]:

$$A_m(k) = c\frac{po(k)}{\pi}\sin(\pi f(k)) \tag{3.10}$$

Here, c is the amplitude of the modulation, assumed here to be pure tone, and $A_m(k)$ is the amplitude of the filtered $m(k)$. The relation between $f(k)$ and the time-varying frequency of the pure tone is $f(k) = F(t_k)po(k)$. It is important to note that the relation between $A_m(k)$ and c depends on the trend of the mean heart period ($po(k)$) and the frequency itself ($f(k)$).

Some results are given in [2] where it has been shown that using the global processing, including frequency tracking and time-varying filtering, the observed heart rate variability magnitude at the respiration frequency is linearly correlated to the ventilation. It is noticeable that the model (3.10) is in agreement with this finding since at the maximum workload the product $f(k) = F(t_k)po(k)$ is low enough to allow the approximation $A_m(k) \approx cpo(k)^2 F(t_k)$. It is expected that the coefficient c is proportional to the tidal volume V_T giving $A_m(k) \approx \alpha\, po(k)^2 F(t_k) V_T \approx \alpha\, po(k)^2 V_{vent}$. The mentioned linearity is thus obtained when V_{vent} stands for the maximal ventilation at t_k corresponding to the maximum workload because $po(k)$ is almost the same for all subjects at that level.

The existence of a model is helpful to relate the information extracted from the observation to the physiological knowledge. Another example is provided in [20] where a very high frequency component is under the scope of the heart rate variability analysis. It has been shown that the presence of this component can be explained by using a model of an oscillatory venous blood flow that returns to the right auricle from the lower limbs. During cycling exercise, this component oscillates at the pedalling rate and could produce a misleading interpretation of the global spectrum because of the aliasing effect. Indeed, the mean heart rate is not high enough during the exercise test to fulfill the Shannon condition when the pedalling rate is greater than 60 rpm.

An illustration of this presence if shown in Fig. 3.5 where three frequency tracks from the time-varying AR model are plotted superimposed to the short-time Fourier transform of a given $m(k)$. The tracks around 0.5 and 0.2 are from the pedaling and the respiration, respectively. When available, the true respiration frequency can be compared to the estimated one, as shown in Fig. 3.6.

The heart rate variability analysis relies on the estimation of the beat time occurrence, providing the RR series. The choice of the estimator is not drastic since the

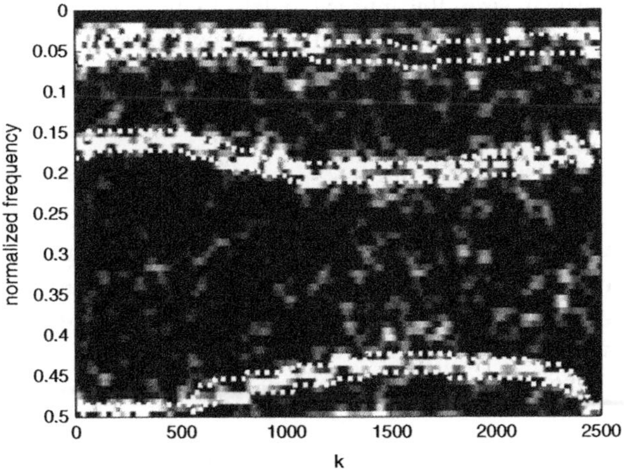

Fig. 3.5 Squared modulus of Short-time Fourier transform of the variability signal. The *gray* range from *black* to white corresponds to low-high range. The estimated three frequency tracks are surrounded by *white dotted* lines

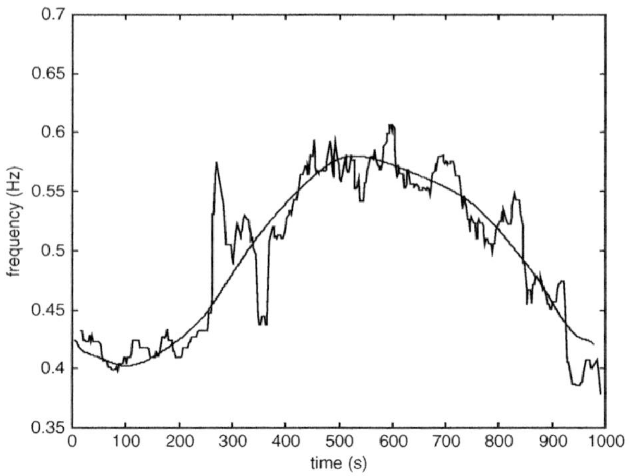

Fig. 3.6 The estimated respiration frequency (*solid line*) compared to the real one (*dashed line*). Note that the high-frequency variability of the measured frequency is artifactual

signal to noise ratio is high when the detection is performed on the R wave. In contrast to the RR estimation, the PR estimation is more tricky as will be shown in the sequel.

3.3.2 PR Interval Estimation

Once the R waves have been located on the time axis, windowing can be performed in order to preserve the P waves with right bound aligned to the R waves fiducial points. In Fig. 3.7, two segment of the ECG aligned on a R wave fiducial point are plotted, at the beginning of the exercise (long T-P) and at the maximum workload (the T and P waves overlap). In the latter case, the windowing could be applied on the interval [400–600 ms]. This effect added to a low signal to noise ratio has limited the analysis of this interval to date. Only very few papers deal with this interval [21] although its analysis could reveal new infomation related to the ANS.

When the entire ECG is processed, all the observations windows indexed by i can be modeled as:

$$x_i(n) = \alpha_i s_{d_i}(n) + \alpha_i T_{d_i}(n;\boldsymbol{\theta}_i) + e_i(n), \quad 1 \leq i \leq I \tag{3.11}$$

where $s_{di}(n)$ stands for the P wave delayed by d_i and $T_{di}(n;\boldsymbol{\theta}_i)$ a parameterized model of the T wave. An amplitude factor is also introduced as α_i. Assuming that the noise $e_i(n)$ is i.i.d. with a normal law and the P wave unknown, the estimation of the variables can be achieved by using a Maximum Likelihood approach [22]. Similarly to the solution of the simpler model [23]

$$x_i(n) = s_{d_i}(n) + e_i(n), \tag{3.12}$$

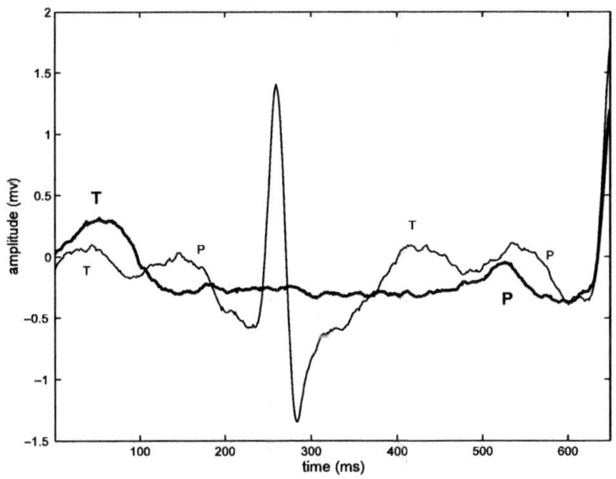

Fig. 3.7 Two superimposed ECG traces. Thick and thin lines correspond to rest and peak exercise periods, respectively. Note that the TP interval has been reduced in accordance to the RR shortening, producing the overlap of the waves

improved in [24], the parameters of (3.11), except *s(n)*, will be estimated by minimizing iteratively the criterion [25]:

$$J = \sum_i \left\| \mathbf{x}_i - \alpha_i T_{d_i}(\boldsymbol{\theta}_i) - \frac{\alpha_i}{I} \sum_k \frac{1}{\alpha_k} \left(\mathbf{x}_{k,d_i-d_k} - \alpha_k T_{d_i-d_k}(\boldsymbol{\theta}_k) \right) \right\|^2 \qquad (3.13)$$

Note that the solution for the d_i's is not unique. Thus the delays will be estimated up to a constant. This limitation is not crucial because the information is more in the variation of the P wave positions than in its true value.

The choice of $T_{di}(n;\boldsymbol{\theta}_i)$ must account for the knowledge of the real T wave features. A gaussian shape has been chosen in [26] but suffers from being a non-linear function of the parameters. The standard acquisition lead II permits us to use a simple feature, namely that the segmented part of the T waves are strictly decreasing in the observation window. Thus polynomial functions or piecewise linear functions [27] can be chosen. The decreasing feature is included in the criteria (3.13) imposing inequality constraints. For a fixed set of α_i's and d_i's, the criteria (3.13) becomes a least square minimization subject to linear-inequality constraints, solved by using a Least Distance Programming [28].

In Fig. 3.8, the d_i's (namely the PR intervals) estimated during exercise test are plotted. It can be noticed that probably because of a strong vagal return at the early stage of the recovery, the PR values may be greater than at rest. This phenomenon has also been described as a hysteresis not due to cardiac memory but to neural balance [29]. To stress the importance of the PR interval richness, we can mention the result given in [27] that uses the slope of I in Fig. 3.8 as a marker to distinguish trained and untrained cyclists. In contrast, the single analysis of the RR series on the same intervals didn't provide such a significant results.

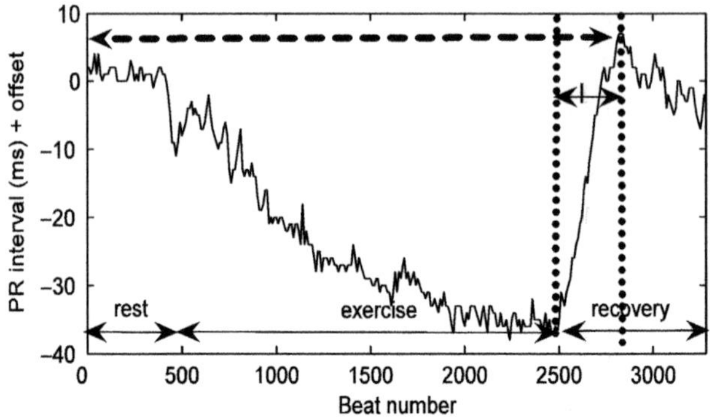

Fig. 3.8 PR intervals during exercise where the periods of rest, exercise and recovery are clearly visible. The interval I defined by the two dashed vertical lines is of interest (see text). Note that in the recovery period the PR attains level higher than at rest

In the introduction it has been mentioned that during exercise, the intrinsic properties of the myocardiac and nodal cells added to the ANS activity should produce changes in the cell action potentials [30]. Expectations are that the model (3.11) would not be valid for any exercise level since $s(n)$ is required to be unique. As it will be shown in the sequel, the shape of the P wave is affected by the exercise but slightly the width. This comment almost validates the assumption of a unique and stable P wave.

The overlap effect of waves can also be encountered in QT analysis. However, thanks to the dedicated lead V_2 for the T wave recording [3] and because the T wave is highly energetic, the problem is not as severe as for the P wave and thus the model (3.12) can be adopted for the observation of the T instead of the P waves. Unfortunately, a side effect is the global stretching of the T wave during exercise [31] in addition to the delays. The model (3.12) is then no longer valid for all the observation windows. An efficient alternative is to split the ensemble of windows into smaller sets where (3.12) is valid. For each set, the d_i's are estimated plus the synchronous averaging that stands for the estimated T wave of this set. Thanks to the averaging process, the offset or other fiducial points of this T wave can be efficiently determined [32] allowing an adjustment of the d_i's for this set. This procedure applied to all the sets will produce a continuous measurement of the QT (or RT) during exercise.

When the problem of shape variation cannot be overcome easily as previously presented, one can turn to more adapted methods. The following section will present such methods in addition to an application to the P wave shape analysis under exercise.

3.4 Shape Variations Estimation and Analysis

Above we have addressed the modification of the rhythm when the electric activity of the heart is recorded during exercise, including the heart rate and its variability. Another feature which may be important to observe is the shape variations of the

ECG waves. In fact, a change in shape must be involved in any dynamic physiological model and also in the estimation of time intervals between two waves since the shape obviously affects each wave in a heart beat.

3.4.1 Curve Registration of P-waves During Exercise

For example, to investigate how the P-wave morphology changes during time-varying effort, Self-Modelling Registration, derived from the Curve Registration (CR) theory has been used [33]. The basic model of curve registration is to assume a generating shape function linked to the set of curves, by increasing functions of time (the warping functions or their inverses) which account for signal variability. These time warping functions represent natural time fluctuations of the generating process [34]. In this application, this hypothesis is coherent with the fact that shape changes of the P-wave, during exercise, are probably due to variations of the atrial depolarization. To perform CR, different methods can be employed, the most famous being Dynamic Time Warping. In our study, we chose a recent CR method, Self-Modelling Registration (SMR). This method estimates the warping functions better than the precedent ones using a semi-parametric model [35]. In the next paragraph, the SMR algorithm is recalled. According to the CR hypothesis, we can suppose that N signals are generated from the shape function *s(t)* as follows:

$$x_i(t) = a_i s(v_i(t)) + \varepsilon_i(t), \tag{3.14}$$

where the non-random function *s(t)* is the shape function, v_i are monotone increasing functions that account for time variability, and a_i and ε_i are random quantities that account for amplitude variability. We can write assuming zero mean process for ε_i and $E\{a\}=1$:

$$E\{x(t)\} = E\{s(v(t))\} \tag{3.15}$$

which can be approximated, for large N, by:

$$\bar{x}(t) = \frac{1}{N}\sum_{i=1}^{N} s(v_i(t)) \tag{3.16}$$

where $\bar{x}(t)$is the classical average and is different, in general, from *s(t)* [35]. Therefore, the objective of the CR operation is to realign or register the signals to *s(t)*. This signal realignment permits to estimate the time warping functions ($v_i^{-1} = w_i$), not directly observable. Then, an estimated shape function or Structural Average (SA) $\mu(t)$ can be obtained as follows [35]:

$$\mu(t) = \frac{1}{N}\sum_{i=1}^{N} x_i(\hat{w}_i(t)), \tag{3.17}$$

where, $\hat{w}_i$ are the estimated warping functions. The signal $\mu(t)$ is unique when the following condition is verified:

$$\frac{1}{N}\sum_{i=1}^{N}\hat{w}_i(t) = t. \tag{3.18}$$

3.4.1.1 Self-Modelling Registration (SMR)

The main idea of the SMR method is to model the warping functions as linear combinations of a small number of functions as in following

$$w_i(t) = t + \sum_{j=1}^{q}\alpha_{ij}\phi_j(t) \tag{3.19}$$

The component functions ϕ_j are estimated from the signals. They are order p linear combinations of cubic B-spline, and then we can write:

$$\phi_j(t) = \sum_{i=1}^{p} c_{ji}\beta_i(t) \tag{3.20}$$

The form of the linear combination of $w_i(t) - t$ given by (3.19) can be viewed as a generalization of the landmark registration. Indeed, imposing the alignment of all the curves at some time positions (landmarks) with the mean landmark, and using linear interpolation for the rest of the curves, leads to the same formula when the ϕ functions are triangles, with peaks at the landmark times. The SMR technique furnishes a more flexible modelling, with bell shaped components, localized around points which may be interpreted as hidden landmarks. Their number, i.e. parameter q, and parameter p are chosen empirically as explained in [35].

The parameters of the signal generation model defined in (3.14) can be estimated by integrated least squares minimization as follows:

$$\min F = \min\sum_{i=1}^{N}\int_0^T \left\{x_i(t) - a_i s\left(v_i(t)\right)\right\}^2 dt \tag{3.21}$$

and in another form:

$$\min F = \min\sum_{i=1}^{N}\int_0^T \left\{x_i(w_i(t)) - a_i s(t)\right\}^2 w_i'(t)dt \tag{3.22}$$

The objective function F is minimized by an iterative algorithm, given the estimated warping functions as follows:

$$\hat{s}(t) = \frac{\sum_{i=1}^{N} \hat{a}_i \hat{w}_i'(t) \hat{x}_i(\hat{w}_i(t))}{\sum_{i=1}^{N} \hat{a}_i^2 \hat{w}_i'(t)} \tag{3.23}$$

In the method, the warping functions are estimated by taking as reference de time axis of $\hat{s}(t)$. In this study, to better show the shape evolution of the P-wave, the time axis reference is changed to the one of another signal (e.g. the beginning of the exercise) $x_1(t)$, so we can write:

$$\hat{w}_{i,1} = \hat{w}_i \circ \hat{w}_1^{-1}, \quad 1 \leq i \leq N \tag{3.24}$$

where $\hat{w}_{i,1}$ are the estimated warping functions related to the time axis of $x_1(t)$. This is an alternative to the solution (3.18) to impose the unicity of $\mu(t)$.

3.4.1.2 Application to Real P-waves

We recall in the following some results given in [33], applying SMR to P-waves coming from ECG signals in exercise conditions.

In Fig. 3.9, 30 selected averaged P-waves (on 10 consecutive beats) going from P-wave 1 to P-wave 30 when the exercise is increasing are shown (Fig. 3.9a) together with the warping curves (Fig. 3.9b) with P-wave 1 as reference. In Fig. 3.9c,d, the same analysis was made during the recovery phase when the effort has been released abruptly. In this case P-wave 1 corresponds to the beginning of the recovery phase. The obtained warping functions show an evolution of the P-wave shape, the important point being that the P-wave duration is not sufficient to characterize its evolution. On another hand, the warping curves of the recovery are not exactly the reflected ones of the exercise phase about the line y = x, suggesting a hysteresis behaviour. To give a physiological explanation of this P-wave morphing, a scenario, linking the separation of the depolarisation signals of the left and right auricles under exercise, was proposed, and validated by simulation. We recall in the following, the simulation study of [33].

3.4.2 Simulation Study

Through the presentation of the following model, we try to give a physiological explanation to the P-wave morphing observed on the precedent results. The model consists in the addition of two Gaussian signals representing both atrial contributions. The evolution of the mean and standard deviation of the gaussians are modeled as affine functions of time. We can write, for the P-wave signal of beat number i, which is the sum of the right (R) and left (L) auricle contributions:

$$P_{i,tot}(t) = A_R G_R(\sigma_{i,R}, m_{i,R}, t) + A_L G_L(\sigma_{i,L}, m_{i,L}, t) \tag{3.25}$$

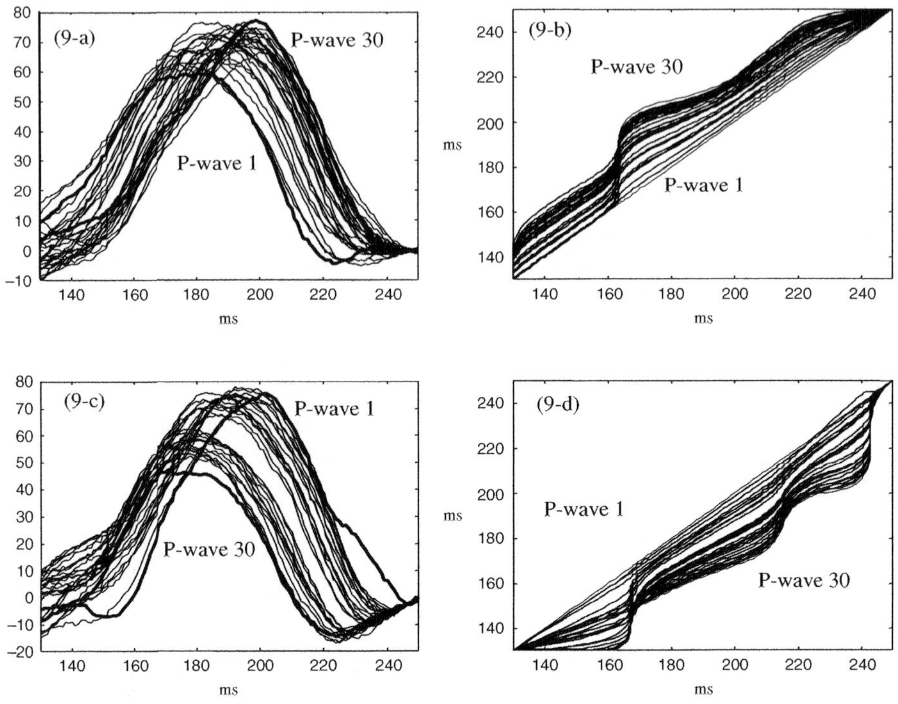

Fig. 3.9 (**a**) 30 selected P-waves in exercise phase, from P-wave 1 to P-wave 30, (**b**) the corresponding warpings of (**a**) with P-wave 1 as reference, (**c**) 30 selected P-waves in recovery phase, from P-wave 1 to P-wave 30, (**d**) the corresponding warpings of (**c**) with P-wave 1 as reference

with:

$$\sigma_{i,R} = \sigma_{0,R} - \alpha_i t_i, \;\; \sigma_{i,L} = \sigma_{0,L} - \alpha_i t_i \tag{3.26}$$

$$m_{i,R} = m_{0,R} + \beta_{i,R} t_i, \;\; m_{i,L} = m_{0,L} + \beta_{i,L} t_i, \;\; 1 \le i \le 30. \tag{3.27}$$

The parameter values of the model are chosen as follows:

$$\begin{cases} A_R = 10, A_L = 9 \\ \alpha_i = 0.16 \\ \sigma_{0,R} = 19, \sigma_{0,L} = 20 \\ m_{0,R} = 105, m_{0,L} = 75 \\ \beta_{i,R} = 0.4, \beta_{i,L} = 0.6 \end{cases} \tag{3.28}$$

Since the signals are selected linearly in beat number but not in time, to generate the time parameter t_i, we used the following formula:

$$t_i = \sqrt{150(i-1)}. \tag{3.29}$$

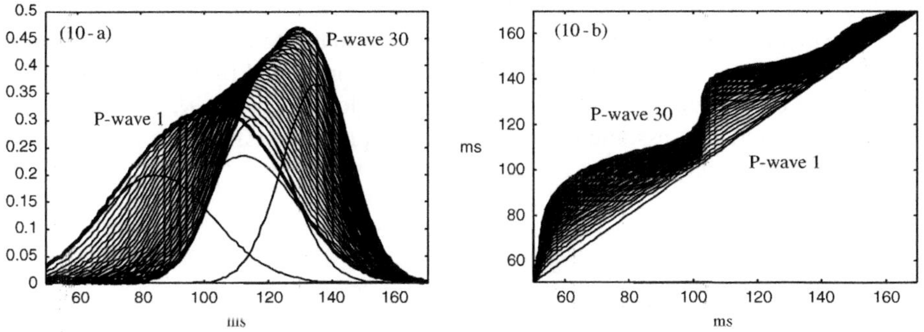

Fig. 3.10 The simulated shapes (**a**) and theirs corresponding warping functions (**b**)

We can see the simulated data on Fig. 3.10.

For the simulation purpose, we supposed that the right atrium contribution is slightly more important with a conduction time distribution more narrow than the left one. During exercise, as the heart rate increases, the conduction rate (represented by $\alpha_i t_i$) increases too. In the same time, the distance between both atria contributions decreases due to this conduction rate increase. These time variations produce, in addition to a time duration reduction, the morphing on Fig. 3.10. As it is shown, the simulated warping functions mimic in a realistic way the ones presented on Fig. 3.9b. The shape evolution at rest can be simulated just by inversing the sense of the shape variation.

3.5 Signal Averaging with Exercise

These non linear warping functions, indicating shape changes, are not in favour of the classical averaging technique to obtain an average signal. In fact, signal averaging has been introduced in HR ECG to increase the signal-to-noise ratio. This technique is optimal when it is possible to model the signal as a repetitive template added to an independent zero mean and stationary noise. Assuming a perfect alignment process of N beats, the standard deviation of the noise is divided by the square root of N on the average. In order to take into account the departures from this ideal hypothesis, we have to distinguish two cases.

3.5.1 Equal Shape Signals

In the first case the averaged signals are not perfectly aligned, but their shape is constant from beat to beat. So, we can assume the following model:

$$s_i(t) = k_i s(\alpha_i t - d_i) + w_i(t) \tag{3.30}$$

The index i is for the beat, k_i and α_i are scaling factors on amplitude and time respectively, d_i is a time delay, representing a residual jitter, $s(t)$ is the signal template and $w_i(t)$ is a zero mean noise. The values represented by k_i, α_i, d_i, $w_i(t)$ are interpreted as realizations of the independent random variables K, A, D and $W(t)$ respectively, and K is assumed to have a mean equal to 1. Averaging over N beats gives the mean signal:

$$\bar{s}(t) = \frac{1}{N}\sum_{i=1}^{N} s_i(t) \tag{3.31}$$

As a matter of fact, this average signal and the template are not the same shape. As shown in [36], when N is sufficiently large, the average (in fact its mathematical expectation) is linked to the template by:

$$\bar{s}(t) = \frac{1}{t}\int_{0}^{\infty} f_A\left(\frac{\tau}{t}\right)(s * f_D)(\tau)d\tau \tag{3.32}$$

where f_D and f_A are respectively the pdfs of the delay D and the time scale factor A. It is important to notice that this result is obtained assuming these random variables are independent in the model used in (3.30). Introducing the time scale in the form:

$$t' = \frac{t+d}{\alpha} \tag{3.33}$$

does not lead to the relation established in (3.32).

A way to obtain a mean signal, preserving the shape of s(t), which is common to all the signals, is to use Integral Shape Averaging (ISA) [36, 37]. Assuming the signals to average are positive on their supports, ISA computes the arithmetic mean of the times associated to a given value y, $0 < y < 1$, of the ordinate of the normalized integral functions of the signals. Plotting y in function of the mean time leads to the normalized integral of the ISA signal. Its derivate is proportional to the Integral Shape Averaged signal. The amplitude factor is easily obtained in computing the average of the areas of all the signals. This mean signal has good properties: it is the same shape as all the averaged signals, and its position, range and height are respectively the average of the positions, ranges and heights of the individual signals. In addition, the integration works as a low pass filter, reducing the noise considerably on the ISA signal. If the signals have both a positive and a negative part, a solution may be to process separately the two parts; another one is to apply ISA to a positive function of the signals, e.g. their square or their absolute value. So the ISA technique appears as an alternative to synchronous averaging, with the aim to reduce noise, but also to preserve the common shape, that is, not to be influenced by jitter and scale fluctuations.

3.5.2 Non Equal Shape Signals

In the second case, that is for example, in exercise ECG, the individual shapes vary from beat to beat. Now the problem is no longer the same. The first question that arises is about the meaning of an average signal, in the classical sense or another like ISA. The variability of the individual signals around the average affects not only the parameters (time delay and time scale) which have no influence over the shape, but also the shape itself. Indeed, nearly all works dealing with curve registration, in the field of functional data analysis, focus on estimating a good representative of a set of signals, while generally making the underlying assumption that its variability must be ignored. Also, The problem of the robustness against noise is generally avoided, assuming a rather good signal to noise ratio. The drawback of using ISA in case of non equal shape signals is the lack of invariance under affine transformations. To overcome this drawback, Corrected Integral Shape Averaging (CISA) has been proposed [38], with an application to the change of P-wave shape due to obstructive sleep apnea. In fact, the CISA signal is invariant when time delays or time scales are applied on the set of curves subjected to be averaged. Of course, CISA coincides with ISA when the signal shapes are equal.

References

1. Berger R, Saul P, Cohen R J (1989) Assessment of autonomic response by broad-band respiration. IEEE Trans Biomed Eng 36(11):1061–1065
2. Blain G, Meste O, Bermon S (2005) Influences of breathing patterns on respiratory sinus arrhythmia in humans during exercise. Am J Physiol 288:H887–H895
3. Malmivuo J, Plonsey R (1995) Bioelectromagnetism. Oxford University Press, New York
4. Mason R, Likar L (1966) A new system of multiple leads exercise electrocardiography. Am Heart J 71(2):196–205
5. Pomeranz B, Macaulay R J B, Caudill M A et al. (1985) Assessment of autonomic function in man by heart rate analysis. Am J Physiol 248:H151–H153
6. Akselrod S, Gordon D, Ubel F A, Shannon D C, Berger A C, Cohen R J (1981) Power spectrum analysis of heart rate fluctuation: a quantitative probe of beat-to-beat cardiovascular control. Science 213:220–222
7. Bernardi L, Salvucci F, Suardi R, Solda P L, Calciati A, Perlini S, Falcone C, Ricciardi L (1990) Evidence for an intrinsic mechanism regulating heart rate variability in the transplanted and the intact heart during submaximal dynamic exercise. Cardiovasc Res 24:969–981
8. Warner M R, DeTarnowsky J M, Whitson C C, Loeb J M (1986) Beat-by-beat modulation of AV conduction. II. Autonomic neural mechanisms. Am J Physiol 251:H1134–H1142
9. Bianchi A, Mainardi L T, Petrucci E, Signorini M, Mainardi M, Cerutti S (1993) Time-variant power spectrum analysis for the detection of transient episodes in HRV signal. IEEE Trans Biomed Eng 40(2):136–144
10. Keselbrener L, Akselrod S (1996) Selective discrete fourier transform algorithm for time-frequency analysis: method and application on simulated and cardiovascular signals. IEEE Trans Biomed Eng 43(8):789–802
11. Toledo E, Gurevitz O, Hod H, Eldar M, Akselrod S (2003) Wavelet analysis of instantaneous heart rate: a study of autonomic control during thrombolysis. Am J Physiol Regul Integr Comp Physiol 284(4):R1079–R1091

12. Mainardi L T, Montano N, Cerutti S (2004) Automatic decomposition of wigner distribution and its application to heart rate variability. Methods Inf Med 43:17–21
13. Meste O, Khaddoumi B, Blain G, Bermon S (2005) Time-varying analysis methods and models for the respiratory and cardiac system coupling in graded exercise. IEEE Trans Biomed Eng 52(11):1921–1930
14. Blain G, Meste O, Bermon S (2005) Assessment of ventilatory threshold during graded and maximal exercise test using time-varying analysis of respiratory arrhythmia. Br J Sports Med 39:448–452
15. Bailon R, Mainardi L T, Laguna P (2006) Time-frequency analysis of heart rate variability during stress testing using a priori information of respiratory frequency. Proc Comput Cardiol 33:169-172
16. Bailon R, Sörnmo L, Laguna P (2006) A robust method for ECG-based estimation of the respiratory frequency during stress testing. IEEE Trans Biomed Eng 53(7):1273–1285
17. Sörnmo L, Laguna P (2005) Bioelectrical signal processing in cardiac and neurological applications. Elsevier Academic Press, New York
18. Mateo J, Laguna P (2000) Improved heart rate variability signal analysis from the beat occurrence times according to the IPFM model. IEEE Trans Biomed Eng 47(8):985–996
19. Brennan M, Malaniswami M, Kamen P (2001) Distortion properties of the interval spectrum of IPFM generated heart beats for the heart rate variability analysis. IEEE Trans Biomed Eng 48(11):1251–1264
20. Meste O, Blain G, Bermon S (2007) Influence of the pedalling frequency on the Heart Rate Variability. Proceedings of the 29th Annual International Conference of the IEEE EMBS 279–282
21. Shouldice R, Heneghan C, Nolan P, Nolan P G, McNicholas W (2002) Modulating effect of respiration on atrioventricular conduction time assessed using PR interval variation. Med Biol Eng Comput 40:609–617
22. Kay S M (1993) Fundamentals of statistical signal processing: estimation theory. Prentice Hall, Englewood Cliffs, NJ
23. Woody C D (1967) Characterization of an adaptative filter for the analysis of variable latency neuroelectric signals. Med Biol Eng Comput 5:539–553
24. Cabasson A, Meste O (2008) Time delay estimation: a new insight into the Woody's method. IEEE Signal Processing Letters 15:1001–1004
25. Cabasson A, Meste O, Blain G, Bermon S (2006) Optimality statement of the woody's method and improvement. Research Report ISRN I3S/RR-2006-28-FR: http://www.i3s.unice.fr/%7Emh/RR/2006/liste-2006.html
26. McSharry P, Clifford G, Tarassenko L, Smith L (2003) A dynamical model for generating synthetic electrocardiogram signals. IEEE Trans Biomed Eng 50:289–294
27. Cabasson A, Meste O (2008) A time delay estimation technique for overlapping signals in electrocardiograms. Proceedings of the 16th European Signal Processing Conference
28. Lawson C L, Hanson R J (1974) Solving least squares problems. Prentice Hall, Englewood Cliffs, NJ, USA
29. Meste O, Blain G, Bermon S (2004) Hysteresis analysis of the PR-PP relation under exercise conditions. Proc Comput Cardiol 31:461–464
30. Klabunde R E (2005) Cardiovascular physiology concepts. Lippincott Williams & Wilkins, Philadelphia, PA USA
31. Langley P, Di Bernardo D, Murray A (2002) Quantification of T wave shape changes following exercise. Pacing Clin Electrophysiol 25(8):1230–1234
32. Zhang Q, Illanes Manriquez A, Medigue C, Papelier Y, Sorine M (2005) Robust and efficient location of T-wave ends in electrocardiogram. Proc Comput Cardiol 32:711–714
33. Boudaoud S, Meste O, Rix H, (2004) Curve registration for study of P-wave morphing during exercise. Comput Cardiol 31:433–436
34. Ramsay J O, Silverman B W (1997) Functional data analysis. Springer series in Statistics, New-York

35. Gervini D, Gasser T (2004) Self-modelling warping functions. J R Stat Soc 66(4):959–971
36. Rix H, Meste O, Muhammad W (2004) Averaging Ssignals with random time shift and time scale fluctuations. Methods Inf Med 43:13–16
37. Boudaoud S, Rix H, Meste O (2005) Integral shape averaging and structural average estimation: a comparative study. IEEE Trans Signal Process 53:3644–3650
38. Boudaoud S, Rix H, Meste O, Heneghan C, O'Brien C (2007) Corrected integral shape averaging applied to obstructive sleep apnea detection from the electrocardiogram. EURASIP J Adv Signal Process, doi:10.1155/2007/32570

Chapter 4
Statistical Models Based ECG Classification

Rodrigo Varejão Andreão, Jérôme Boudy, Bernadette Dorizzi,
Jean-Marc Boucher and Salim Graja

Abstract This chapter gives a comprehensible description of two statistical approaches successfully employed to the problem of beat modeling and classification: hidden Markov models and hidden Markov trees. The HMM is a stochastic state machine which models a beat sequence as a cyclostationary Markovian process. It offers the advantage of performing both beat modeling and classification through a unique statistical approach. The HMT exploits the persistence property of the wavelet transform by associating to each wavelet coefficient a state and the states are connected across scales to form a probabilistic graph. This method can also be used for signal segmentation.

4.1 Introduction

The automatic analysis of the electrocardiogram (ECG) has been the subject of intense research during the last three decades. The particular interest for ECG analysis comes from its role as an efficient non-invasive investigative method which provides useful information for the detection, diagnosis and treatment of cardiac diseases [22]. The first step in the analysis of an ECG signal consists in the segmentation of the signal into the beats and in the elementary waves of which each beat is composed (see Fig. 4.1 and see also Chap. 1). The study of each wave relationship with particular heart diseases such as atrial or ventricular fibrillation and arrhythmias is therefore possible.

Although a great variety of approaches has been proposed in order to perform that task, the majority of them have some features in common: signal processing techniques, parameter extraction, heart beat modeling and classification of subwaves. The core of such approaches is the heart beat modeling strategy. The later can be built directly with the help of an expertise, whose knowledge is transformed into a set of rules. This strategy was present in the first ECG analysis systems and it is

R.V. Andreão (✉)
CEFETES, Coord. Eletrotécnica, Av. Vitória, 1729, Jucutuquara, Vitória – ES, Brazil
e-mail: rodrigova@cefetes.br

A. Naït-Ali (ed.), *Advanced Biosignal Processing*,

DOI 10.1007/978-3-540-89506-0_4, © Springer-Verlag Berlin Heidelberg 2009

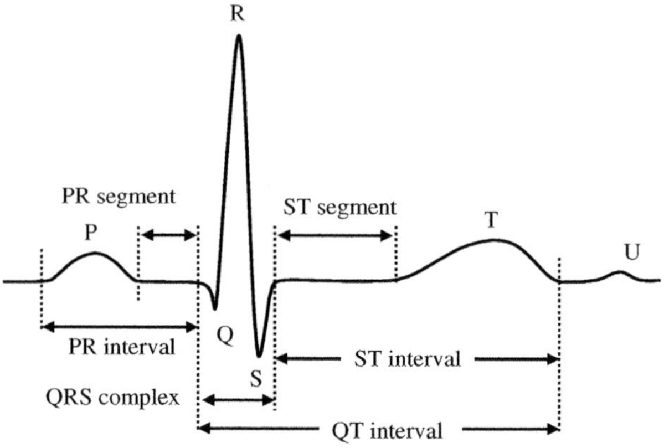

Fig. 4.1 Heart beat observed on an ECG with its elementary waveforms and intervals identified

used so far. Another strategy became popular during the nineties with the introduction of the neural networks and other advanced statistical approaches in the problem of heart beat modeling. By the construction of labeled ECG databases, statistical approaches learn the way the expert classifies the ECG without the need of direct explanation of the expertise since his knowledge about the signal is coded through manually made labels.

Heart beat modeling can be related to entire beat cycle but also to the elementary events which compose the cycle, what we'll call here the elementary waveforms. The two types of modeling are linked: if one wants to identify the elementary waveforms, he has first to detect each beat in the ECG signal. Moreover, to classify the beats, the knowledge of the onset and offset of a complete cycle is necessary, starting from a P wave (if it isn't present, the cycle starts from the QRS) and finishing at the offset of the T wave.

From the scientific literature in the field of computer ECG analysis, we find that most works employ a set of heuristic rules to model and classify the heart beats. The procedure consists first in segmenting each heart beat automatically from the ECG signal after performing a suitable signal processing technique [23, 25, 26, 28, 30, 36, 38, 39]. Then, the elementary beat waveforms of each beat are properly identified by another combination of signal processing techniques and heuristic rules [26, 28, 30, 36]. With the information of each beat, the classification is finally performed [21, 13, 19].

One of the drawbacks of these rule-based segmentation methods is their difficulty to adapt to the variability that can be encountered when dealing with several persons and different types of anomalies. Statistical methods, relying on learning processes, can therefore be envisaged to provide a solution to these problems.

In practice, most of the statistical approaches have been proposed to carry out only the beat classification part [24, 12, 8, 19, 3, 15]. The segmentation itself is realized thanks to the ECG beat morphological information obtained through heuristic rules. Nevertheless, the Coast's pioneer work [11, 12] based on hidden Markov models (HMM) performed both beat modeling and classification through a unique statistical approach. The HMM replaced the heuristic rules commonly used for beat modeling, which generally requires thresholds. Based on this idea and on the fact that P wave is a low amplitude signal often disturbed by noise, a particular application of HMM modeling for P-wave delineation was also proposed [10]. Delineation means waveform segmentation with the purpose of identifying precisely the point representing the onset and offset of the waveform.

A more recent system introduced in [2] proposes the modeling of each elementary wave by a specific HMM, allowing this way to take into account the morphological diversity of each elementary beat waveform. Moreover the authors have also introduced an incremental HMM training for model adaptation to each individual [2].

Contrary to the segmentation approach proposed above which explicitly takes into account the temporal nature of the signal, multiresolution analysis with continuous or discrete wavelets [1, 37, 28] was used for this segmentation (or delineation) purpose, where waves are seen at different scales, and where their transitions can be observed as high values of wavelet coefficients. This method has its own advantages and drawbacks. It is accurate, but sensitive to noise. Wavelets and statistical methods, which are more robust to noise, can be used complementarily for wave delineation associating local and global segmentation. By combining wavelet analysis and Markov models, Hidden Markov Trees were developed [14] and their application to ECG delineation [18] showed a robust and accurate behavior.

The aim of this chapter is to give a comprehensive description of the HMM adapted to the problem of beat modeling and classification and the HMT for ECG delineation.

4.2 Hidden Markov Models

4.2.1 Overview

The basic theory of hidden Markov models was developed by Baum et al. at the end of the sixties [6] and was introduced in the field of automatic speech recognition during the seventies. The description given below is a short overview, and for more details please refers to Rabiner et al. [33].

A hidden Markov model is a stochastic state machine, characterized by the following parameter set:

$$\lambda = (\mathbf{A}, \mathbf{B}, \pi) \tag{4.1}$$

where **A** is the matrix of state-transition probabilities, **B** is the observation probability and π is the initial state probability.

The HMM models a sequence of events or observations by combining two main properties:

- the observation $\mathbf{o}_t$ is generated by a stochastic process whose state is hidden to the observer;
- we suppose that the hidden state qt satisfies a Markov process $P(q_t = j|q_{t-1} = i, q_{t-2} = h, \ldots) = P(q_t = j|q_{t-1} = i)$, which means that the current state q_t depends only on the previous one q_{t-1}.

One way to characterize the topology of the HMM is by the structure of the transition matrix $\mathbf{A}$:

$$\mathbf{A} = \begin{bmatrix} a_{11} & \cdots & a_{1N} \\ \vdots & \ddots & \cdots \\ a_{N1} & \cdots & a_{NN} \end{bmatrix} \tag{4.2}$$

where

$$\sum_{j=1}^{N} a_{ij} = 1, \quad \forall i, j \in \mathrm{N}. \tag{4.3}$$

It can be fully connected (ergodic or $a_{ij} \neq 0 \forall i, j$) as shown in Fig. 4.2. However, the left-right structure is more appropriated to model the ECG signal. One such structure is shown in Fig. 4.3.

The HMM represents the observation probability distribution of an observation sequence $\mathbf{O} = (\mathbf{o}_1\mathbf{o}_2 \cdots \mathbf{o}_T)$, where $\mathbf{o}_t$ can be a symbol from a discrete alphabet or a real or an integer number. We consider that the observations are sampled by discrete and regular intervals, where t represents a discrete time instant. Moreover, the observations are considered independent and identically distributed (also called iid process).

The observation probabilities are assigned to each model state as follows:

$$b_j(\mathbf{o}_t) = P(\mathbf{o}_t|q_t = j) \tag{4.4}$$

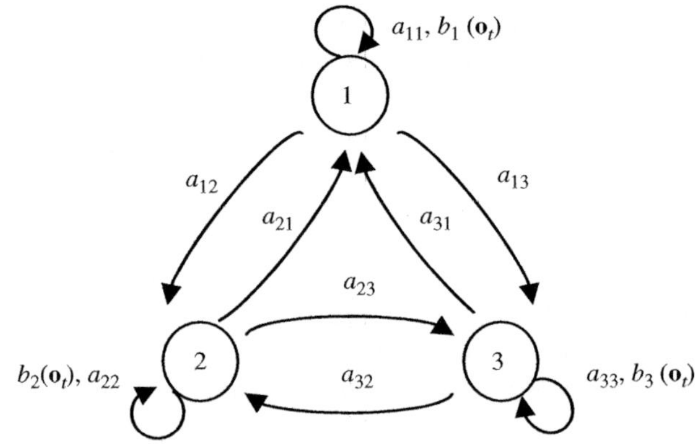

Fig. 4.2 Graphical representation of an ergodic hidden Markov model

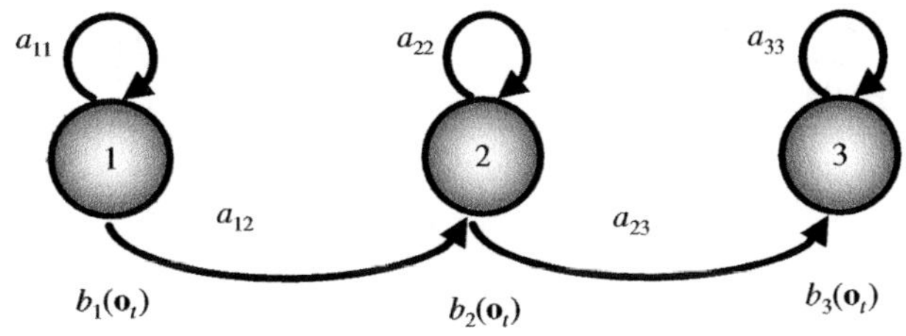

Fig. 4.3 An example of a 3-state *left-right* HMM [4]

where q_t is the model state at time t and $b_j(\mathbf{o}_t)$ is the probability of observation $\mathbf{o}_t$ given the state q_t. We assume that $b_j(\mathbf{o}_t)$ is independent from states and observations of other time instants. If we consider that $\mathbf{O} = (\mathbf{o}_1\mathbf{o}_2\cdots\mathbf{o}_T)$ are continuous signal representations (signal features), modeled by a Gaussian probability density function, then

$$b_j(\mathbf{o}_t) = \frac{1}{\sqrt{2\pi|\mathbf{U}_j|}} \exp\left\{-\frac{1}{2}(\mathbf{o}_t - \boldsymbol{\mu}_j)^T \mathbf{U}_j^{-1}(\mathbf{o}_j - \boldsymbol{\mu}_j)\right\} \tag{4.5}$$

where $\mathbf{o}_t$ is the observation vector at time t, $\boldsymbol{\mu}_j$ is the mean vector and $\mathbf{U}_j$ is the covariance matrix at state j. The size of the observation vector $\mathbf{o}_t$ is related to the number of distinct observation symbols used to represent the signal.

Considering the fact that the HMM is a stochastic model, its parameter set $\boldsymbol{\lambda}$ is estimated with the purpose of maximizing the modeling of the observations, i.e., maximizing locally the likelihood $P(\mathbf{O}|\boldsymbol{\lambda})$ of the model $\boldsymbol{\lambda}$ using an iterative procedure such as the Baum-Welch method (which is a particular case of the expectation-maximization method) or using gradient techniques [33]. In the speech recognition field, the Baum-Welch method, also called forward-backward algorithm, is considered as a standard for HMM training. Furthermore, in biomedical applications using HMMs, this method is also widely employed [13].

With the parameter set $\boldsymbol{\lambda}$ already estimated, the HMM is ready to be used to measure the likelihood of the observations given the model $P(\mathbf{O}|\boldsymbol{\lambda})$. To perform this task, two methods are available: Viterbi algorithm and forward algorithm [33].

4.2.2 Heart Beat Modeling

As discussed earlier, a heart beat can be seen as a waveform sequence, separated by isoelectric segments (PQ and ST segments) (see Fig. 4.1). Moreover, these waveforms are produced cyclically. Therefore, it is reasonable to consider each waveform or segment as a left-right Markov model (see Fig. 4.4). As a result, by connecting each elementary HMM, we obtain the beat model. If we extend that to the ECG signal, connected beat models will represent the whole ECG.

The relation presented above and shown in Fig. 4.4 works for a normal beat model but it isn't general enough to take into account different beat types. Indeed, if we take a beat characterized by an atrial activity not conducted (i.e., a heart beat

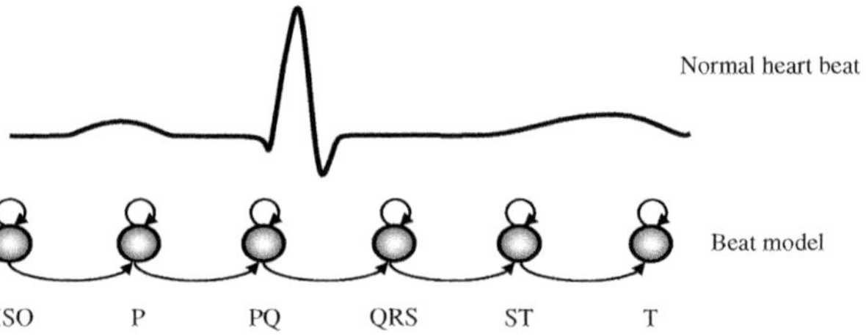

Fig. 4.4 A beat model composed of connected HMM of each beat waveform and segment

where the P wave is not followed by a QRS complex), it is easy to verify that the beat model assumes that the P wave is necessarily followed by a state transition towards the QRS complex. In order to make the beat model more general, it is necessary to introduce new arcs or transitions among the waveform models, as follows:

- Transition from P wave model to ISO model: this transition represents P waves not conducted by a ventricular activity (typical symptom of bundle block [20]);
- Transition from ISO to QRS model, skipping P wave model: in this case, ventricular and supraventricular arrhythmias without visible P wave can be modeled [20];
- Transition from T wave to ISO model: it allows modeling a sequence of beats.

The final beat model is presented in Fig. 4.5. It is important to point out that this model is consistent with the constraints of the heart electrical activity.

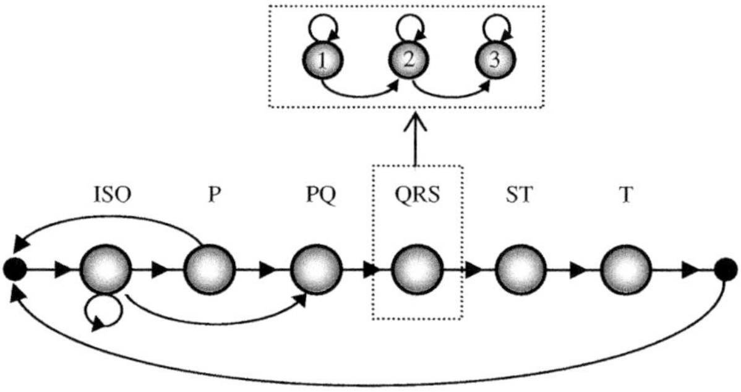

Fig. 4.5 Beat model composed of connected HMM of each beat waveform and segment. The transition from P to ISO models ECG signals with P waves not conducted by a ventricular activity. The transition from ISO to PQ models ECG signals with supraventricular arrhythmias without visible P wave. (© 2006 IEEE)

4.2.3 Beat Classification

The beat classification problem using HMMs can be performed directly by the HMM [12]. The main idea is to consider beat models as specific to each beat type. To understand the method, let's take as an example the premature ventricular contraction beat (PVC). It is know that PVC beats have two well defined features which are sufficient to distinguish this abnormality from the other ones. The first feature is related to beat prematurity. It means that the PVC beat is characterized by an R-R

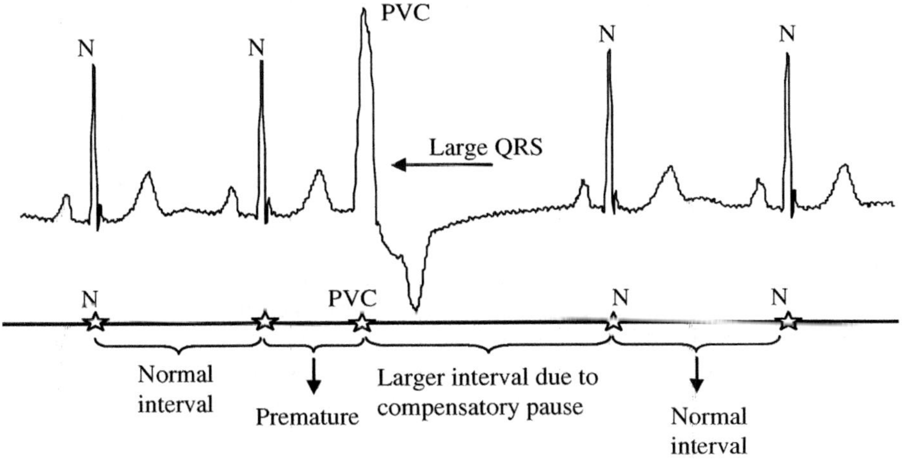

Fig. 4.6 Premature ventricular beat characterized by: (i) a interval shorter than the previous one; (ii) a compensatory pause; and (iii) QRS-complex wider than the one of the normal beats (N)

interval shorter than the previous one (see Fig. 4.6) [20]. In most of the cases, a PVC beat is also followed by a compensatory pause. The second feature concerns the QRS-complex morphology. In fact, a PVC beat is also one type of ventricular beat, which is characterized by a QRS-complex wider than the normal beats (N) (see Fig. 4.6) [20]. Thus, a PVC beat is a premature and a ventricular beat at the same time.

From the information about the beat type, the beat model is constructed. In the case of the PVC beat, the model will not have a state assigned to the P wave. Moreover, a state must be introduced after the last state to take into account the time spent by compensatory pause. It is important to remark that interval durations are modeled by the state transitions.

Finally, to classify the whole beat sequence of Fig. 4.6, two beat models are needed: one for normal beats and one for PVC beats. Figure 4.7 shows the beat sequence modeled by HMMs. It is important to note that the model of Fig. 4.7 may represent either the states of one HMM as proposed in [12] or connected HMMs of each waveform as described in Sect. 4.2.2.

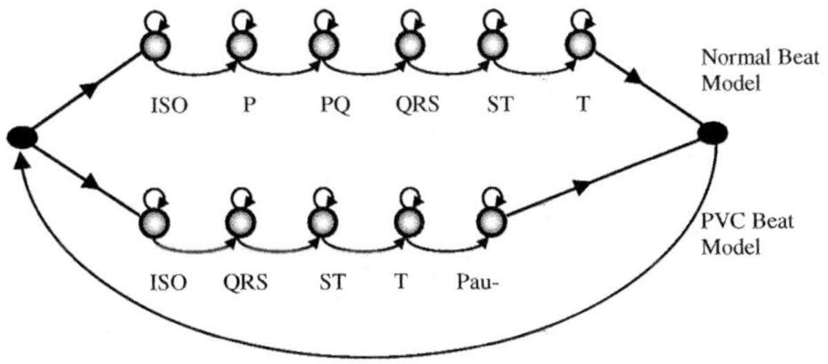

Fig. 4.7 Graphical representation of two HMMs connected to model a beat sequence composed of Normal and PVC beats

4.2.4 HMM for Beat Segmentation and Classification

4.2.4.1 Parameter Extraction

Parameter extraction is the front-end of statistical modeling. Features must be extracted from the signal in order to provide relevant information which compactly represents the signal. When dealing with HMMs, the information extracted from the signal corresponds to the observation sequence.

The ECG signal has some particularities which must be considered during this phase: amplitude offset, i.e., the isoelectric line isn't really placed at 0 mV (signal not centered); noise affecting typically the isoelectric line and P wave. To face that, the parameter extraction strategy must act as a band-pass filter, removing the DC offset (i.e., 0 Hz) and the noise. The strategy which has been successfully applied for that is based on the wavelet transform.[1] Indeed, it implements:

- A multiresolution analysis: the signal is decomposed in different scales, corresponding to different frequency bands. Thus, regarding the signal spectrum contents, it allows to take into account only the scales where the useful information is present and the signal-to-noise ratio is larger.
- A localized and transitory event analysis: time-frequency methods are suitable to analyze possible time evolution of the signal spectrum contents [16].

Figure 4.8 shows the ECG signal transformed into three different signals, each one corresponding to a particular Mexican Hat wavelet scale (second derivative of the Gaussian function). In fact, the transformation consists in band-pass filtering the signal using the Mexican Hat function at different scales as follows:

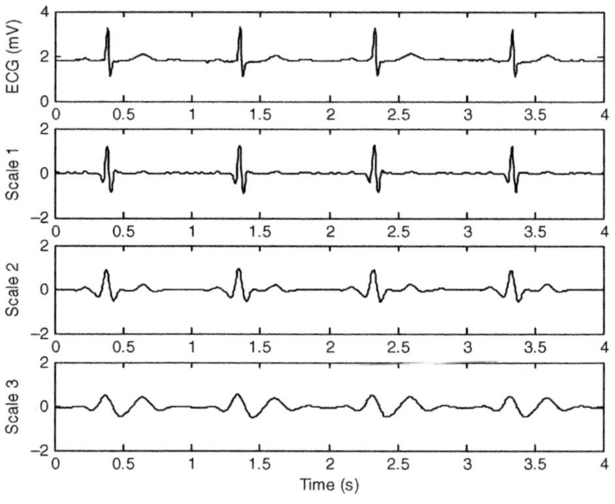

Fig. 4.8 The ECG signal and its representation at 3 dyadic scales (scale $s = 2^j$, where $j = 1,2$ and 3) using Mexican Hat wavelet transform [4]

[1]For more details on wavelet transforms, the reader may refer to [27].

$$Wf(n,j) = \sum_{m=0}^{M-1} f[m] \times \overline{\psi}_j[m-n] \tag{4.6}$$

$$\overline{\psi}_j[n] = \frac{1}{\sqrt{2^j}} \frac{2}{\sqrt{3}\pi^{1/4}} \left[1 - \left(\frac{n}{2^j}\right)^2\right] \exp\left[\frac{1}{2} \times \left(\frac{n}{2^j}\right)^2\right] \tag{4.7}$$

where f is the sampled signal composed of M samples, $\overline{\psi}_j$ is the Mexican Hat wavelet function at the dyadic scale j for $j \in \mathrm{N}$ (dyadic means power of two), and $-5 \leq n \leq 5$ for $n \in \mathrm{Z}$.

For the example above, the observation sequence generated after the parameter extraction is of the form $\mathbf{O} = (\mathbf{o}_1 \mathbf{o}_2 \cdots \mathbf{o}_T)$, where T is the signal length in number of samples and each observation $\mathbf{o}_t$ is a vector whose size means the number of scales. It is important to point out that the wavelet scales have the same time resolution as the original signal.

Some other wavelet mother shapes were retained for comparison to the Mexican Hat, namely first derivative of Gaussian, Paul, Morlet and B-Spline wavelets [4]. They gave interesting results, but Mexican Hat appeared as the best trade-off for various sub-beat frequencies (P, T of low-frequencies versus QRS of higher frequency). Also combined or extended schemes of static and dynamic extraction parameters (such as 1st derivative and Mexican Hat) are deeply detailed in [4].

4.2.4.2 Training HMMs

The HMM training consists of estimating the model parameter set $\boldsymbol{\lambda} = (\mathbf{A}, \mathbf{B}, \boldsymbol{\pi})$ from the observation sequence $\mathbf{O}$. The observation sequence can be a beat waveform (P or T waves, QRS complex, PQ and ST segment) or a beat type (normal or PVC beat). The goal is to build a generic system which works independently of the nature of the ECG signal. Certainly, considering the ECG signal diversity of different individuals, we cannot expect from such a system an optimal result for each individual.

The HMM parameter estimation using the Baum-Welch method (expectation-maximization method) requires labeled datasets [33, 7]. It consists in a reverse problem: to find the stochastic process (HMM) which generated the observation sequences of the training set. Each HMM is adapted to its respective set of patterns or morphologies, as illustrated in Fig. 4.9.

The learning procedure starts with the parameters initialization. The matrix $\mathbf{A}$ is initialized using a uniform distribution. As regards to the vector $\boldsymbol{\pi}$, the first state probability π_1 is one while the remaining states have probability π_i zero. Considering that the observations are real numbers, observation probability parameter $\mathbf{B}$ is given by probability density functions (pdf). Gaussian pdf is typically used. However, for a suitable modeling, it is necessary to study the probability distribution of the observation set at hand. The study consists in segmenting the observation sequence uniformly by the number of states. After repeating that procedure for all examples from the training set, the statistical behavior of the observations at each

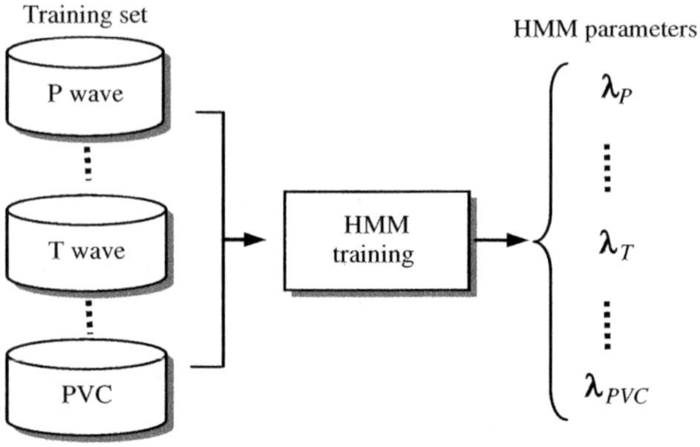

Fig. 4.9 HMM training. Each model is trained on its respective training set composed of patterns or morphologies from several individuals

state is obtained through histograms. If a Gaussian pdf is a good fit for the histograms, then the parameters $\boldsymbol{\mu}_j$ and $\mathbf{U}_j$ at each state j can be directly estimated. Finally, starting from the initial model $\boldsymbol{\lambda}_{Initial}$, the model parameters are adjusted during a training phase in order to maximize the likelihood $P(\mathbf{O}|\boldsymbol{\lambda})$ of the observation sequence given the model $\boldsymbol{\lambda}$. The training phase stops (i.e., the convergence is achieved) either when the maximum number of iterations is reached or when the likelihood between two consecutive iterations is below a threshold [33].

In the work of Andreão et al. [2], it was proposed the use of multiple HMMs for each pattern (beat waveform or beat type). Actually, the number of models for each pattern depends on the variety of morphologies present in the training set. This issue increases the model complexity. It can be easily noticed that among the beat waveforms, the QRS-complex is the one which has the largest variability. As a result, the number of models used to represent the QRS-complex is greater than the number of models of the other waveforms. The training algorithm is called HMM likelihood clustering, and was firstly applied by Rabiner to the speech recognition problem [34].

4.2.4.3 Classifying Patterns

The classification step can be seen as the decoding procedure of an observation sequence in terms of beat waveforms and beat types. The main point of the decoding procedure when dealing with HMMs is the use of the one-pass algorithm [33], which was originally conceived to perform online decoding when working with connected HMMs. This method has been widely employed in the speech recognition field [33].

The one-pass method reduces significantly the complexity of the decoding problem. It works on two dimensions: time and level. In the speech recognition problem, each level corresponds to a word in a sentence (or an utterance in a word). For the problem of beat modeling, we have associated the level to the waveform position in the beat model as shown in Fig. 4.10. Hence, level 1 represents the isoelectric line or ISO model, level 2 the P wave model, and so on until level 6 which represents the T wave model. The same association is carried out when working with beat

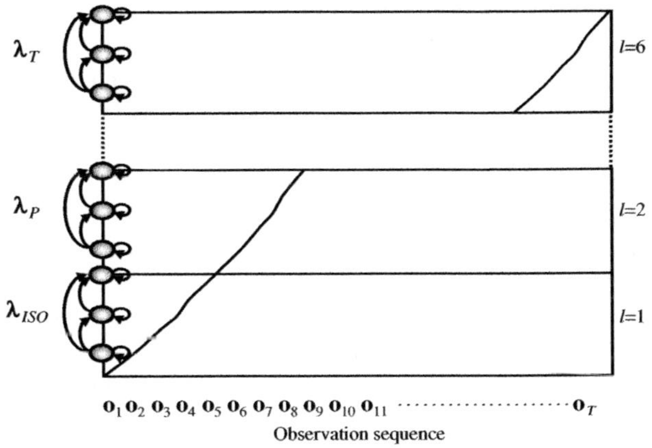

Fig. 4.10 Observation sequence decoding via the one-pass method. The most likely HMM is associated to its respective observation sequence (*diagonal line*), which represents one specific waveform (or beat type). The number of levels *l* corresponds to the number of beat waveforms (or beat types)

classification. In this case, each level of Fig. 4.10 will be assigned to a beat type modeled by HMMs.

The main idea of the method is to perform a time warping between the observation sequence and the connected HMMs through a Viterbi decoding procedure. However, to pass from one level *l* to another level *l*+1, we only consider the most likely model from level *l*. Hence, we avoid a time consuming procedure which tries unnecessary combinations of models. Its efficiency is still more significant when multiple models are employed to represent the same pattern (beat waveform or beat type) at each level.

4.2.5 HMM Adaptation

Using the concepts described above, we obtain a model, called generic, trained on a large set composed of examples from several individuals. This model is able to provide waveform segmentation or beat classification of a given beat sequence no matter the individual, even if the individual is not present in the training set.

However, the performance of the system in terms of segmentation precision (particularly on the P wave) decreases when working on signals which are very different from those present in the training set. It is clear that the P wave segmentation requires a very accurate modeling due to its low amplitude. Additionally, when performing online monitoring of the ECG of an individual, system adaptation becomes a very important tool, for tracking signal variations caused by any change in the state of a patient along time. For this reason, it is expected a significant performance improvement after adapting the generic model to the individual's specific waveforms.

The generic model adaptation corresponds to the re-estimation of each generic HMM on a new training set (specific to the individual) via a training method. In fact, adaptation is performed each time a representative training set of one waveform is available. The training set is built from the segmentation and labeling of the ECG signal in an unsupervised way by the one-pass algorithm.

Nevertheless, the classical approaches for HMM training, namely expectation-maximization (EM), segmental k-means and Maximum a Posteriori (MAP), make the training schemes of the HMM computationally intensive, because the corresponding algorithms require multiple iterations to get model convergence. Furthermore, it is necessary to pass through all the data to re-estimate the models, because the training set is composed of the examples used to build the generic model and the examples specific to the individual. To reduce the complexity of the re-estimation procedure, incremental HMM training approaches have been proposed [32, 5] to improve the speed of convergence, compared to the classical training methods, and to reduce the computational requirements, thus making them suitable for online applications.

Indeed, Andreão et al. [5] have adapted the idea of incremental HMM training originally proposed for speech recognition to the problem of beat segmentation. The adaptation works in the following way: each time an elementary waveform is segmented, the corresponding observation vector is placed in the waveform receipt (see Fig. 4.11), and as soon as 30 observation vectors per Gaussian pdf are available for one waveform, the corresponding model parameters are re-estimated through an incremental training algorithm. The training methods are incremental versions of the expectation-maximization (EM), segmental k-means and Maximum a Posteriori (MAP) algorithms.

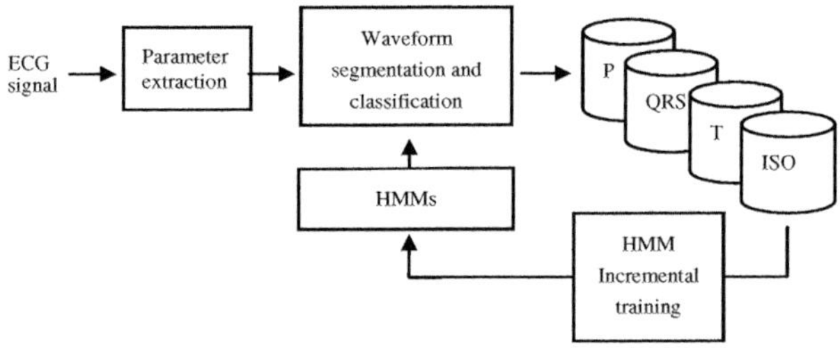

Fig. 4.11 General block diagram of the procedure for HMM adaptation

4.2.6 Discussion

The HMM approach described above has been successfully tested for the problem of ECG segmentation. According to Andreão et al. [2], it reaches similar performances to those considered as the best one in the literature on the QT database [30, 27]. In some particular cases where the P wave is detected with a great difficulty, it is convenient to perform manual labeling of these waves yielding to a suitable P-wave model re-estimation. Indeed, the patient model results showed that a small number of examples is enough to train appropriately the HMM. All these considerations are not valuable for the heuristic approaches. It is still possible to improve this HMM system by modeling explicitly the waveform duration [34]. As a result, we could avoid segmenting the waveforms whose duration deviates too much from the estimated one.

The HMMs are statistical tools efficient for robust ECG signal segmentation able to face the variability of the waveform shapes and allowing the replacement of heuristic rules commonly employed to waveform detection. Moreover, the HMM can be adapted during online segmentation (Sect. 4.2.5) in order to track small signal variations caused by any change in the state of a patient along time. Nevertheless, transitory changes on the waveforms are not correctly processed by statistical approaches. This can be observed in particular for individuals suffering from ischemia. Indeed, the ST-T complex amplitude changes in time during an ischemic episode. In order to overcome this problem, Andreão et al. added some rules dependent on the QT-interval duration [35].

4.3 Hidden Markov Trees

4.3.1 Overview

The Hidden Markov Tree (HMT) model is based on the persistence property of the wavelet transform: the large/small values of wavelet coefficients tend to propagate across scales. By associating to each wavelet coefficient, which will be the observation process, a state that measures its energy, the states can be connected by a probabilistic graph. This gives the hidden Markov tree (see Fig. 4.12). In the case of two hidden states, with high (H) and low (L) energy, this tree is characterized by the following transition matrix:

$$A_i = \begin{bmatrix} p_i^{L \to L} & p_i^{H \to L} \\ p_i^{L \to H} & p_i^{H \to H} \end{bmatrix} \tag{4.8}$$

where $p_i^{x \to y}$ is the probability of the energy state change from x to y.

The observation distribution will be modeled by a mixture of Gaussian distributions. For example, in the case of two hidden states, this distribution will be:

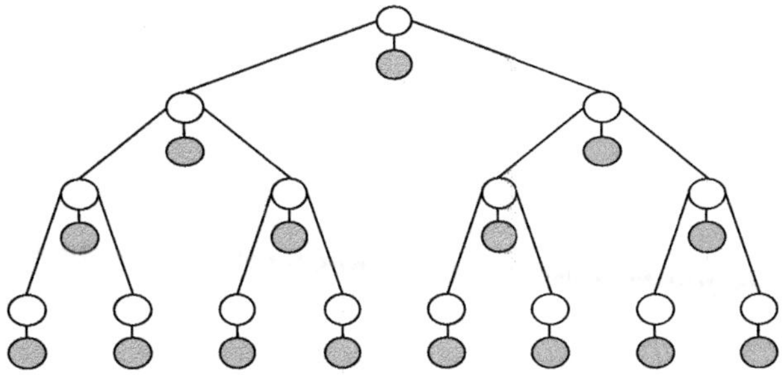

Fig. 4.12 Wavelet tree. *White dots*: hidden state variable; *black dot*: wavelet coefficient

$$f(w_i) = p_i^H \times f(w_i|S_i = H) + p_i^L \times f(w_i|S_i = L) \tag{4.9}$$

where:

- p_i^H (resp. p_i^L) are the *prior* probabilities that w_i is in the high (resp. low) state of energy.
- $f(w_i|S_i = H) = \mathcal{N}(\mu_{i,H}, \sigma_{i,H})$ and $f(w_i|S_i = L) = \mathcal{N}(\mu_{i,L}, \sigma_{i,L})$ are conditional distribution of the wavelet coefficients in function of the state and where $\mathcal{N}(\mu, \sigma)$ is the normal low with mean μ and standard deviation σ.

The HMT parameters are:

- $p_{S_i}(m)$: the prior probability of the state S_i.
- $\varepsilon_{i,\rho(i)}^{mr}$: the transition probabilities between states in the tree structure.
- $\mu_{i,m}, \sigma_{i,m}$: the mean and the variance of the Gaussian distribution.

All these parameters are grouped in a vector $\theta = \{p_{S_i}(m), \varepsilon_{i,\rho(i)}^{mr}, \mu_{i,m}, \sigma_{i,m}\}$. In the wavelet tree, the parent of w_i is its immediate ancestor and is denoted by $w_{\rho(i)}$.

4.3.2 Electrocardiogram Delineation by HMT

The ECG delineation by HMT was proposed by Graja and Boucher [18]. The cardiac cycle is decomposed into five classes (see Fig. 4.13) corresponding to the main waves or isoelectric segments. The idea is to characterize each class by an HMT model. So the three following steps are used for ECG segmentation:

- Model training: estimation of the parameter vector $\boldsymbol{\theta}$ for each class. The training data are wavelet coefficients of one class C_l. For parameter estimation the maximum likelihood (ML) criterion is approached by an EM algorithm because it is an incomplete problem case.

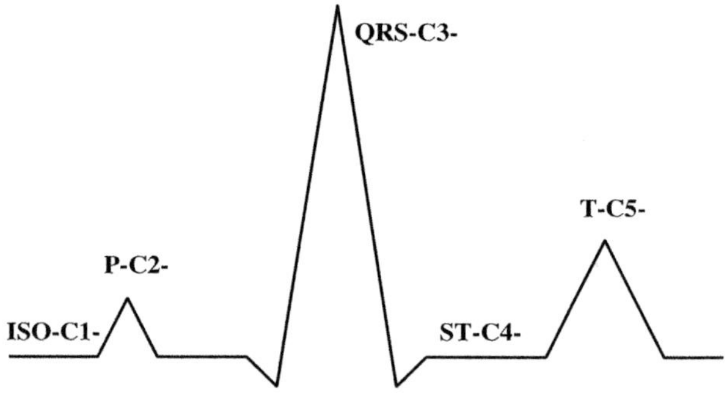

Fig. 4.13 ECG class description

- Scale segmentation: identification of the limit of each class. The wavelet coefficients are classified by ML at each scale: $c_i^{ML} = \underset{l}{ArgMax}\,[f(w_i|\boldsymbol{\theta}_l)]$.
- Inter-scale fusion: to exploit the time dependency between wavelet coefficients, called context. This step is used to refine the scale segmentation.

4.3.2.1 Model Training

For each class, the HMT parameter vector might be estimated by ML:

$$\hat{\boldsymbol{\theta}}_{ML} = \underset{\boldsymbol{\theta}_l}{ArgMax}\,[f(\mathbf{W}, \mathbf{S}\,|\boldsymbol{\theta}_l)] \tag{4.10}$$

where $\mathbf{W}$ is the Haar wavelet coefficient vector of the class and $\mathbf{S}$ the hidden states vector. Since $\mathbf{S}$ is unobserved then the direct ML estimation cannot be done. This problem is solved by the following EM algorithm [14].

- **E step**: estimate the hidden state's probabilities of each wavelet coefficient by propagating the hidden state information once up the tree and then once down the tree. For this, the up down variables β_i and α_i defined as follows are introduced:

$$\beta_i(m) = f(\mathbf{T}_i\,|S_i = m, \boldsymbol{\theta}) \tag{4.11}$$

$$\alpha_i(m) = f\left(S_i = m, \mathbf{T}_{1/i}\,|\boldsymbol{\theta}\right) \tag{4.12}$$

where: $\mathbf{T}_i$ is defined to be the subtree of observed wavelet coefficients with root at node i and $\mathbf{T}_{i/j}$ to be the set of wavelet coefficients obtained by removing the subtree $\mathbf{T}_j$ from $\mathbf{T}_i$.

Finally the hidden state probability estimations are given by:

$$p(S_i = m|\mathbf{W}, \boldsymbol{\theta}) = \frac{\alpha_i\,(m)\,\beta_i\,(m)}{\sum_{n=1}^{M} \alpha_i\,(n)\,\beta_i\,(n)} \tag{4.13}$$

$$p\left(S_i = m, S_{\rho(i)} = n\,|\mathbf{W}, \boldsymbol{\theta}\right) = \frac{\alpha_{\rho(i)}\,(n)\,\beta_{\rho(i)/i}\,(n)\,\beta_i\,(m)\,\varepsilon_{i,\rho(i)}^{nm}}{\sum_{n=1}^{M} \alpha_i\,(n)\,\beta_i\,(n)} \tag{4.14}$$

- **M step**: update the HMT parameters vector so that the likelihood function is maximized.

$$p_{S_i}(m) = \frac{1}{K}\sum_{k=1}^{K} p\left(S_i^k = m\middle|\mathbf{W}^k, \boldsymbol{\theta}\right) \tag{4.15}$$

$$\varepsilon_{i,\rho(i)}^{nm} = \frac{1}{Kp_{S_{\rho(i)}}(m)} \sum_{k=1}^{K} p\big(S_i^k = n,\, S_{\rho(i)}^k = m \big| \mathbf{W}^k, \boldsymbol{\theta}\big) \tag{4.16}$$

$$\mu_{i,m} = \frac{1}{Kp_{S_i}(m)} \sum_{k=1}^{K} w_i^k p\big(S_i^k = m \big| \mathbf{W}^k, \boldsymbol{\theta}\big) \tag{4.17}$$

$$\sigma_{i,m}^2 = \frac{1}{Kp_{S_i}(m)} \sum_{k=1}^{K} \left(w_i^k - \mu_{i,m}\right)^2 p\big(S_i^k = m | \mathbf{W}, \boldsymbol{\theta}\big) \tag{4.18}$$

From a database created by the cardiology unit at Brest Hospital for the study of atrial fibrillation risks [10], lead II of healthy and ill patients is extracted and used, after sampling at 1 KHz. Based on a manual delineation made by cardiologists, the training base consists of 10 ECG including 10 beats, which give one hundred waves for each class. Then we carry out signal decomposition by an orthogonal Haar wavelet transform until the third scale level is reached. The statistical distribution of the coefficients of each scale corresponds to the mixture of three Gaussian distributions.

4.3.2.2 Scale Segmentation

The aim of this step is to determine the class limits at each scale. As the transition between the ECG waves correspond to high values of these coefficients, a good classification of their values can be used for beat delineation. So, the Haar wavelet transform is applied to each ECG beat then the wavelet coefficients are classified by ML at each scale:

$$c_i^{ML} = \underset{l}{ArgMax}\left[f(w_i\, |\boldsymbol{\theta}_l\,)\right] \tag{4.19}$$

The wavelet distribution at each scale is a mixture of three Gaussian distributions, so we can write:

$$f(w_i\, |\boldsymbol{\theta}_l\,) = \sum_{m=1}^{3} f(w_i\, |S_i = m,\, \boldsymbol{\theta}_l\,)\, p_{S_i}(m) \tag{4.20}$$

$f(w_i|S_i = m, \boldsymbol{\theta}_l)$ is the β_i variable computed at the E step of the EM algorithm, and $p_{S_i}(m)$ is the state prior probability obtained from the HMT training step.

In addition, because of the wave shape variability of an ECG, a normalization step is necessary. The R peak is then detected and three windows (central, left and right) are opened to select parts of the ECG beat. The Central window processes the QRS complex and isoelectric line, the right window processes the T wave and the isoelectric line and the left window processes the P wave and the isoelectric line.

Figure 4.14 shows an example of P-wave segmentation on the three first scales. At scale 3, results are more robust than those on scales below. However, the temporal

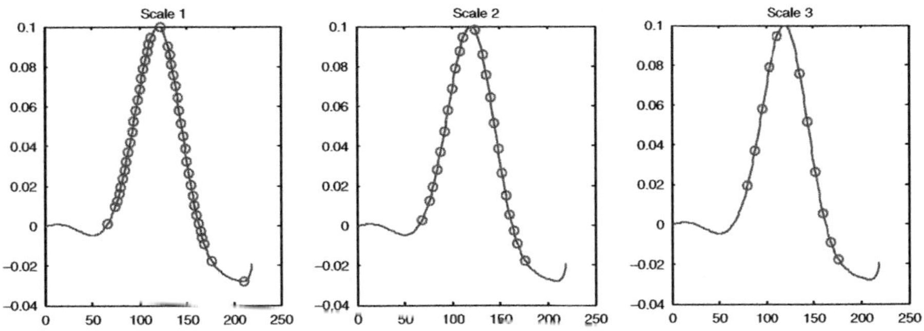

Fig. 4.14 Delineation of P wave at the first three scales

resolution at this scale is greater. In scale 1 and 2 we improve the temporal resolution but we loose in robustness. To obtain high quality segmentation, the multiscale results must be combined to benefit from the robustness of the coarse scale and the resolution of the finer scales. To this end, a fusion procedure between scales including context is undertaken.

4.3.2.3 Inter-Scale Fusion

This idea was firstly proposed by Choi and Baraniuk [9]. To improve the scale segmentation, Graja and Boucher [18] fuse the decisions taken at the different scales by introducing the concept of context. Context is defined as the states of a temporal father's neighbors in the tree. Its definition must be chosen carefully; otherwise it could damage the segmentation instead of improving it. Let denote by $\mathbf{D}_i^j$ the sub tree to position i in scale j. In [18], the context of $\mathbf{D}_i^j$ is defined as a length-2 vector $\mathbf{V}_i^j$. The first element $\mathbf{V}_{i,1}^j$ is the class label of the parent $\mathbf{D}_{\rho(i)}^{j+1}$. The second element $\mathbf{V}_{i,2}^j$ is the majority labels of the subset formed by $D_{\rho(i)}^{j+1}$ and its two neighbors (right and left). Figure 4.15 illustrates the definition of this context.

To include information supplied by the context, the Maximum a Posteriori (MAP) criterion is used:

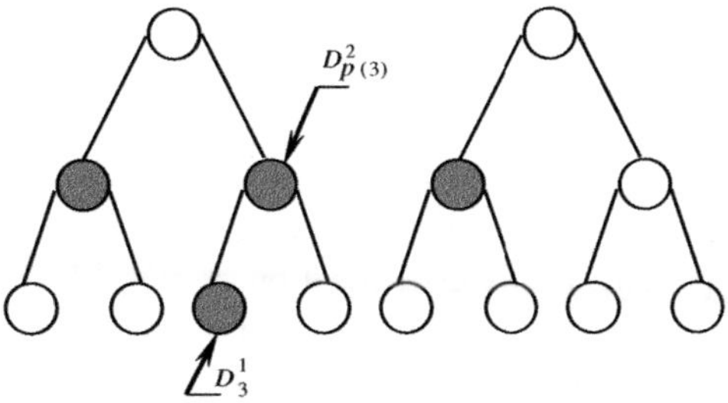

Fig. 4.15 Context definition for the subtreeD_3^1

$$c_i^{MAP} = \underset{c\in\{1,2,...,5\}}{ArgMax}\, p\left(c_i \left| \mathbf{D}_i^j, \mathbf{V}_i^j \right.\right) \quad (4.21)$$

The posterior distribution is defined by:

$$p\left(c_i \left| \mathbf{D}_i^j, \mathbf{V}_i^j \right.\right) = \frac{f\left(\mathbf{D}_i^j \left| c_i, \mathbf{V}_i^j \right.\right) p\left(c_i \left| \mathbf{V}_i^j \right.\right)}{f\left(\mathbf{D}_i^j \left| \mathbf{V}_i^j \right.\right)} \quad (4.22)$$

Assuming that $\mathbf{D}_i^j$ giving c_i is independent of $\mathbf{V}_i^j$, allows to simplify the MAP criterion:

$$c_i^{MAP} = \underset{c\in\{1,2,...,5\}}{ArgMax}\, f\left(\mathbf{D}_i^j | c_i\right) p\left(c_i \left| \mathbf{V}_i^j \right.\right) \quad (4.23)$$

$f\left(\mathbf{D}_i^j | c_i\right)$ is the likelihood calculated in scale segmentation. So there only remains the calculation of $p\left(c_i \left| \mathbf{V}_i^j \right.\right)$.

The Bayes rule permits to write:

$$p\left(c_i \left| \mathbf{V}_i^j \right.\right) = \frac{p\left(\mathbf{V}_i^j | c_i\right) p(c_i)}{p\left(\mathbf{V}_i^j\right)} \quad (4.24)$$

There only remains to determine the couple $\left\{p\left(\mathbf{V}_i^j | c_i\right), p\left(c_i\right)\right\}$. Another EM algorithm proposed by [14] permits to compute these two probabilities.

Finally, the decision criterion is:

$$c_i^{MAP} = \underset{c\in\{1,2,...,5\}}{ArgMax}\, f\left(\mathbf{D}_i^j | c_i\right) p\left(\mathbf{V}_i^j | c_i\right) p(c_i) \quad (4.25)$$

Figure 4.16 shows an example of P-wave segmentation with and without the inter-scale fusion step. By comparing to scale segmentation (see Fig. 4.14) we note that the use of context improves the segmentation by taking into account the robustness of the coarsest scale (scale 3) and the precision of the finest scale (scale 1). This leads to higher classification homogeneity and to an elimination of the fragmentary aspect of previous decisions.

For delineation performance measurement, a comparison was carried out with previous published results. Wavelet-based delineation (WT) is a recently proposed method [26, 30], which gives good performance and has been tested on standard databases such as MIT-BIH Arrhythmia, QT, European ST-T and the CSE database. The whole procedure described in this chapter was applied to a test base of 100 ECGs of 10 beats, registered for the automatic P-wave analysis of patients of patients for ECG delineation. A manual delineation was performed by a cardiologist on this test base, providing a reference. Automatic wave delineation was comparable to the manual one with similar mean values on onset and offset. As an error less

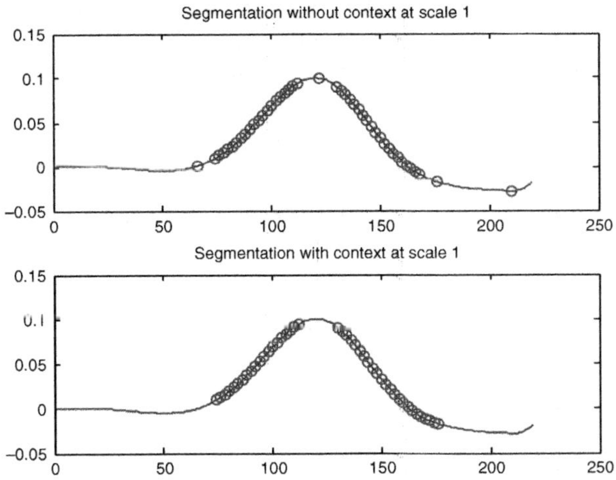

Fig. 4.16 P-wave segmentation with and without inter-scale fusion

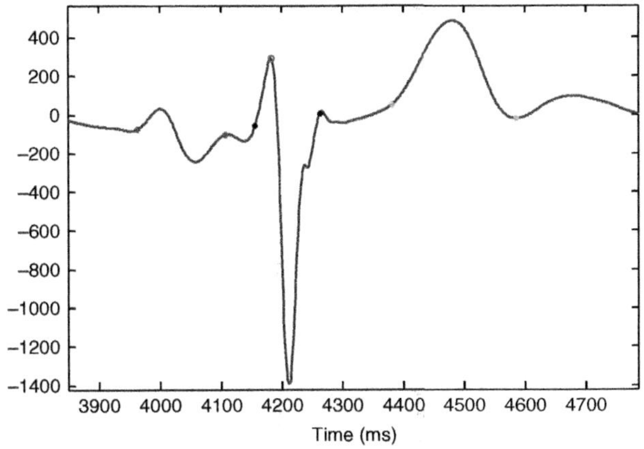

Fig. 4.17 Examples of ECG segmentation with diphasic P-wave

than 25 ms on the P-wave length is acceptable according to the cardiologist measurement norm, 13% of misclassifications for the P-wave, 11% for the QRS and 1% for the T-wave end were obtained. Figure 4.17 shows an example of ECG wave delineation.

4.3.2.4 Discussion

It can be seen that the standard deviation on the onset and end of the P-wave and QRS complex seems lower with the HMT method than with the method based on WT, which leads to the idea that the HMT method presents properties of robustness and low variance for ECG delineation.

A present drawback of the HMT method lies in the fact that it needs the detection of the R peak and the opening of a time window on the right and left of the R peak for the delineation of waves. As many algorithms [30, 17] give excellent results for the detection of the R peak, this can be viewed as a minor drawback.

4.3.3 Association of HMM and HMT

As previously mentioned, a drawback in using HMT for ECG delineation is the necessity of R peak detection and the use of the windows. To become independent of such a processing, an HMM can be introduced to model the temporal evolution of the waves into the beat. This can be done at each scale. As described at 4.2.1, in the learning phase and segmentation phase of HMM, one must compute the conditional distribution of the observation $b_j(\mathbf{o}_t) = P(\mathbf{o}_t|q_t = j)$. In this case, it is only the likelihood function of the wavelet coefficient of state q_j, which can be given by HMT, and this fact associates both algorithms. Each state of HMM is modeled by an HMT as shown on Fig. 4.18. It is important to notice that the HMMs combined here is just a classical version of the HMM described in Sect. 4.2, and it doesn't take into account all the possibilities of the HMM. Good results are obtained when signals of the learning base and beats to be processed have similar amplitude levels. Indeed this method is sensitive to energy variation in the ECG signal, and can lead to bad segmentation. For example, when the T wave has a small amplitude and after normalization by the beat energy the limit of this wave cannot be detected. But, by normalizing by 25% of the energy beat a good T wave delineation is obtained. A solution of this problem is to process each wave alone but this leads to the use of

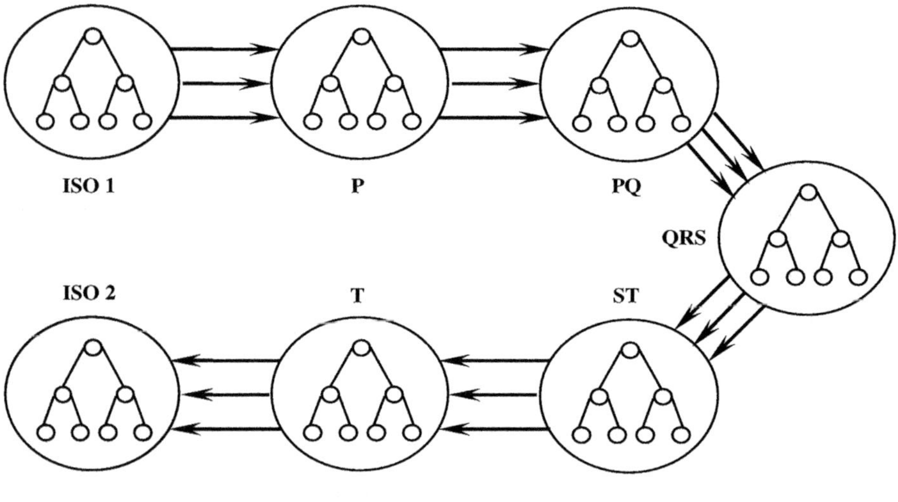

Fig. 4.18 Association of HMM and HMT models

windows. On conclusion, the association between HMT and HMM is still interesting for applications which do not require a local normalization.

4.4 Conclusions

This chapter described two Markov process-based statistical approaches for the problem of beat modeling and classification. The main difference between the two systems is that one (the HMM described in Sect. 4.2) takes into account the dynamics of the ECG, modeling successfully the beat sequence as a cyclostationary Markov process. The other approach (the HMT described in Sect. 4.3) performs the segmentation through each isolated beat, exploiting the persistence property of the wavelet transform: the large/small values of wavelet coefficients tend to propagate across scales, according to a Markov process. Consequently, the state isn't related to time but to the wavelet coefficient energy instead. Note that a transition between waves is associated to high values of the wavelet coefficients.

For both HMM and HMT approaches, experimental assessments have been performed for various pathological ECG contexts such as arrhythmia, ischemia and fibrillation detection or prediction.

The perspective that could be explored is to go towards the combination of these approaches aiming at obtaining a more robust modeling of the ECG. As was underlined before, the signal amplitude, which may vary in some recordings (like ambulatory ECG), must be carefully controlled since the statistical models depends on the signal amplitude used for training. Moreover, the use of rich databases and multiple models can help significantly the modeling.

Finally, both approaches were described for single lead configuration, and the lead combination was just performed in a post-processing phase. Nevertheless, the HMM and HMT can be improved in order to take into account the observations of each lead directly at the input of the model. In this case, it will be necessary to study more complex topologies for the HMM and HMT [31].

References

1. Addison PS (2005) Wavelet transform and the ECG: a review. Physiol Meas 26(1):155–199
2. Andreão RV, Dorizzi B and Boudy J (2006) ECG Signal Analysis through Hidden Markov-Models. IEEE Trans Biom Eng 53(8):1541–1549
3. Andreão RV, Dorizzi B; Cortez PC and Mota JCM (2002) Efficient ECG multi-level wavelet classification through neural network dimensionality reduction. In: Proc. IEEE Workshop on Neural Network for Signal Processing, Martigny, Suisse
4. Andreão RV and Boudy J (2007) Combining wavelet transform and hidden Markov models for ECG segmentation. EURASIP J Appl Signal Process. doi:10.1155/2007/56215
5. Andreão RV, Muller SMT, Boudy J et al. (2008) Incremental HMM Training Applied to ECG Signal Analysis. Comput Biol Med 38(6):659–667

6. Baum LE and Petrie T (1966) Statistical Inference for probabilistic functions of finite state Markov chains. Ann Math Stat 37:1554–1563
7. Boite R, Bourlard H et al. (2000) Traitement de la parole. Presses polytechniques et universitaires romandes, Lausanne
8. Bortolan G, Degani R and Willems JL (1990) Neural Networks for ECG classification. In: Computers in Cardiology, Chicago, USA
9. Choi H and Baraniuk RG (2001) Multiscale image segmentation using wavelet-domain hidden Markov models. IEEE Trans Image Process 10(9):1309–1321
10. Clavier L, Boucher JM, Lepage R et al. (2002) Automatic P-wave analysis of patients prone to atrial fibrillation. Med Biol Eng Comput 40(1):63–78
11. Coast DA and Cano GG (1989) QRS detection based on hidden Markov modeling. In: Proc. of the Annual International Conference of IEEE EMBS, Seattle, WA, USA
12. Coast DA, Stern RM et al. (1990) An approach to cardiac arrhythmia analysis using hidden markov models. IEEE Trans Biom Eng 37(9):826–836
13. Cohen A (1998) Hidden Markov models in biomedical signal processing. In: Proc. of the Annual International Conference of IEEE EMBS, Hong Kong, China
14. Crouse MS, Novak RD and Baraniuk RG (1998) Wavelet-Based statistical signal processing using Hidden Markov Models. IEEE Trans Signal Process 46(4):886–902
15. Elghazzawi Z and Gehed F (1996) A Knowledge-Based System for Arrhythmia Detection. In: Computers in Cardiology, Indianapolis, IN USA
16. Flandrin P (1998) Temps-Fréquence. Hermes, Paris
17. Friesen GM et al. (1990) A comparison of the noise sensitivity of nine QRS detection algorithm. IEEE Trans Biomed Eng 37(1):85–98
18. Graja S and Boucher J-M (2005) Hidden Markov Tree Model Applied to ECG Delineation. IEEE Trans Instrum Meas 54(6):2163 –2168
19. Hamilton P (2002) Open Source ECG Analysis. Comput Cardiol 29(1):101–104
20. Houghton A R and Gray D (2000) Maîtriser l'ECG de la théorie à la clinique. Masson, Paris
21. Jager F, Mark RG et al. (1991) Analysis of transient ST segment changes during ambulatory monitoring using the Karhunen-Loève transform. In: Computers in Cardiology, Durham, USA
22. Kadish A et al. (2001) ACC/AHA Clinical Competence Statement on Electrocardiography and Ambulatory Electrocardiography. J Am Coll Cardiol 38(7):2091–2100
23. Köhler B-U, Hening C and Orglmeister R (2002) The principles of software QRS detection. IEEE Trans Biom Eng 21(1):42–57
24. Kors JA and van Bemmel JH (1990) Classification Methods for Computerized Interpretation of the Electrocardiogram. Meth Inform Med 29(1):330-336
25. Koski A, Juhola M and Meriste M (1995) Syntactic Recognition of ECG Signals by Attributed Finite Automata. Pattern Recognit 28(12):1927–1940
26. Laguna P, Jané R and Caminal P (1994) Automatic detection of wave boundaries in multilead ECG signals. Validation with the CSE database. Comput Biomed Res 27(1):45–60
27. Laguna P, Mark RG et al. (1997) A Database for evaluation of algorithms for measurement of QT and other waveform intervals in the ECG. In: Computers in Cardiology, Lund, Sweden
28. Li C, Zheng C and Tai C (1995) Detection of ECG characteristic points using wavelet transforms. IEEE Trans Biom Eng 42(1):21–28
29. Mallat S. (1998) A Wavelet Tour of Signal Processing. Academic Press, San Diego, CA
30. Martínez JP, Almeida R et al. (2004) A Wavelet-Based ECG Delineator: Evaluation on Standard Databases. IEEE Trans Biom Eng 51(4):570–581
31. Murphy KP (2002) Dynamic Bayesian Networks: Representation, Inference and Learning. PhD Thesis, University of California
32. Neal RM and Hinton GE (1998) A new view of the EM algorithm that justifies incremental, sparse and other variants. In: M. I. Jordan (ed) Learning in Graphical Models, Kluwer Academic Publishers, Dodrecht
33. Rabiner LR and Juang BH (1993) Fundamentals of Speech Recognition. Prentice Hall, Englewood Cliffs, NJ

34. Rabiner LR, Lee CH, Juang BH and Wilpon JG (1989) HMM Clustering for Connected Word Recognition. In: Proc. ICASSP, Glasgow, UK
35. Rautaharju PM, Zhou SH et al. (1993) Function Characteristics of QT Prediction Formulas. The Concepts of QTmax and QT Rate Sensitivity. Comput Biomed Res 26(2):188–204
36. Sahambi JS, Tandon SN and Bhatt RKP (1997) Using wavelet transforms for ECG characterization. An on-line digital signal processing system. IEEE Eng Med Biol 16(1):77–83
37. Senhadji L, Bellanger JJ, Carrault G and Coatrieux JL (1990) Wavelet analysis of ECG signals. In: Proc. of the Twelfth Annual International Conference of the IEEE EMBS
38. Vullings HJLM, Verhaegen MHG and Verbruggen HB (1998) Automated ECG segmentation with Dynamic Time Warping. In: Proc. 20th Annual Conf. IEEE EMBS, Hong Kong, China
39. Willems JL, Zywietz C et al. (1987) Influence of noise on wave boundary recognition by ECG measurement programs. Comput Biomed Res 20(6):543–562

Chapter 5
Heart Rate Variability Time-Frequency Analysis for Newborn Seizure Detection

Mostefa Mesbah, Boualem Boashash, Malarvili Balakrishnan and Paul B. Coldiz

Abstract The identification of newborn seizures requires the processing of a number of physiological signals routinely recorded from patients, including the EEG and ECG, as well as EOG and respiration signals. Most existing studies have focused on using the EEG as the sole information source in automatic seizure detection. Some of these studies concluded that the information obtained from the EEG should be supplemented by other information obtained from other recorded physiological signals. This chapter documents an approach that uses the ECG as the basis for seizure detection and explores how such approach could be combined with the EEG based methodologies to achieve a robust automatic seizure detector.

5.1 Identification of Newborn Seizures Using EEG, ECG and HRV Signals

This chapter presents the issue of automatic seizure detection in the newborn using a number of advanced signal processing methods, including time-frequency signal analysis. A number of physiological signals, such as the electroencephalogram (EEG), electrocardiogram (ECG), electro-oculogram (EOG), and the electromyogram (EMG) may be recorded and monitored in neonatal intensive care units in babies at clinical risk of neurological dysfunction. The core signals measured from the newborn for the application described in this chapter include the EEG and the ECG from which we extract the Heart Rate Variability (HRV). The intent of this chapter is to document the applicability of using information from the time-frequency representation of the HRV to supplement EEG based seizure detection methodologies and focus on the additional gain made by incorporating the information obtained from the analysis of HRV.

B. Boashash (✉)
Perinatal Research Centre, University of Queensland Centre for Clinical Research, The University of Queensland, Australia & College of Engineering, University of Sharjah, Sharjah, UAE
e-mail: b.boashash@uq.edu.au

A. Naït-Ali (ed.), *Advanced Biosignal Processing*,
DOI 10.1007/978-3-540-89506-0_5,

5.1.1 Origin of Seizures

Neonatal seizures are the most common manifestation of neurological dysfunction in the newborn [19]. Although the causes of seizure are many and varied, in most neonates the seizure is triggered by acute illnesses such as hypoxic-ischemic encephalopathy (HIE), intracerebral birth injuries, central nervous system (CNS) infections and metabolic disturbances [44]. The incidence of seizure is higher in the neonatal period (i.e., the first 4 weeks after birth) than in any other period of life [15]. The reported incidence of neonatal seizures varies widely from 3% to 25%, reflecting the difficulties in diagnosis [14].

5.1.2 Seizure Manifestation

A seizure is a paroxysmal behaviour caused by hyper-synchronous discharge of a large group of neurons. It manifests itself through clinical and/or electrical signs. The clinical manifestations, when present, involve some stereotypical physical behaviors whilst the electrical ones are identified by a number of EEG patterns. In the adult, the clinical signs include repeated tonic-clonic spasms and/or may involve changes in the patient's state of consciousness and behavior, such as increased agitation, frightened or confused behavior, visual sensations and amnesia. In the newborn, the signs are more subtle and may include sustained eye opening with ocular fixation, repetitive blinking or fluttering of the eyelids, drooling, sucking and other slight facial movements; tonic-clonic activity is commonly absent [51]. These characteristics may also be part of the repertoire of normal behavior in newborns. Autonomic nervous system manifestations or associations with seizures may result in changes in heart rate, blood pressure and skin perfusion. The fact that motor phenomena may be absent as well as the difficulties in distinguishing any seizure signs from normal behavior mean that it is imperative to use physiological signals for seizure detection.

5.1.3 The Need for Early Seizure Detection

Seizures are among the most common and important signs of acute neonatal encephalopathy (degenerative disease of the brain) and are a major risk factor leading to subsequent neurological disability or even death. Neonatal seizures are associated with increased rates of long-term chronic neurological morbidity and neonatal mortality. Infants with neonatal seizures are 55–70 times more likely to have severe cerebral palsy and 18 times more likely to have epilepsy than those without seizures [4]. Once seizures are recognized, they can be treated with anticonvulsant drugs and ongoing monitoring is necessary to detect the response to the anticonvulsant medication. The early detection of seizure in the newborn is a significant responsibility faced by society in order to prevent long term neurological damage in the population.

5.1.4 Seizure Monitoring Through EEG

Several physiological signals are normally recorded and monitored in neonatal intensive care units. Over the last few decades, the EEG has been used as a primary tool in monitoring and diagnosing seizures. The EEG represents a continuous time-varying activity that reflects on-going synaptic activity in the brain and, therefore, reflects the status of the nervous system at any point in time. This justifies its use as the primary tool for seizure detection.

The other recorded signals are currently mainly used to detect artifacts in order to prevent misinterpretation of the EEG data. Since visual monitoring of the EEG has limitations [34], attention has shifted to the development of automatic computer based methods for newborn EEG seizure.

Considerable work, funded by the Australian Research Council, was performed by the authors in the decade 1997–2007 to analyze the newborn EEG signals in a joint time-frequency domain and develop automatic methods based on the findings of these investigations, and a concise summary appears in [12, 23, 24, 42, 43]. The approach that was followed is based in essence on the material described in Sect. 5.2 of this chapter. A key outcome of the analysis was the modeling of the newborn EEG as a piece-wise linear multi-component FM signal in noise, well suited for a quadratic time-frequency analysis. Several methods were developed and one of the most recent one was reported in [41]. An international patent on this subject was also taken.

5.1.5 Limitations of EEG-Based Seizure Identification

The complexity of the EEG signals led to a number of different approaches for automatic detection of the seizure based solely on the EEG, each approach making different approximations and assumptions [11]. Difficulties abound, especially in the case of scalp-recorded EEG, because of the large number of artifacts usually present alongside the EEG [31]. Additional information from other physiological sources was needed to filter out other undesirable components in the signal and ensure the development of a robust automated newborn EEG seizure detection. For example, EOG signals provide a measure of the movements of the eye and allow for pre-processing techniques to remove the effect of artifacts. Other signals such as ECG and respiration signals may also be used for this purpose.

5.1.6 ECG-Based Seizure Identification

The ECG also contains information relevant to the seizure. In the presence of a seizure, the heart rate is altered because of a correlation of epileptic activity with cardiovascular phenomena [50]. Most studies that have used EEG and ECG recordings simultaneously reported an increase in heart rate (tachycardia) in the presence of seizure. Zijlmans et al. [53] studied 281 seizures in 81 patients mostly of temporal lobe origin and found an increase in heart rate (tachycardia) of at least 10 beats/min in 73% of seizures, and 7% showed a decrease (bradycardia) of at least 10 beats/min. The authors also found that in 23% of seizures, the heart rate change preceded the onset both electrical and clinical manifestations of the seizure. Most of these studies have been done on animal models or human adults and only very few investigators have studied the cardio-regulatory mechanisms in children with epilepsy. In [36], the authors found tachycardia in 98% of children suffering complex partial seizures of temporal lobe origin, more frequent than in adults.

5.1.7 Combining EEG and ECG for Robust Seizure Identification

Heart rate and rhythm are largely under the control of the autonomic nervous system (ANS), with sympathetic stimulation causing an increase in heart rate and parasympathetic stimulation causing a reduction in heart rate. Heart rate variability (HRV), the changes in the heart's beat-to-beat interval can be calculated from ECG (see also Chap. 3). The estimation of the HRV before, during, and after a seizure provides an indication of the sum of sympathetic and parasympathetic inputs to the heart. HRV is a well established non-invasive tool for assessing ANS regulation [30]. It should naturally provide the additional information needed for the detection of seizure in newborns. It is then possible to combine both HRV and EEG analysis and processing in a way that leads to the development of an improved and robust algorithm that can automatically detect the presence of seizure in the newborn. To achieve this, we need first to perform an accurate and detailed analysis of the HRV signal.

5.1.8 The Need for Time-Frequency Signal Processing

Like EEG signals, HRV signals have been studied in the time or frequency domains using both linear and nonlinear methods [48]. These methods do not show information about the time-varying structure of these signals. In the use of the frequency domain, by far the most widely used domain, it is observed that the instantaneous frequency changes of the signal content, typical in physiological signals, are actually smeared out or appear as wideband noise. Therefore, it is common practice to restrict the analysis to an area that is a "reasonably stationary" part of the signal that is identified and analyzed [28]. Any precise spectral estimation, however, is dependent on the chosen observation window, and the lack of such a tedious adaptation of the parameters would consequently lead to an erroneous or limited interpretability of the results.

The more appropriate approach for such non-stationary signals, and therefore both the EEG and the HRV, is to apply the concept of time-frequency distribution (TFD) [1, 11]. The TFD is a two-dimensional function that describes the instantaneous frequency content of the signal in the joint time-frequency domain. The use of a TFD for biological signals is reported in [2, 32].

The chapter is organized as follows. Next, in the second section, we briefly introduce the time-frequency signal analysis tools needed to develop the automatic seizure detector. The third section illustrates how the ECG is processed to obtain the HRV and to extract the discriminating features from the HRV. In the fourth section, we introduce the automatic seizure detection algorithm using the features extracted from HRV, assess its performance, and discuss the results obtained.

5.2 Time-Frequency Signal Analysis

5.2.1 Addressing the Non-Stationarity Issue

The frequency content of many biomedical signals such as EEG and ECG are known to vary with time and it is well documented that this variation may be crucial in the important tasks of detection, diagnosis and classification as well as other applications [11]. It was widely reported that conventional spectral analysis techniques based on the Fourier transform are unable to adequately analyze these nonstationary signals. By mapping a one dimensional signal to a two-dimensional domain, time-frequency representations are able to localize the signal energy in both time and frequency. In order to illustrate the inherent limitations of the classical representation of a non-stationary signal, consider two different linear frequency modulated (LFM) signals, signal 1 and signal 2, with length $N = 100$ samples and a sampling frequency $f_s = 1$ Hz. Signal 1 has linearly increasing frequency from 0.1 Hz to 0.4 Hz while signal 2 has linearly decreasing frequency from 0.4 Hz to 0.1 Hz. Fig. 5.1a shows the time domain and frequency domain representations of two different LFM signals. As can be seen in this figure, both signals have similar spectral representations. The spectral representation shows the spread of the power within the whole length of signal but lacks any time localization. On the other hand, the time-frequency representation, shown in Fig. 5.1b, allows an accurate time localization of the spectral energy and reveals how it is progresses over time. It is this progression that may be the critical factor in many applications.

One class of methods for time-frequency representations that has gained wide acceptance in the analysis of biomedical signals is the class of quadratic time-frequency distributions [11]. There is a large number of TFDs that can be used to represent a signal in a time-frequency domain within this class. The choice of a suitable TFD depends on both the characteristics of the signal under analysis and the application at hand.

5.2.2 Formulation of TFDs

For a given real-valued signal $x(t)$, TFDs can be expressed as the two dimensional convolution of the Wigner-Ville Distribution, $W_z(t,f)$, with a two dimensional time-frequency filter $g(t,f)$ [11].

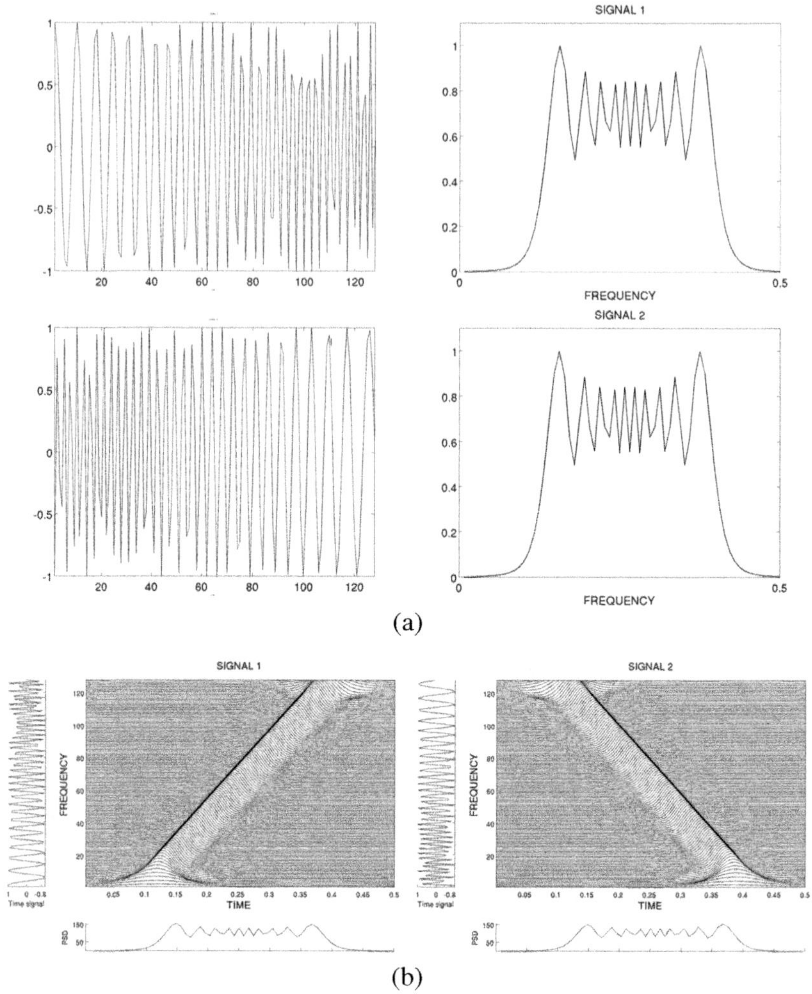

Fig. 5.1 The time-domain and frequency representations of signal 1 and signal 2, (**b**) The time-frequency representation of signal 1 and signal 2 shown in 1 (**a**)

$$\rho_z(t, f) = W_z(t, f) * *g(t, f) \tag{5.1}$$

Where $**$ indicates a convolution operation in both t and f variables and $W_z(t,f)$ is given by:

$$W_z(t, f) = \int_{-\infty}^{\infty} z(t - \tau/2)\bar{z}(t - \tau/2)e^{-j2\pi f\tau}d\tau \tag{5.2}$$

$\bar{z}(t)$ stands for the complex conjugate of $z(t)$, which is the analytic associate of $x(t)$, given by

$$z(t) = x(t) + jH[x(t)] \tag{5.3}$$

H is the Hilbert operator defined as

$$H[x(t)] = \frac{1}{\pi} PV \int_{-\infty}^{\infty} \frac{x(u)}{t-u} du \tag{5.4}$$

In the above equation, PV represents the "principal value" operator. For computation, it is more effective to express the above equation in terms of the time-lag kernel $G(t,\tau)$. This leads to the expression [11]:

$$\rho_z(t, f) = \int_{-\infty}^{\infty} \int_{-\infty}^{\infty} G(t-u, \tau) z(u + \tau/2) \bar{z}(u + \tau/2) e^{-j2\pi f \tau} du d\tau \tag{5.5}$$

The time-lag kernel, $G(t, \tau)$, is a function chosen to satisfy some desired properties of the TFD, in particular resolution and reduction of the cross-terms introduced by the quadratic nature of the transformation. The most widely studied member of the quadratic class of TFDs is the Wigner-Ville distribution (WVD). It is a core member of the quadratic class of TFDs as all other members are smoothed versions of WVD. Table 5.1 shows a number of quadratic TFDs along with their corresponding kernels. The first four are widely used TFDs. The last one, a modified version of the b distribution [5], is a recent addition to the quadratic class of TFDs and has shown to achieve high time-frequency resolution and significant cross-term reduction when applied to different types of nonstationary signals [27].

With the time-lag kernel $\delta(t)$, the WVD provides a high resolution time-frequency representation. It is the ideal representation for the class of mono-component linear frequency modulated signals and satisfies all mathematical desirable properties in a distribution except the non-negativity property that is rarely needed in practice [11]. A disadvantage is that the WVD suffers from cross-terms which appear midway between true signal components for multi-component sig-

Table 5.1 Selected quadratic TFDs and their corresponding time-lag kernels [11]

TFDs	Kernel ($G(t, \tau)$)
WVD	$\delta(t)$
Smoothed Pseudo WVD (SPWVD)	$h(\tau)g(t)$; $h(\tau)$ and $g(t)$ are two window functions
Spectrogram (SP)	$w(t + \tau/2)w(t - \tau/2)$; $w(t)$ is an analysis window function
Choi-Williams (CWD)	$\sqrt{\pi\sigma}/\lvert\tau\rvert e^{-\pi^2 t^2 \sigma/\tau^2}$, σ is a design parameter
Modified B distribution (MBD)	$\dfrac{\cosh^{-2\beta} t}{\int_{-\infty}^{\infty} \cosh^{-2\beta} \xi d\xi}$, β is a design parameter

nals and nonlinear frequency modulated signals. The presence of these cross-terms may make the interpretation of the time-frequency representation difficult in some situations but can be beneficial in others [8].

5.2.3 Trade-Off Resolution Versus Cross-Terms Trade-Off

Considerable effort has been deployed in an attempt to design TFDs which reduce the cross-terms while preserving as many of the desirable properties enjoyed by the WVD as possible. In effect, this is mainly done by smoothing the WVD. One example of such TFD is the smoothed pseudo Wigner-Ville distribution (SPWVD) [8, 11, 27]. The SPWVD has a separable kernel, where the $g(t)$ is the smoothing window and $h(\tau)$ is the analysis window. These two windows are chosen to suppress spurious peaks in both frequency and time directions. The suppression of cross-terms is improved with shorter windows. This, however, results in the undesirable loss of resolution and hence smearing of characteristics such as the instantaneous frequency. Some commonly used windows for $g(t)$ and $h(\tau)$ in the literature dealing with adult HRV signal representation are the unit rectangular window and the Gaussian window respectively [39]. The kernels of the Choi-William distribution (CWD) and Modified-B distribution (MBD) are low pass filters that are designed in the ambiguity domain (the dual of the time-frequency domain). These TFDs, as with most quadratic TFDs, exploit the property that in this domain the auto-terms (also known as the signal terms) are concentrated around the origin while the cross-terms are situated away from it [8]. Unlike the CWD kernel, the MBD kernel is lag-independent which means that the filtering is performed only in the time direction. The design parameters, β and σ, are positive numbers that control the trade-off between cross-term reduction and loss of resolution introduced by the smoothing operations.

5.2.4 The Signal Instantaneous Frequency (IF)

One of the most important pieces of information embedded in the time-frequency representation of nonstationary signals is the instantaneous frequency (IF). The IF is a time-varying parameter which defines the location of the signal's spectral peak as it varies with time. It was originally defined for mono-component signals, where there is only one frequency or a narrow range of frequencies varying as a function of time.

The IF of a real-valued signal $x(t)$ is determined in terms of the phase of its analytic associate $z(t) = a(t)e^{j\phi(t)}$ through the following equation [9]

$$f_i(t) = \frac{1}{2\pi}\frac{d\varphi(t)}{dt} \tag{5.6}$$

where $a(t)$ is referred to as the signal envelope and $\phi(t)$ is the instantaneous phase. Two major approaches for IF estimation are parametric and non-parametric. A review can be found in [10]. A widely used non-parametric IF estimation technique is based on the peak of the TFD. It uses the property that TFDs have their maxima around the IF of signal. The accuracy of the estimate depends on the resolution of the TFDs. Another non-parametric approach for estimating the IF uses the first-order moment of a TFD, $\rho(t, f)$, with respect to the frequency is expresses as [7]

$$f_c(t) = \frac{\int_{-\infty}^{\infty} f\rho(t, f)df}{\int_{-\infty}^{\infty} \rho(t, f)df} \quad (5.7)$$

This first-order moment, sometimes called the central frequency $f_c(t)$, is equal to the IF for the case of TFDs whose Doppler-lag kernel satisfies the IF property; namely $g(\nu, \tau)|_{\tau=0}$ = constant and $\partial g(\nu, \tau)/\partial\tau|_{\tau=0} = 0$ for all ν, where $g(\nu, \tau)$ is obtained from the time-lag kernel $G(t, \tau)$ through a Fourier transform with respect to t [7]. The WVD and CWD are two such TFDs. The first-order moment provides an estimate of the IF for the TFDs that do not satisfy the IF property.

5.2.5 Multi-Component IF Estimation

As mentioned above, for multicomponent signal, the notion of a single-valued IF becomes ill defined, and a break-down into its components is needed. Consider a signal composed of an M monocomponent signals, such that

$$z(t) = \sum_{i=1}^{M} a_i(t)e^{j\varphi_i(t)} \quad (5.8)$$

where $a_i(t)$ and $\phi_i(t)$ are the amplitude envelope and instantaneous phase of the ith signal component respectively. To extract the IF of each component, the signal is first decomposed into a number of monocomponents. Once this is done, the IF of each component is extracted using existing methods. Decomposing the signal into a number of monocomponents can be done in several ways depending on the nature of the signal under analysis. One way is by using the empirical mode decomposition (EMD) [26]. The EMD is an adaptive nonlinear decomposition technique developed to decompose nonstationary signals into a set of monocomponent signals. This technique is thoroughly described in Chap. 16. It is also evoked in Chap. 6 and evaluated in [47]. Another approach for decomposing the multicomponent signals is time-frequency filtering [13, 25]. The main disadvantage of this approach is that prior knowledge about the location of the different spectral components in the time-frequency domain is needed. This restriction prevents using this method in a fully automated detection algorithm. A simpler version adopted in this chapter is to use a band-pass filters to isolate the different monocomponents prior to mapping them into the time-frequency domain for IF extraction. This method assumes that the components are separated in frequency, a condition mostly satisfied by HRV signals. Another IF estimation method specifically designed for multicomponent signals [41] combines a local peak detector and an image processing technique, known as component linking, to estimate the number of TF components and extract their individual IFs. This IF estimation method has the advantage that it does not require prior information about the signal to be decomposed. It uses instead the energy peaks in the time-frequency domain.

5.2.6 TFD as a Density Function

Density functions are basic tools in many fields of science, mathematics and engineering. In many cases, the density function can be sufficiently characterized by some of its lower order moments such as the mean and the variance. In signal analysis, instantaneous power and the spectral density have been used by analogy with density functions. This notion has been extended for the case of nonstationary signal through the joint time-frequency density [8, 16], and this led to some developments such as time-frequency filtering [13]. Continuing the analogy with densities in general, these signal densities may also be characterized by their low-order moments. For the case of the quadratic joint time-frequency density, the first and the second conditional moments are the mean frequency at a particular time given by (5.7) and the variance given by

$$IB(t) = \frac{\int_{-\infty}^{\infty} (f - f_c(t))^2 \rho(t, f) df}{\int_{-\infty}^{\infty} \rho(t, f) df} \tag{5.9}$$

Proceeding further with the analogy, the first and the second conditional moments have been called instantaneous frequency and instantaneous bandwidth (IB) respectively although it is recognized that most of the quadratic TFDs are not positive definite and some of them do not even satisfy the marginal conditions [13, 38]. This problem becomes clearly apparent in the case of multi-component signals as such signals require a breakdown of the signal in its primary components before applying the analogy [7, 8].

5.3 ECG Pre-Processing and HRV Extraction

5.3.1 Data Acquisition

The newborn EEG and ECG data and signals presented in this chapter were recorded from newborns admitted to the Royal Brisbane and Women's Hospital in Brisbane, Australia. A single ECG channel was recorded simultaneously along with 20 channels of EEG. These EEG channels were formed from 14 electrodes using a bipolar montage. The 14 electrodes were placed over the newborn's scalp according to the international 10–20 system. The data were recorded using the Medelec Profile System (Medelec, Oxford Instruments, Old Woking, UK) including video. The EEG seizure and non-seizure segments were identified and annotated by a qualified paediatric neurologist from Royal Children's Hospital, Brisbane, Australia. The start and end of the seizures were identified based on EEG evidence with video surveillance as necessary. The raw ECG and EEG were sampled at 256 Hz. A 50 Hz notch filter was also used to remove power line interference. In the present study 6 seizure-related and 4 non-seizure-related non-overlapping ECG epochs of 64 s collected from 5 newborns were used.

5.3.2 Extracting HRV from ECG

This section presents the different steps required to obtain the HRV from the raw ECG. These steps are illustrated in Fig. 5.2.

5.3.2.1 The ECG and QRS Wave Detection

The contractions of the heart are initiated by an electrical impulse. The formation and propagation of the electrical impulse through the heart muscle results in

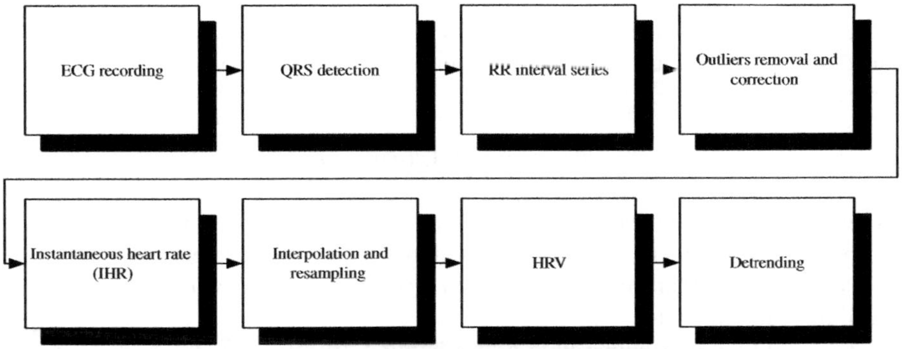

Fig. 5.2 Pre-processing steps to extract HRV from ECG

a time-varying potential on the surface of the body known as the ECG. The impulse propagates from one node to the other resulting in a P wave. After a pause, the impulse enters the bundle branches resulting in the contraction of the ventricular walls known as QRS complex. Then, the ventricular muscles regain their shape and size, resulting in the T wave (the reader can also refer to Chap. 1).

A QRS detection algorithm is used to detect the QRS complexes and localize the R wave. This is the most sensitive parameter in obtaining accurate RR intervals. The raw ECG is initially filtered using a 6th order band-pass finite impulse response (FIR) filter using a Hamming window with a lower and upper cut-off frequency of 8 Hz and 18 Hz respectively. This filter allows the frequencies related to the QRS complex to pass while rejecting noise, artifacts and non-QRS waveforms in the ECG signal such as the P and T waves. The lower cut-off frequency is chosen so to minimize the influence of large amplitude of P and T waves while still accentuating ectopic beats and QRS waveforms. The upper cut-off frequency is chosen to suppress motion artifacts but not to affect the QRS complexes. This process enhances the QRS waveform of the digitised ECG to allow for more efficient detection of the QRS onset and offset.

Approaches to the problem of QRS detection include template matching, linear prediction, wavelet transforms, and nonlinear transformation with thresholding [21]. The last approach was selected for its computational efficiency, ease of implementation, and reliability in recognizing the QRS waveforms [52]. Specifically, the algorithm for QRS detection uses the smoothed nonlinear energy operator (SNEO) proposed in [35, 37] for the detection of spikes in signals. For discrete time signals, $x(n)$, the NEO operator, Ψ, is defined as [29]:

$$\Psi[x(n)] = x^2(n) - x(n+1)x(n-1) \tag{5.10}$$

NEO is also known as the energy-tracking operator. Equation (5.10) indicates that only three samples are required for energy computation at each time instant. This gives good time resolution in capturing the energy fluctuations instantaneously and enables the NEO to accentuate the high frequency content of a signal. A spike with a high magnitude and a high gradient produces a high peak at the output of the NEO. This method also generates non-warranted cross-terms due to its quadratic property, and requires smoothing to reduce these interferences. This is expressed as:

$$\Psi_s[x(n)] = \Psi[x(n)] * w(n) \tag{5.11}$$

where $*$ denotes the convolution operator, and $w(n)$ is a time domain window whose properties, shape and its width are selected to achieve a good interference reduction while preserving high time resolution. A five point Bartlett window was used to fulfil this requirement

The SNEO is used along with a thresholding operation to extract the R points which are treated as spikes in the ECG signal. The threshold value selected is signal

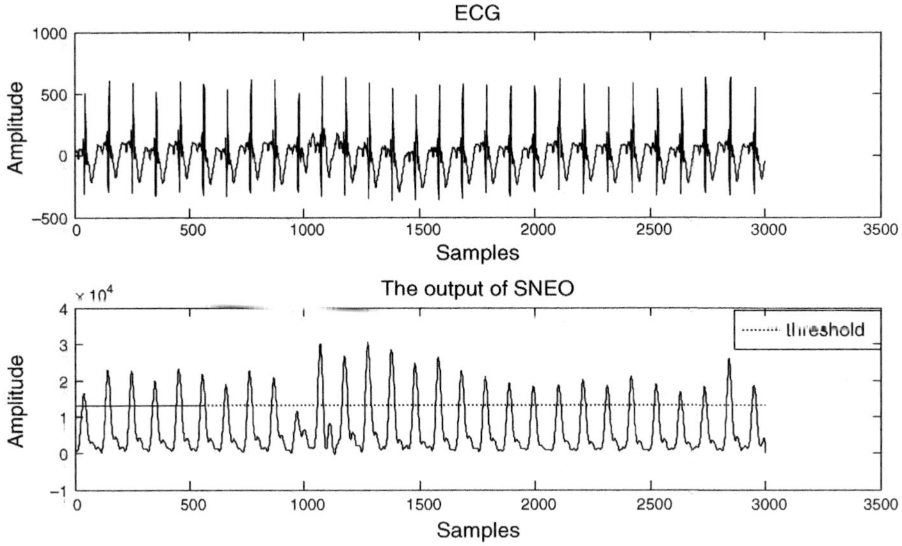

Fig. 5.3 (**a**) a segment of ECG (**b**) the output of SNEO and the threshold used in the detection of the QRS complex

dependent and is obtained using the histogram of the output of the SNEO. The size of bin, W, for the histogram is determined as in [20]:

$$W = 3.49\sigma_x N^{-1/3} \tag{5.12}$$

where σ_x and N are the standard deviation and the length of the signal $x(n)$ respectively. The value of the SNEO output which belongs to the second highest bin is taken as the threshold value. Figure 5.3 shows the result of applying the SNEO on newborn ECG. The peaks in Fig. 5.3b above the threshold are taken as the locations of the R points (the maximum of the R wave). The time duration between consecutive R points is used to represent the heart's beat-to-beat interval. This series is known as the RR interval time series, *RRi*, or tachogram.

5.3.2.2 Removal of Outliers

The next step in the pre-processing stage is the removal of outliers from the *RRi* data. Any *RRi* values that contain artifacts due to QRS missed detections, false detections, ectopic beats, or other random-like physiological disturbances are known to skew the data and adversely affect the parameters estimated. The outliers, therefore, are removed from the *RRi* before further analysis.

Outliers are defined as the *RRi* values which are not within a specified limited interval. These values are considered as statistically inconsistent with the time series and removed. In this work, outliers are defined as [49]:

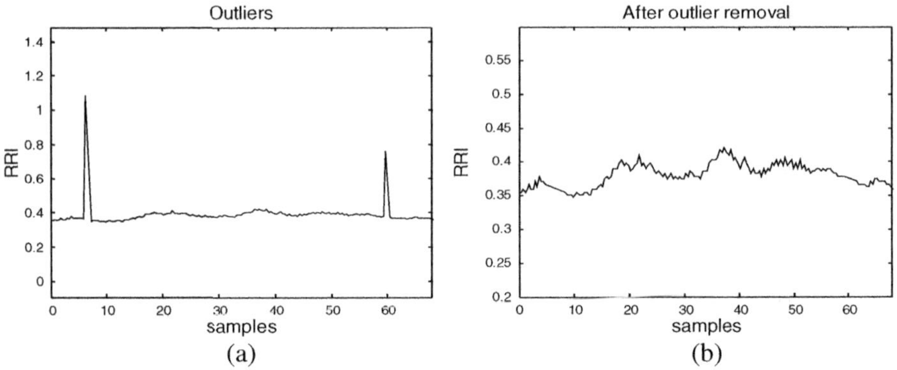

Fig. 5.4 *RRi* (**a**) with outliers (**b**) after outliers removed

For $0 < n \leq \text{length}\,(RRi)$

$$Outlier\ (n) = \begin{cases} RRi(n) & \text{if } RRi(n) < 1^{st}\ \text{quartile}(RRi) \\ & \quad - \text{interquartile\_range}(RRi) \times C, \text{ or} \\ RRi(n) & \text{if } RRi(n) > 3^{rd}\ \text{quartile}(RRi) \\ & \quad + \text{interquartile\_range}(RRi) \times C \end{cases} \tag{5.13}$$

where C is a constant. In [49], the authors suggested a value of 1.5 for C. We found, however, that a value of 3.5 is more suitable for our newborn database. In the case of normally distributed *RRi*, the outliers are identified as the *RRi* values that are more than 3 standard deviations from the mean.

Once the outliers are removed, the resulting series is linearly interpolated to replace the missing data. Linear interpolation is used as it has been reported to be a better choice when faced with runs of ectopic beats which is a common phenomenon in newborns [40]. Figure 5.4a shows one particular case with 2 outliers in the *RRi* epoch and Fig. 5.4b shows the same epoch after the outlier has been removed.

5.3.2.3 Quantification of HRV

An instantaneous heart rate (IHR) signal is obtained by taking the inverse of the *RRi*. The IHR is a non-uniformly time-sampled signal. This is not an issue in time domain analysis but it is for the case of frequency domain or time-frequency analyses, which implicitly assume the signal to be uniformly sampled. This irregularity generates additional harmonic components and artifacts [6, 22]. The IHR, therefore, needs to be processed to produce a uniformly sampled signal suitable for TF analysis. A performance comparison of such methods in [22] concluded that the technique based on cubic spline interpolation and resampling of IHR is efficient, fast, and is simply implemented. This method is used here to interpolate the IHR to obtain a uniformly sampled rate of 4 Hz. The resulting signal constitutes the HRV signal

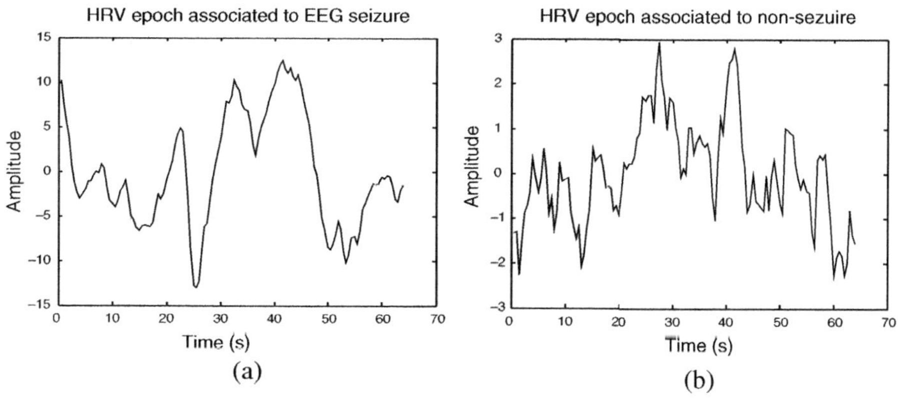

Fig. 5.5 The HRV related to (**a**) EEG seizure (**b**) EEG background

which is used for seizure detection. Figure 5.5 shows examples of HRV signals coinciding with the EEG seizure and EEG background (non-seizure) of the same newborn.

5.3.2.4 Detrending

Finally, average heart rate and trends are removed from the HRV. Detrending is achieved by subtracting the linear trend (straight line obtained by linear regression) from the signal.

5.4 HRV Time-Frequency Feature Extraction

5.4.1 HRV Spectral Components

Investigators usually divide the HRV power spectrum into different spectral bands. Pioneering works established three major spectral peaks in the short-term HRV power spectrum of the adult [3, 46]. A high-frequency (HF) spectral peak appears generally between 0.15 and 0.5 Hz, a low-frequency (LF) peak occurs around 0.1 Hz (generally between 0.04 and 0.15 Hz), and a very low-frequency (VLF) peak is found below 0.04 Hz. As the neonatal heart rate oscillations differ from that of the adult, studies utilizing frequency analysis of newborn HRV show somewhat different spectral divisions. Currently, the most commonly recommended frequency bands for short-term newborn HRV are [0.01–0.05] Hz for LF, [0.05–0.2] Hz for MF, and [0.2–1] Hz for HF [33]. The spectral peaks in the HRV power spectrum were shown to reflect the amplitude of heart rate fluctuations present at different oscillation frequencies [45]. In newborns, sympathetic activities manifest themselves in the LF band ascribed to baroreceptor reflex and vasomotor activity. The MF band is known to be both parasympathetically and sympathetically mediated.

The HF band correlates with respiratory fluctuations mediated by parasympathetic activities [33, 45].

5.4.2 Selection of a TFD for HRV Analysis

5.4.2.1 Performance Comparison of Relevant TFDs for HRV Analysis

In order to extract features, the HRV signals are processed using a number of quadratic TFDs so as to provide the most accurate time-frequency representation (TFR). The quadratic TFDs considered to be a priori appropriate for this task are those which meet the criteria defined earlier in terms of accuracy, resolution and cross-terms resolution; these are the smoothed pseudo Wigner-Ville distribution (SPWVD), the spectrogram (SP), the Choi-William distribution (CWD), and the Modified-B distribution (MBD) [11]. The performance analysis is carried using only two events (one representing seizure and the other representing non-seizure) out of the 10 events studied. These two events were found to be representative of the general characteristics observed. The TFDs of the HRV related to the seizure and non-seizure signals are shown in Figs. 5.6 and 5.7 respectively. In all plots in Figs. 5.7 and 5.8, the left plot is the HRV time representation, the centre figure shows the TFD, and the lower plot is the frequency representation. The sequence

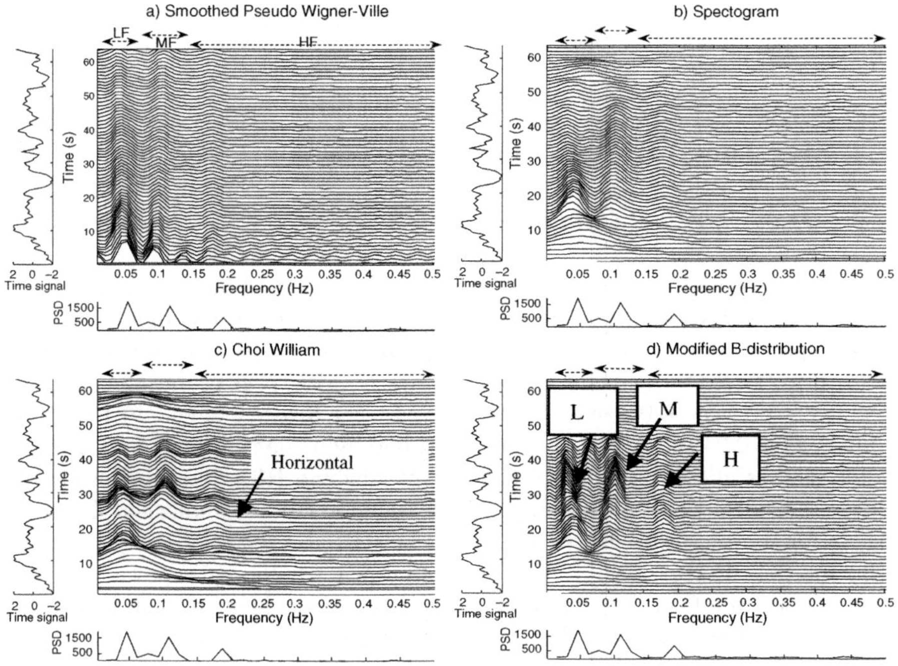

Fig. 5.6 TFDs for HRV seizure case: (**a**) SPWVD (**b**) SP (**c**) CWD (**d**) MBD

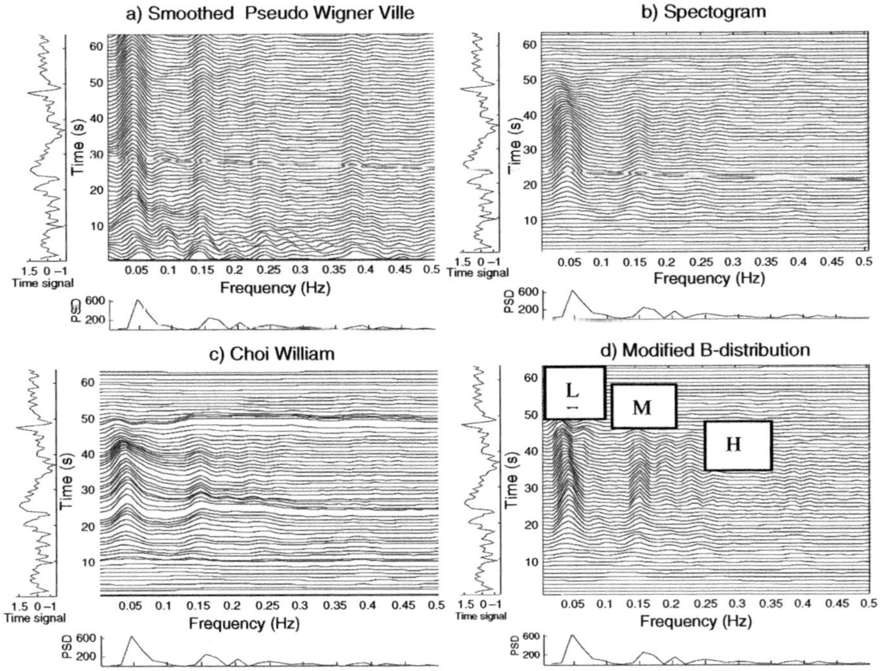

Fig. 5.7 TFDs for HRV non-seizure case: (**a**) SPWVD (**b**) SP (**c**) CWD (**d**) MBD

of plots labeled with (a), (b), (c), and (d) corresponds to the TFDs of the SPWVD, SP, CWD, and MBD, respectively. For clarity of illustration, the relevant frequency bands are labeled with LF, MF, and HF. The arrows are indicated in Fig. 5.6 because the relative position of those frequencies prevails in all the sequence of figures. The optimal parameters for the SPWVD, SP, CW and MBD are chosen so each TFD achieves the best compromise between TF resolution and the cross-terms suppression. The parameters were selected by comparing the TF plots of the signals visually for different values of parameters.

For the SPWVD, a Gaussian window of 121 samples was chosen for $h(\tau)$ and a rectangular window of 63 samples was selected for $g(t)$. In Fig. 5.6a, the dominant frequency content can be observed in the LF, MF and HF bands. The frequency resolution is satisfactory and the TFD is cross-terms free.

A Hamming window of 111 samples was selected for the SP. Figure 5.6b shows that the SP lacks time resolution which makes the TF components smeared. The SP suppresses all interference terms at the expense of the resolution of the signal components.

For the CWD, a kernel parameter σ of 0.4 was chosen. It can be seen that the TFD is almost cross-terms free but the existence of horizontal makes the TF components smeared (especially the LF and MF). This is due to the shape of the kernel which accentuates components that are parallel to the time and frequency axes.

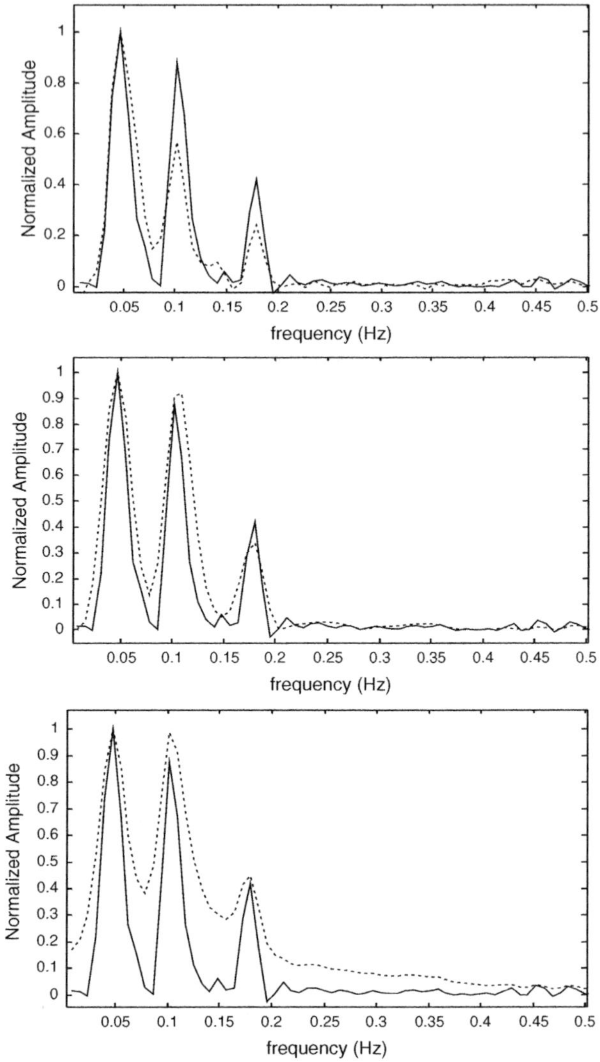

Fig. 5.8 Normalized slices (*dashed*) of SPWVD (*top*), SP (*middle*), and CWD (*bottom*). All plots are compared against the MBD (*solid*)

5.4.2.2 The Choice of the Modified B Distribution for HRV Analysis

The MBD's parameter, β was selected as 0.01. We observe that the TFD is cross-term free and have better TF resolution compared to the SP and CWD. This better resolution facilitates the identification/interpretation of the frequency components of the HRV in seizure newborn. The dominant frequency content can be observed in the LF, MF and HF bands. The results indicate that MBD provides the best compromise in terms of high TF resolution and effective cross-terms reduction for the signals considered.

The results of the TFD analysis of the HRV signal for a non-seizure epoch are presented in Fig. 5.7. A similar conclusion is reached regarding the time-frequency resolution and suppression of cross-term interference, as the case of seizure HRV. To better visualize the performance of the different TFDs, we compared the frequency resolution using a time slice of the TFDs, taken at specific time, t_0. A normalized slice at time interval, $t_0 = 23$ s is taken and displayed in Fig. 5.8 for each TFD of the seizure HRV epoch represented in Fig. 5.6.

The SPWVD shows performance similar to the MBD in cross-terms suppression but is outperformed in terms of resolution. Also compared to the MBD, the SP shows a poor TF resolution, and the CWD exhibits an unsatisfactory cross-term reduction. The MBD provides the best compromise between cross-term reduction and high resolution in both cases of seizure and non-seizure; it is, therefore, selected to represent HRV in the time-frequency domain.

5.4.3 *Feature Selection in the Time-Frequency Domain*

The features used to classify the HRV as corresponding to seizure or non-seizure are the first and second joint conditional moments related to each of the 3 spectral components discussed above.

The feature extraction procedure is made of the following three steps:

Band-pass filtering: Based on the information provided by the time-frequency analysis of the HRV, FIR band-pass filters are applied to isolate the three frequency bands mentioned earlier. This step results in three sub-signals corresponding to the LF, MF, and HF components.

Time-frequency mapping: The three sub-signals are represented in the time-frequency domain using the MBD. This step results in three different time-frequency representations corresponding to the LF, MF and HF components.

Moment Estimation: The parameters $f_c(t)$ and $IB(t)$ are computed for each of the three signal TFDs obtained in step 2. Examples of the parameters $f_c(t)$ and $IB(t)$ related to LF, MF and HF are shown in Figs. 5.9 and 5.10 respectively.

5.5 Performance Evaluation and Discussion of the Results

5.5.1 *Performance of the Classifier*

Two common methods used to estimate the classifier performance are holdout and cross-validation. The holdout method, used when the available data is large, splits the data into two separate sets: the training set and the test set. It is not selected here as it makes inefficient use of the data (using only a part of it to train the classifier) and gives a biased error estimate [17].

The cross-validation approach has several variants [17]. Examples are ten-fold cross-validation, leave-one-out and bootstrap. The difference between these three

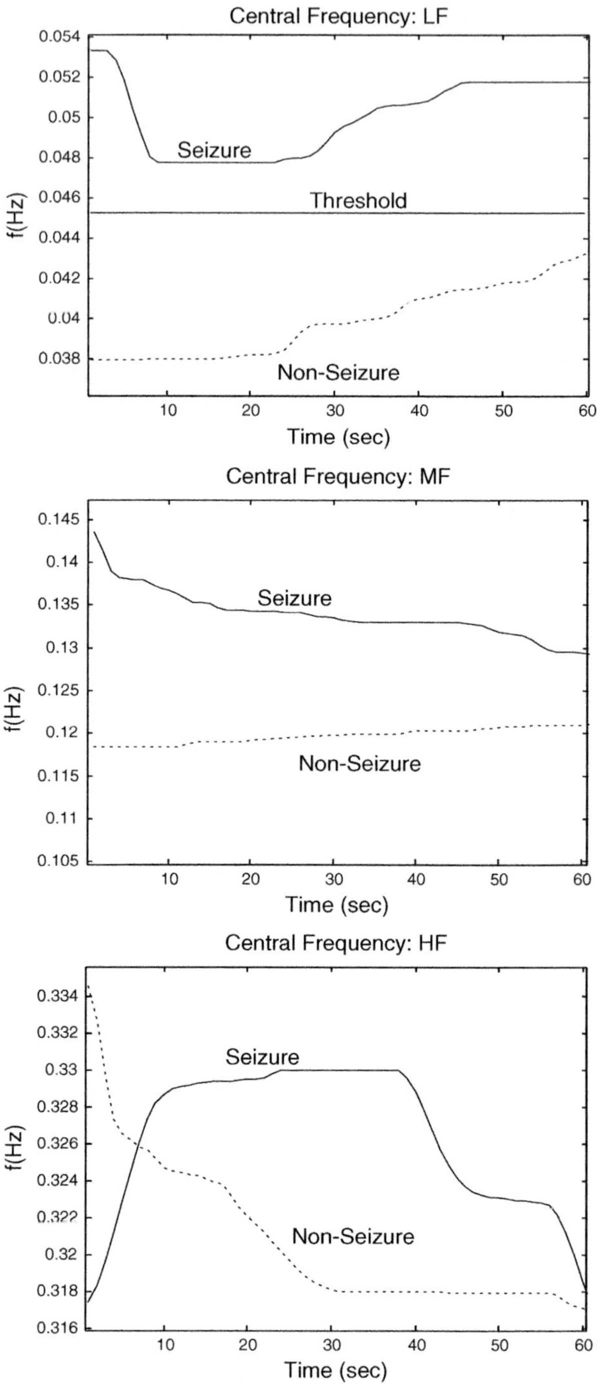

Fig. 5.9 The central frequency of the LF, MF and HF of the HRV

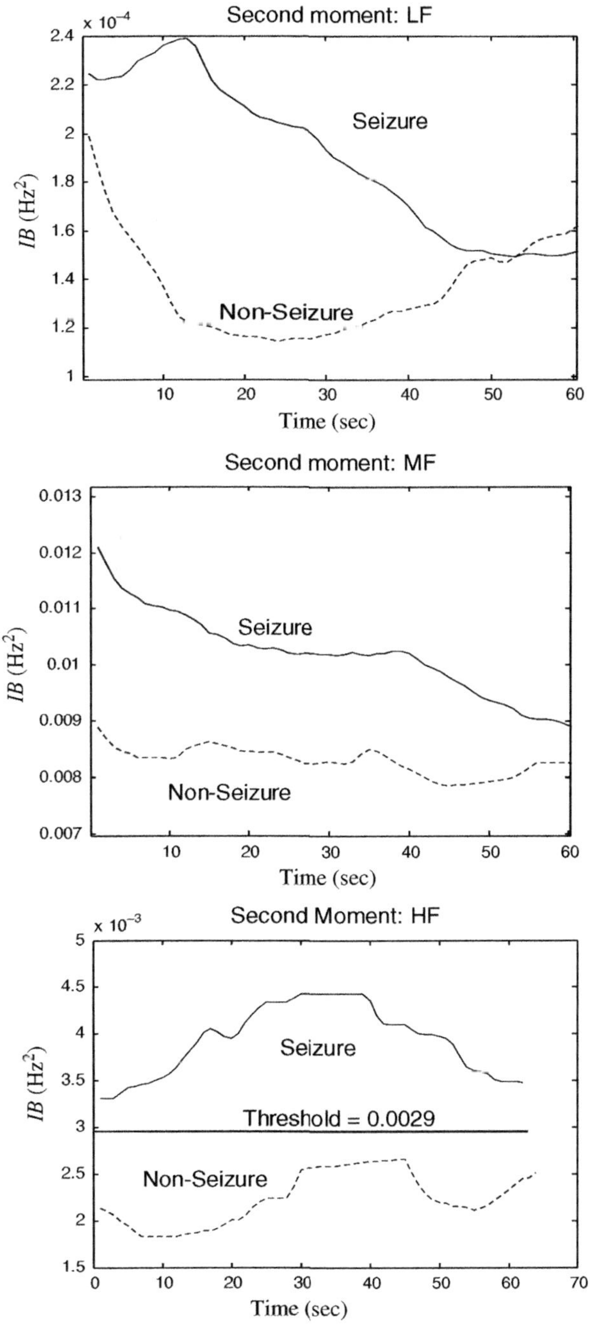

Fig. 5.10 The variance of the LF, MF and HF of the HRV

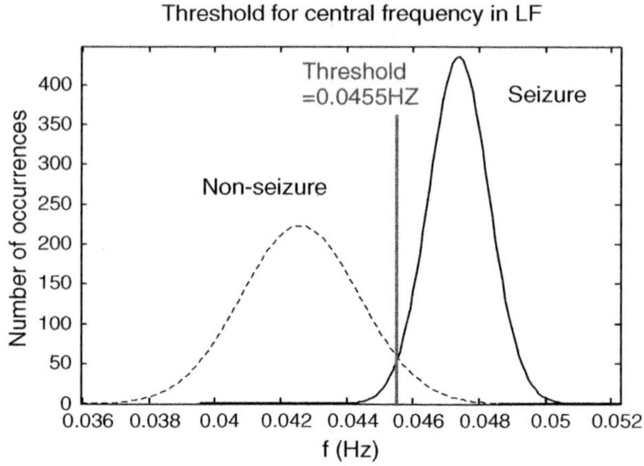

Fig. 5.11 The distributions used to determine threshold for central frequency in LF

types of cross-validation is in the way that data is partitioned. Leave-one-out is also known as n-fold cross-validation, where n stands for the number of subsets or folds. The process is performed splitting the data set D into n mutually exclusive subsets $D_1, D_2, \ldots, D_n$. The classifier is trained and tested n times; each time $k = 1, \ldots, n$, it is trained on $D \setminus D_k$ and tested on D_k. As the leave-one-out variant is suitable when the size of the data is small, it is adopted here, as it fits the circumstances.

For the 10 available events, 9 events (seizure and non-seizure) were used for training at a time. Each time, the $f_c(t)$ and $IB(t)$ obtained from seizure-related epochs are compared with those from non-seizure related, and thresholds are chosen that best differentiated the two groups. In the selection of the threshold, it is assume that both $f_c(t)$ and $IB(t)$ are normally distributed. Figures 5.11 and 5.12 illustrate how the threshold is obtained.

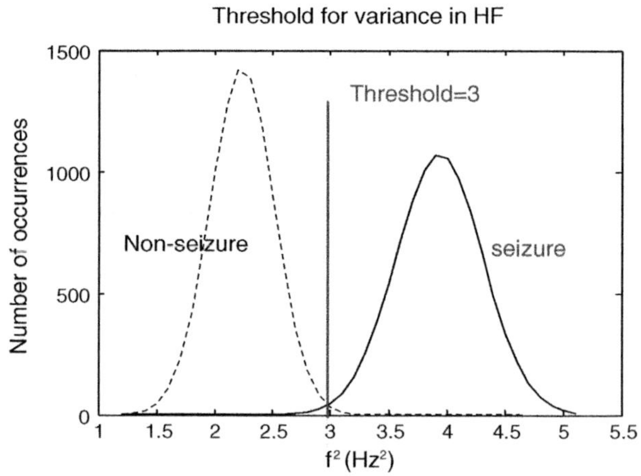

Fig. 5.12 The distributions used to determine threshold for variance in HF

The $f_c(t)$ and IB(t) parameters which was not included in the training set was then compared against the obtained thresholds and the decisions were recorded. The procedure was repeated 10 times for the $f_c(t)$ and IB(t) related to the three frequency bands. It was observed that the thresholds selected were newborn-dependent. The decisions recorded from the different tests were used to calculate the sensitivity and the specificity.

The sensitivity, R_{sn}, and specificity, R_{sp}, are defined as:

$$R_{sn} = \frac{TP}{TP + FN}; \quad R_{sp} = \frac{TN}{TN + FP} \tag{5.14}$$

where *TP, TN, FN*, and *FP* represent the number of true positive, true negatives, false negatives and false positives respectively. The sensitivity is the proportion of seizure events correctly recognized by the test (the seizure detection rate) while specificity is the proportion of non-seizure events correctly recognized by the test (the non-seizure detection rate).

5.6 Interpretation and Discussion of the Results

Table 5.2 illustrate the results obtained for the case of $f_c(t)$ while Table 5.3 shows the results for the case of $IB(t)$.

The tables indicate that seizures can be best discriminated from the non-seizure using $f_c(t)$ in the LF band (83.33% of sensitivity and 100% of specificity). The average threshold was found to be 0.0453 Hz. These results tend to indicate that newborn seizure manifests itself mostly in the LF component (sympathetic activity) of the HRV. The MF component was more affected than the HF component since it is both parasympathetically and sympathetically mediated. The $f_c(t)$ parameter related to the HF band shows the worst performance. This suggests that seizures have the least effect on the parasympathetic activity.

Table 5.2 Results for the central frequency

Frequency band	R_{sn} (%)	R_{sp} (%)
LF	**83.33**	**100.00**
MF	83.33	66.67
HF	50.00	16.67

Table 5.3 Results for the variance

Frequency band	R_{sn} (%)	R_{sp} (%)
LF	66.67	66.67
MF	83.33	66.67
HF	**83.33**	**100.00**

Table 5.3 indicates that non-seizure can be well discriminated from seizure in the HF band (83.33% of sensitivity and 100% of specificity). The averaged threshold found was 0.0026 Hz^2. These results show that the parameter $IB(t)$ related to the HF has been affected greatly during seizure compared to those from the LF and MF. The HF band is mediated by the respiration rate. So, these results suggest that seizure tends to increase the respiration variation in newborns. In addition, while the $f_c(t)$ parameter in the HF is less affected by seizure, the spread of the frequency in this band shows significant difference. The $IB(t)$ parameter obtained from the LF and MF bands did not show considerable changes, indicating that those features then do not seem to be good discriminating features.

5.7 Conclusions and Perspectives

This chapter has described a general time-frequency based methodology for processing the ECG for the purpose of automatic newborn seizure detection. The specific focus is to test the hypothesis that the information extracted from the HRV can be used to detect seizures and could, therefore, be used to improve the performance of EEG-based seizure detectors and classifiers by combining the information gained separately from the EEG and ECG.

The results in the previous chapters show that the first and second order moments of the TFD applied to filtered versions of the HRV signals provide good features for an automatic seizure/non-seizure classification process; so, the hypothesis has been validated, and this is an important step to plan the further improvement of automatic seizure detection methods. The performance of the method presented will be further assessed and refined using a larger ECG database, and improved methodologies should result.

The preliminary results of using the information from the HRV and combining it with the one from the EEG first at both the feature level (feature fusion) and the classifier level (classifier combination) look promising and more details of related ongoing experiments will appear elsewhere.

The long term aim of this study is to fuse information from different biological signals, including but not limited to EEGs, ECGs, EOGs, to design an accurate and robust automatic seizure detector. Specifically, it is intended to combine the methods described in this chapter with other approaches that focus on the time-frequency approach to seizure detection using solely the EEG, as mentioned in Sect. 5.1.4. Several key questions will arise and will need to be investigated. How best to combine and fuse the information obtained from the various signals? What relative weight needs to be given to each separate signal? How best to calibrate the novel procedures with currently practiced methodologies? Which method is best implemented online in real time, and what is the trade-off speed versus detection performance? As we approach towards a general resolution of the problem of automatic newborn seizure detection, there is a need to review all previous approaches with the new constraints and criteria provided by the need of speed and the need for an efficient and robust classification/fusion of critical features originating from a time-frequency representation of multiple physiological signals. Additional details about the most relevant of these approaches can be found in [12, 23, 24, 42, 43] and references therein.

This body of work has important implications for clinical management of newborns with seizures. Accurate automatic detection of seizures over periods of days is an essential basis for establishing, in clinical trials, the most effective managements for babies at risk of adverse neurodevelopmental outcomes. A robust system will also underpin investigations into the efficacy of new anticonvulsant medications that can be "designed" to incorporate new knowledge about seizure mechanisms in the developing brain.

Acknowledgement The signals presented in this chapter were collected and labeled with the assistance of Dr Chris Burke, paediatric neurologist, and Ms. Jane Richmond from the Royal Children's Hospital in Brisbane. This research was funded by a grant from the Australian Research Council (ARC).

References

1. Abeysekera SS, Boashash B (1991) Methods of signal classification using the images produced by the Wigner-Ville distribution. *Patter Recognit Lett* 12(11):717–729
2. Akay M (ed.) (1998) Time-frequency and wavelets in biomedical signal analysis. IEEE Press, New York
3. Akselrod S, Gordon D, Ubel FA, Shannon DC, Barger AC, Cohen RJ (1981) Power spectrum analysis of heart rate fluctuation: a quantitative probe of beat to beat cardiovascular control. *Science* 213:220–222
4. Aylward GP (1989) Outcome of the high-risk infant: fact versus fiction. In: Gottlieb MI, Williams JE (Eds.) Developmental-behavioral disorders: selected topics, Volume 2, Springer, New York
5. Barkat B, Boashash B (2001) A high-resolution quadratic time-frequency distribution for multicomponent signals analysis. *IEEE Trans Signal Process* 49(10):2232–2239
6. Berger RD, Akselrod S, Gordon D, Cohen RJ (1986) An efficient algorithm for spectral analysis of heart rate variability. *IEEE Trans Biomed Eng* 33(9):384–387
7. Boashash B (1990) Time-frequency signal analysis. In Haykin S (ed.) Advances in spectrum estimation and array processing, Prentice-Hall, Englewood Cliffs, NJ
8. Boashash B (ed.) (1992) Time-frequency signal analysis. Longman Cheshire, Melbourne
9. Boashash B (1992) Estimating and interpreting the instantaneous frequency of a signal – Part 1: Fundamentals. *Proc IEEE* 80(4):520–538
10. Boashash B (1992) Estimating and interpreting the instantaneous frequency of a signal – Part 2: Algorithms and Applications. *Proc IEEE,* 80(4):540–568
11. Boashash B (ed.) (2003) Time frequency signal analysis and processing: a comprehensive reference. Elsevier, Oxford
12. Boashash B, Mesbah M (2001) A time-frequency approach for newborn seizure detection, *IEEE Eng Med and Biol Magazine* 20(5):54–64
13. Boashash B, Mesbah M (2004) Signal enhancement by time-frequency peak filtering. *IEEE Trans Signal Process* 52(4):929–937
14. Bromfield EB, Cavazos JE, Sirven JI (eds.) (2006) An Introduction to Epilepsy. Bethesda (MD): Am. Epilepsy soc. http://www.ncbi.nlm.nih.gov/books/bv.fcgi?rid=epilepsy.chapter.107. Accessed 17 May 2008
15. Clancy RR (2006) The newborn drug development initiative workshop: summary proceedings from the neurology group on neonatal seizures. *Clin Ther* 28(9):1342–1352

16. Davidson KL, Loughlin (2000) Instantaneous spectral moments. *J Franklin Institute*, 337(4):421–436
17. Devijver PA, Kittler I (1982) Pattern recognition: a statistical approach. Prentice-Hall, Englewood Cliffs, NJ
18. Faul S, Boylan G, Connolly S, Liam M, Gordon L (2005) An evaluation of automated neonatal seizure detection methods. *Clin Neurophysiol* 116:1533–1541
19. Fisher RS, Boas W, Blume W, Elger C, Genton P, Lee P, Engel J Jr (2005) Epileptic seizures and epilepsy: definitions proposed by the International League Against Epilepsy (ILAE) and the International Bureau for Epilepsy (IBE). *Epilepsia* 46(4):470–472
20. Freedman D, Diaconis P (1981) On the histogram as a density estimator: L_2 theory. Wahrscheinlischkeittheorie Verw. *Gebiete* 57:453–476
21. Friesen GM, Jannett TC, Jadallah MA, Yates SL, Quint SR, Nagle HT (1990) A comparison of the noise sensitivity of nine QRS detection algorithms. *IEEE Trans Biomed Eng* 37:85–98
22. Guimarães HN, Santos RA (1998) A comparative analysis of preprocessing techniques of cardiac event series for the study of heart rhythm variability using simulated signals. *Braz J Med and Biol Res* 31:421–430
23. Hassanpour H, Mesbah M, Boashash B (2004) Time-frequency feature extraction of newborn EEG seizure using SVD-based techniques. *EURASIP J Appl Signal Process* 16:1–11
24. Hassanpour H, Mesbah M, Boashash B (2004) Time–frequency based newborn EEG seizure detection using low and high frequency signatures. *Physiol Meas* 25:934–944
25. Hlawatsch F (1998) Time-frequency analysis and synthesis of linear signal spaces. Springer, Norwell, Massachusetts
26. Huang NE, Shen Z, Long SR, Wu ML, Shih HH, Zheng Q, Yen NC, Tung CC, Liu HH (1998) The empirical mode decomposition and Hilbert spectrum for nonlinear and nonstationary time series analysis. *Proc Roy Soc London, Ser* A 454:903–995
27. Hussain ZM, Boashash B (2002) Adaptive instantaneous frequency estimation of multicomponent FM signals using quadratic time-frequency distributions. *IEEE Trans Signal Process* 50(8):1866–1876
28. Jurystaa F, Van de Borneb P, Migeottec PF, Dumontd M, Lanquarta JP, Degauteb JP, and Linkowskia P (2003) A study of the dynamic interactions between sleep EEG and heart rate variability in healthy young men. *Clin Neurophysiol* 14:2146–2155
29. Kaiser JF (1990) On a simple algorithm to calculate the energy of a signal. *IEEE International Conference Acoustic Speech and Signal Processing* 381–384, Albuquerque, USA
30. Kobayashi H, Ishibashi K, Noguchi H (1999) Heart rate variability: an index for monitoring and analyzing human autonomic activities. *J Physiol Anthropol Appl Hum Sci* 18:53–59
31. Koszer S (2007) Visual analysis of neonatal EEG. *eMedecine* http://www.emedicine.com/neuro/topic493.htm Accessed 21 May 2007
32. Lin Z, Chen J (1996) Advances in time-frequency analysis of biomedical signals. *Critical Rev Biomed Eng* 24 (1):1–72
33. Longin E, Schaible T, Lenz T, Konig S (2005) Short term heart rate variability in healthy neonates: normative data and physiological observations. *Early Hum Dev* 81:663–71
34. Lombroso CT, Holmes GL (1993) Value of the EEG in neonatal seizures. *J Epilepsy* 6:39–70
35. Malarvili MB, Hassanpour H, Mesbah M, Boashash B (2005) A histogram-based electroencephalogram spike detection. *International Symposium on Signal Processing and its applications* 207–210, Sydney
36. Mayer H, Benninger F, Urak L, Plattner B, Geldner J, Feucht M (2004) EKG abnormalities in children and adolescents with symptomatic temporal lobe epilepsy. *Neurol* 63(2):324–328
37. Mukhopadhyay S, Ray GC (1998) A new interpretation of nonlinear energy operator and its efficiency in spike detection. *IEEE Trans Biomed Eng* 45(2):180–187
38. Nho W, Loughlin P (1999) When is instantaneous frequency the average frequency at each time? *IEEE Signal Process Lett* 6(4):78–80
39. Novak P, Novak V (1993) Time-frequency mapping of the heart rate, blood pressure and respiratory signals. *Med Biol Eng Comput* 31:103–110

40. Ramanathan A, Myers GA (1996) Data preprocessing in spectral analysis of heart rate variability: handling trends, ectopy and electrical noise. *J Electrocardiol* 29(1):45–47
41. Rankine L, Mesbah M, Boashash B (2007) IF Estimation for multicomponent signals using image processing techniques in the time-frequency domain. *Signal Process* 87(6):1234–1250
42. Rankine L, Mesbah M, Boashash B (2007) A matching pursuit-based signal complexity measure for the analysis of newborn EEG. *Med Biol Eng Comput* 45(3):251–260
43. Rankine L, Stevenson N, Mesbah M, Boashash B (2007) A nonstationary model of newborn EEG, *IEEE Trans Biomed Eng* 54(1):19–29
44. Rennie JM (1997) Neonatal seizures. *Eur J Pediatr* 156:83–87
45. Rosenstock EG, Cassuto Y, Zmora E (1999) Heart rate variability in the neonate and infant: analytical methods, physiological and clinical observations. *Acta Paediatrica* 88:477–482
46. Sayers BM (1973) Analysis of heart rate variability. *Ergonomics* 16(1):17–32
47. Stevenson N, Mesbah M, Boashash B, Whitehouse HJ (2007) A joint time-frequency empirical mode decomposition for nonstationary signal separation. *Int Symposium of Signal Processing and its Applications* (CD ROM), Sharjah, UAE
48. Task Force of the European Society of Cardiology and the North American Society of Pacing and Electrophysiology (1996) Heart rate variability: standards of measurement, physiological interpretation, and clinical use. *Eur Heart J* 17:354–381
49. Tukey JW (1977) Exploratory Data Analysis. Addison-Wesley, Reading, MA
50. Van Buren JM (1958) Some autonomic concomitant of ictal automatism. *Brain* 81:505–522
51. Volpe JJ (1989) Neonatal seizures: current concepts and revised classification. *Pediatr* 84:422–428
52. Wan H, Cammarota JP, Akin A, Sun HH (1997) Comparison of QRS peak detection algorithms in extracting HRV signal *International Conference of the IEEE Engineering in Medicine and Engineering*, 302–305, Chicago
53. Zijlmans M, Flanagan D, Gotman J (2002) Heart rate change and ECG abnormalities during epileptic seizures: prevalence and definition of an objective clinical sign. *Epilepsia* 43(8):847–854

Chapter 6
Adaptive Tracking of EEG Frequency Components

Laurent Uldry, Cédric Duchêne, Yann Prudat, Micah M. Murray
and Jean-Marc Vesin

Abstract In this chapter, we propose a novel method for tracking oscillatory components in EEG signals by means of an adaptive filter bank. The specific utility of our tracking algorithm is to maximize the oscillatory behavior of its output rather than its spectral power, which shows interesting properties for the observation of neuronal oscillations. In addition, the structure of the filter bank allows for efficiently tracking multiple frequency components perturbed by noise, therefore providing a good framework for EEG spectral analysis. Moreover, our algorithm can be generalized to multivariate data analysis, allowing the simultaneous investigation of several EEG sensors. Thus, a more precise extraction of spectral information can be obtained from the EEG signal under study. After a short introduction, we present our algorithm as well as synthetic examples illustrating its potential. Then, the performance of the method on real EEG signals is presented for the tracking of both a single oscillatory component and multiple components. Finally, future lines of improvement as well as areas of applications are discussed.

6.1 Motivation

6.1.1 Oscillatory Activity as a Key Neuronal Mechanism

Oscillatory phenomena have gained increasing attention in the field of neuroscience, particularly because improvements in analysis methods have revealed how oscillatory activity is an information-rich signal. Neuronal oscillations represent a major component of the fundamental paradigm shift that has occurred across different fields of neuroscience [36, 9], wherein the brain actively and selectively processes external stimuli under the control of top-down influences, rather than simply treating sensory information in a largely bottom-up, passive, and serial manner. Perception

J.-M. Vesin (✉)
Swiss Federal Institute of Technology Lausanne (EPFL), Institute of Electrical Engineering (IEL), EPFL-STI-IEL-LTS1, Station 11, 1015 Lausanne, Switzerland
e-mail: jean-marc.vesin@epfl.ch

A. Naït-Ali (ed.), *Advanced Biosignal Processing*,
DOI 10.1007/978-3-540-89506-0_6, © Springer-Verlag Berlin Heidelberg 2009

and behavior are thus strongly modulated by top-down internal states, such as expectations derived from past experience or general knowledge [30], selective attention [12, 38], awareness [8], emotional states or planned actions. Oscillatory activity is considered a key component for the top-down control of perception, because modulation of sensory inputs might manifest through the temporal structure of both stimulus-evoked and ongoing activity, and could be expressed through the modulation of synchronization between multiple areas, and through the large scale coherence of different neuronal populations [37]. The fundamental role of oscillatory activity in brain responses is evident in recent findings that oscillations can make neurons transiently sensitive to inputs by shifting their membrane potential [2] and that the phase of such oscillations is linked to activity of single neurons [17]. It has furthermore been shown that the ongoing oscillatory state of the brain before a given stimulus can predict parameters of numerous behavioral responses in both motor and sensory tasks [23, 39]; thus, brain oscillations could have a crucial effect on behavioral outcome.

The role of network oscillations and their relation to specific brain functions or behaviors, however, remain poorly understood, though several hypotheses have been formulated. For example, the binding-by-synchronization (BBS) hypothesis proposes that oscillatory activity provides a tag that binds neurons representing the same perceptual object [32, 8]. According to this theory, synchrony enhances the saliency of specific neural patterns in response to a given sensory input, and objects are then represented through the binding of distributed synchronous neural assemblies, coding for each of the different object features. This representational code based on synchronization patterns has been widely described in visual object perception. Recently, another model for the role of neuronal oscillations has been proposed and is referred to as the 'communication-through-coherence' (CTC) hypothesis [11]. According to this model, the fixed anatomical structure of the brain, in order to allow effective and selective neuronal communication, requires a flexible communication structure, which is mechanistically implemented by the patterns of coherence between interacting neuronal populations. The CTC hypothesis considers only the mechanistic aspect of oscillations for neuronal communication and has been successfully tested in cortico-spinal communication through cortico-spinal coherence [31]. Finally, Salinas and Sejnowski [29] proposed a model of neuronal oscillations where correlations could be controlled independently of firing rate and would serve to regulate the strength of information flow rather than its meaning.

Collectively, these mutually non-exclusive models show that neuronal oscillations could either serve as a representational code for object perception, or as a mechanistic substrate for optimized neuronal communication, or maybe both; in addition, they provide a framework for understanding the neurophysiologic underpinnings and functional consequences of brain rhythms. Consequently, the field of neuroscience would clearly benefit from the development of sophisticated methods for analyzing neuronal oscillatory phenomena.

6.1.2 Exploring the Oscillatory Content of EEG

Electroencephalography (EEG) is a functional neuroimaging technique with millisecond time resolution. Acquired on the surface of the head, scalp EEG is non-invasive, and since intracranial EEG are only used for patients undergoing presurgical evaluation, scalp EEG has become a widely approved technique for the investigation of oscillatory phenomena in healthy humans.[1] The EEG signals are potential variations detected at the scalp, resulting from the joint electrical activity of active neurons in the brain. In addition, the different resistances between the neural tissue, cerebral spinal fluid, skull/skin, etc. create a low-pass filtering on the source signal, resulting in a noisy signal. Thus, robust signal processing methods are required in order to extract physiologically meaningful information about the oscillatory content of the neuronal sources. So far, several lines of research have been investigated in order to observe the principal spectral components of an EEG signal and describe their evolution over time. Each of these existing methods has its advantages and drawbacks, since a perfect trade-off between time resolution and spectral resolution cannot be found objectively due to the Heisenberg uncertainty principle. It is worth mentioning these methods, in order to provide a clear framework for the presentation of our algorithm.

6.1.2.1 Time-Frequency Analysis

The most standard way of analyzing the spectral behavior of a signal over time is time-frequency analysis [6]. There exist several methods to perform time-frequency analysis, for instance short-term Fourier transforms, Cohen's class distribution functions, or Wavelets [24]. Through the mapping of the investigated signal into a two-dimensional time-frequency representation, this kind of analysis provides a global view of the frequency components of the signal evolving over time. Moreover, these methods do not require extensive parameter tuning and can be quickly applied, resulting in a direct observation of the spectro-temporal content of the signal. Time-frequency analysis has proved to be very fruitful in the analysis of biomedical signals, and particularly EEG signals [34].

However, time-frequency analysis can sometimes be incomplete when dealing with non-stationary signals like EEGs. Indeed, a typical EEG signal contains numerous frequency components of different magnitudes competing with each other over time; a given oscillation may transiently have a low magnitude, although it does not disappear and may be of physiological importance. Time-frequency analysis only focuses on the detection of time-varying spectral power. Consequently, changes in oscillations could be partially ignored by the above time-frequency analysis, thereby potentially failing to provide a continuous description of each of the simultaneously evolving oscillations. Moreover, rapid transitions in mental state or responses to

[1] In this chapter, the term EEG will implicitly refer to scalp EEG.

external stimuli that are characterized by quick changes of finite duration in the oscillatory components of the EEG are difficult to track precisely and do not always appear clearly in time-frequency representations at single-trial level. An example of these problematic situations is shown with the time-frequency analysis at a parieto-occipital EEG electrode following visual stimulation (Fig. 6.1; see [25, 26] for methodological details). In this experiment, a visual stimulus consisting in an illusory contour is presented at time zero, immediately provoking a cascade of oscillatory responses at different frequencies. For analysis purposes, the raw EEG signal has been band-pass filtered between 20 Hz and 100 Hz (top panel), in order to make observation of the high-beta (20–30 Hz) and gamma (30–80 Hz) frequency bands easier. Then, a smoothed pseudo Wigner-Ville distribution [10] was applied on this filtered signal (bottom panel). At stimulus onset (green dashed line), there is clearly increased power at 40 Hz. By contrast, it is more difficult to interpret phenomena occurring at 60 Hz, where it appears that there is a decrease in power immediately after stimulus onset (red dashed circle), although this transient response is likely of strong physiological relevance. A similar limitation of this approach is seen in

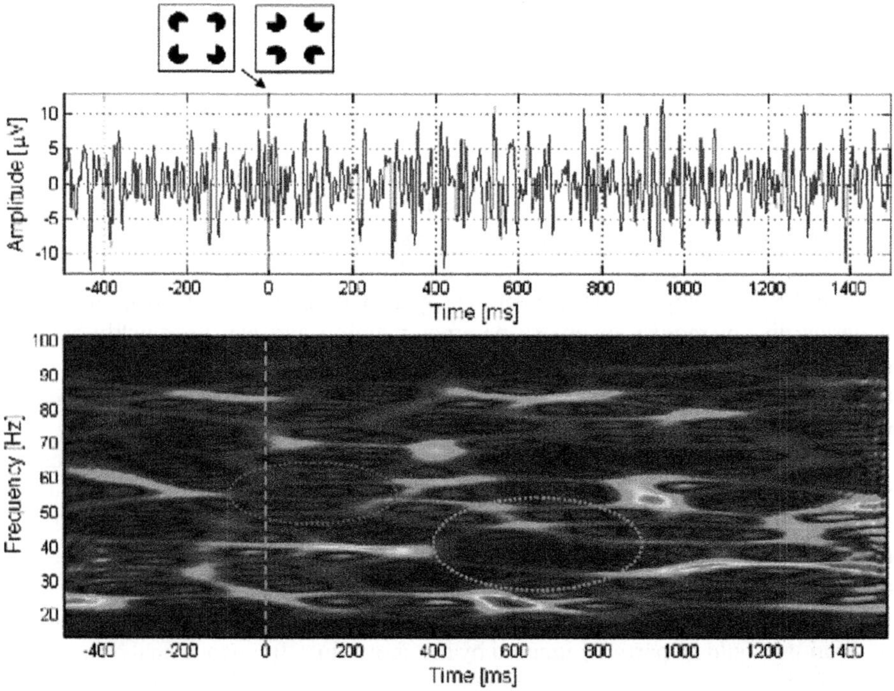

Fig. 6.1 Smoothed Pseudo Wigner-Ville time-frequency analysis of an EEG single-trial during visual stimulation. *Top panel*: band-pass filtered (20–100 Hz) single-trial EEG at a parieto-occipital electrode (PO4) during visual stimulation. Stimulus presentation occurs at time zero (*dashed green line*). *Bottom panel*: smoothed pseudo Wigner-Ville distribution of the filtered signal in higher panel

the temporal evolution of 40 Hz oscillations over the 400 to 900 ms post-stimulus period, which could either be increasing to 55 Hz or decreasing to 30 Hz. Which of these alternatives is correct would, of course, alter the interpretation of the data both in absolute as well as in neurophysiologic terms. This example shows that more precision and continuity are needed in the tracking of the oscillatory components of EEG traces in order to allow a finer physiological interpretation of the phenomena under study. For these purposes, time-frequency analysis would clearly benefit from additional information provided by a robust and continuous tracking of oscillatory components.

6.1.2.2 Filter Bank Decomposition

A well-accepted solution to the problem caused by the broad frequency range of EEG is to first decompose the raw signal into distinct frequency components by means of a narrow filter bank, and then to apply the desired spectral analysis on the resulting filtered oscillations. Thanks to this pre-processing step, one can eliminate differences in magnitudes between the oscillations of different frequencies. Because every output of the filter bank is processed independently, several intrinsic properties of neuronal oscillations as well as interdependencies between oscillations of different frequencies can be revealed in this manner. This process has been widely used in the field of both intracranial and scalp-recorded EEG [3, 7].

The main drawback of this pre-processing step is that the cut-off frequencies of each band-pass filter must be pre-defined and are assumed to remain constant during the whole neurophysiologic process under investigation. This constraint can produce physiologically misleading results, in the case of an oscillating component crossing the cut-off frequency limit of a filter. For instance, the 40 Hz oscillation in Fig. 6.1, which changes its frequency towards either 30 Hz or 55 Hz at ~500 ms post-stimulus onset (green dashed circle), is a typical problematic case. In such a situation, a narrow band-pass filter around 40 Hz could not have described the evolution of this oscillatory component in a precise manner, maybe missing a crucial stage of the global neurophysiologic phenomenon. Thus, in order to perform rigorous descriptions of neuronal processes, new methods are needed that allow for an adaptive tracking of narrow-band EEG oscillations over time.

6.1.2.3 Empirical Mode Decomposition

The so-called empirical mode decomposition (EMD) was introduced in 1998 as a method for nonlinear and non-stationary time series analysis [15]. By means of a so-called sifting process, the EMD decomposes the raw input signal into narrow-band oscillatory components representing the physical time scales intrinsic to the data. These intrinsic mode functions (IMFs) are extracted according to geometrical criteria describing the structure of the signal itself. Because the method is directly and exclusively based on the structure of the signal and considers oscillations at a very local level, it can be described as an automatic and adaptive signal-dependent filter. The EMD has already proven to be powerful in the field of applied neuroscience;

the method was shown to successfully decompose the local field potentials of a macaque monkey in the standard alpha, beta and gamma frequency bands [21], and provided the basis for complex synchronization detection schemes in EEG [33].

Nevertheless, some drawbacks related to the use of EMD for filtering EEG signals should be mentioned. First, the EMD method is influenced by sampling rate, and requires high sampling rates in order to be optimally applied. Moreover, the physiological meaning of the IMFs extracted from the raw signal is still under debate. In order to prove the efficiency of the EMD method and its physiological meaning, it would be useful to propose alternative methods allowing adaptive tracking of narrow spectral components in EEG signals. We present such a method in the following section.

6.2 Adaptive Frequency Tracking

The extraction of oscillatory components in a noisy signal is a classical task in the field of signal processing. The information of interest is the time evolution of these components and of their instantaneous frequencies. Several algorithms have been proposed in the literature to track a single frequency component using either an adaptive band-pass (BP) filter or an adaptive notch filter [16, 4]. The general approach is to maximize (minimize) the energy of the BP filter (notch filter) output. Later algorithms for multiple frequency tracking, often based on those for single frequency tracking, have been proposed [18, 35].

In this chapter we first propose an improved single frequency tracking scheme based on Liao [22]. Then, we extend it to the multiple frequency case by using and refining the scheme in Rao and Kumaresan [28]. Finally, we propose an approach inspired from Prudat and Vesin [27] that uses simultaneously the frequency information possibly present in several signals.

Note first that while the scheme in Liao [22] is designed for real-valued signals, here we use the complex signal framework more employed in the relevant literature.[2] Of course, in practice, the signals acquired are real-valued. Using the Hilbert transform [13] one obtains the so-called analytic representation whose real part is the original signal. This permits working more simply in the complex domain and then to extract real-valued results.

6.2.1 Single Frequency Tracking

In this section we propose a solution based on a real-valued scheme [22] working for a single component situation. This simple tracking algorithm is composed of

[2]In the real case, the simplest band-pass filter requires two poles. Only one pole is needed in the complex case. In this chapter j stands for the pure imaginary complex number of modulus one and argument $\pi/2$.

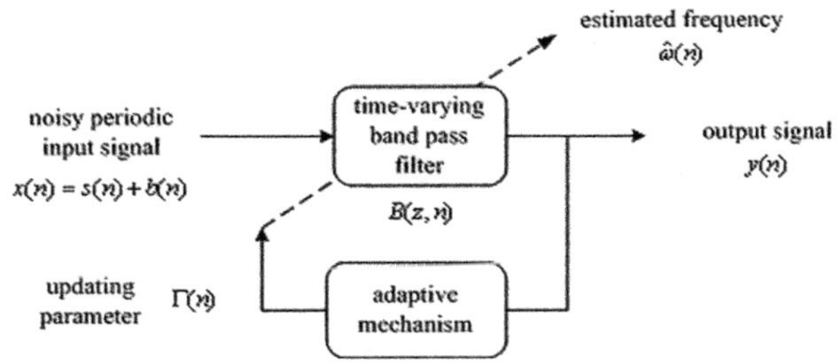

Fig. 6.2 Simple frequency tracking structure composed of a time-varying band-pass filter and an adaptive mechanism. Modified from [22]

two parts: (1) a single pole band-pass (BP) filter with an adaptive transfer function B(z,n), and (2) a feedback mechanism that adapts the central frequency of the BP filter based on some measurement of its output. Together, these two parts form a time-varying scheme able to track frequency changes in the input signal x(n). This structure bears some resemblance with that of the phase-locked loop (PLL) systems used in communications.

At time index n, the transfer function of the BP filter with a central normalized frequency $\omega(n)$ is expressed as follows:

$$B(z,n) = \frac{1-\beta}{1-\beta\Gamma(n)z^{-1}} \tag{6.1}$$

with a pole at $\beta\Gamma(n)$, $\Gamma(n) = e^{j\omega(n)}$. The parameter β, $0 << \beta < 1$, controls the bandwidth of the filter. The closer β is to unity, the more selective the BP filter is (see Fig. 6.3). Note also that (6.1) defines a filter with a unit gain and a zero phase at $\omega(n)$.

Let us assume first that the input signal is an oscillatory signal (cisoid) $x(n) = e^{j\omega_x n}$. Then $x(n)$ must obey the equation for a discrete complex oscillator, that is:

$$x(n) = e^{j\omega_x} x(n-1) \tag{6.2}$$

If the central frequency of the BP filter is constant at $\omega(n) = \omega_x$ then the output signal $y(n)$ computed with:

$$y(n) = \beta\Gamma(n)y(n-1) + (1-\beta)x(n) \tag{6.3}$$

should obey the same equation:

$$y(n) = e^{j\omega_x} y(n-1) \tag{6.4}$$

Thus the coefficient of $y(n{-}1)$ in (6.4) defines the pole of the BP filter. Now if the input signal is corrupted by a complex interference $b(n)$ then the filter output is perturbed. The Eq. (6.4) is not satisfied, all the more so if additionally the frequency of the input $\omega_x(n)$ changes with time. In order to reduce the mismatch between the

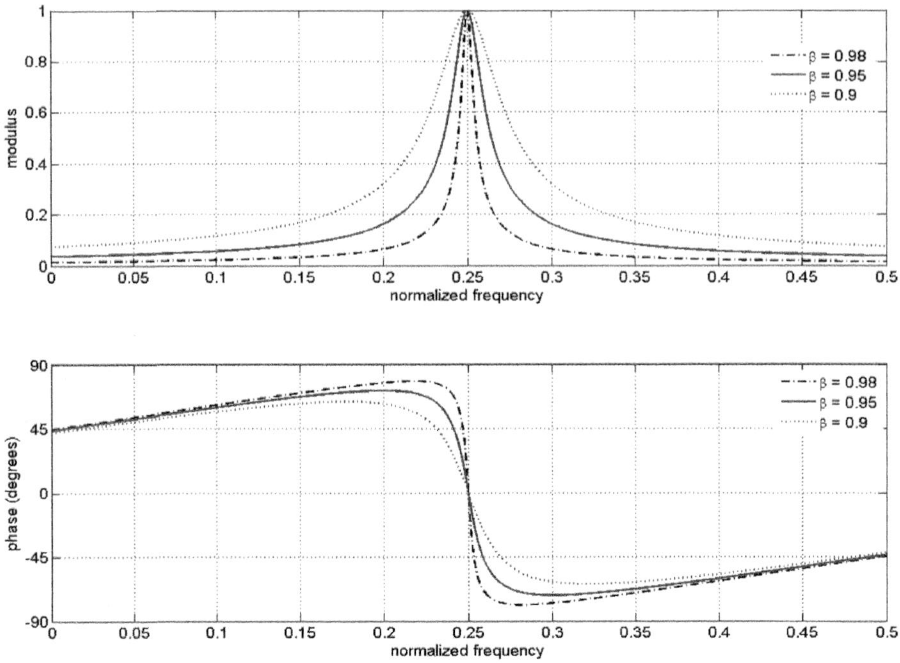

Fig. 6.3 Example of frequency response of the filter $B(z,n)$ for different values of β and for $\Gamma(n) = e^{j\omega n}$ with $\omega(n) = \pi/2$

BP filter central frequency and signal frequency, the idea consists in writing $y(n)$ as:

$$y(n) = \theta(n)y(n-1) + e(n) \tag{6.5}$$

where the error term $e(n)$ represents all the imperfections due to $b(n)$ and frequency maladjustment. A suitable criterion to estimate $\theta(n)$ is the classical mean square error $\mathrm{E}[|e(n)|^2]$. Minimizing it leads to the following expression for $\theta(n)$:

$$\theta(n) = \frac{\mathrm{E}\left[y(n)\bar{y}(n-1)\right]}{\mathrm{E}\left[|y(n-1)|^2\right]}$$

where the upper bar stands for complex conjugation. The expectations of the numerator and the denominator cannot be computed in practice. In the recursive least square solution [14] these quantities are estimated as follows:

$$\hat{\theta}(n) = \frac{Q(n)}{P(n)} = \frac{\delta Q(n-1) + (1-\delta)y(n)\bar{y}(n-1)}{\delta P(n-1) + (1-\delta)\,|y(n-1)|^2} \tag{6.7}$$

where the forgetting factor δ, $0 << \delta < 1$, controls the convergence rate of the estimation. By analogy between (6.4) and (6.5), the feedback mechanism translates as:

$$\Gamma(n+1) = \frac{\hat{\theta}(n)}{|\hat{\theta}(n)|} = e^{j\hat{\omega}_x(n)} \tag{6.8}$$

Finally the estimate $\hat{\omega}_x(n)$ of the instantaneous frequency at time n is obtained as the argument of $\Gamma(n+1)$.

It can be shown that this tracking strategy, referred from now on as single frequency tracker (SFT), leads to an unbiased estimator of the instantaneous frequency.[3]

In order to illustrate the frequency tracking ability of the SFT we test it in two different non-stationary cases. In the first example, the signal processed is a cisoid with an abrupt normalized frequency change from 0.2 to 0.35 at time index 200 corrupted by a complex, Gaussian, and independent noise. The top panel of Fig. 6.4 displays the time evolution of the estimate of the instantaneous frequency for a signal-to-noise ratio (SNR) of 1 dB. This estimate is quite close to the true value

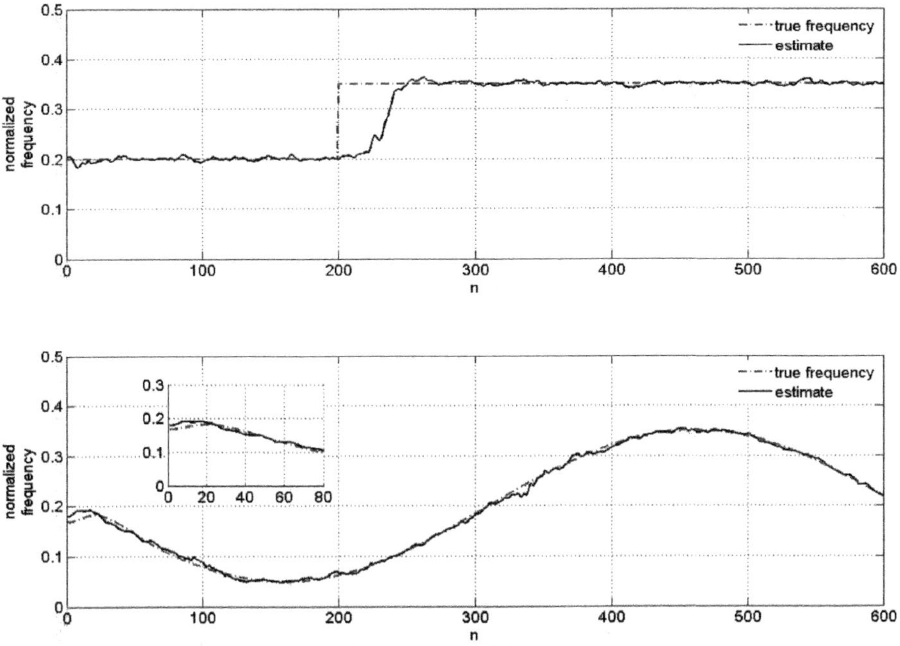

Fig. 6.4 Estimates of the instantaneous frequency of cisoids in two non-stationary situations. *Top panel*: abrupt normalized frequency change between 0.2 and 0.35 for an SNR of 1 dB. *Bottom panel*: sinusoidal frequency modulation between 0.05 and 0.35, modulation period of 600 samples. The update parameter values are: $\beta = \delta = 0.9$

[3]Not presented in this chapter. See Liao [22] for a demonstration in the real-valued case.

(dash-dotted line). After about 75 samples the estimate stabilizes around the final value of the true instantaneous frequency.

In the second example, the frequency tracking ability of the SFT is tested on a cisoid with a sinusoidal frequency modulation between 0.05 and 0.35 (modulation period of 600 samples). The same type of noise is added with the same SNR of 1 dB. The result is represented on the bottom panel of Fig. 6.4. The true profile of the instantaneous frequency is indicated by the dash-dotted line. The estimate of the instantaneous frequency is the solid line. The two curves are quite close. The method converges to the true frequency modulation profile after 40 samples.

6.2.2 A Multiple Frequency Tracking Solution

Now we present an extension of the previous method (SFT) in the multi-component case, i.e., we suppose that we observe a multi-component signal defined by:

$$x(n) = \sum_{k=1}^{p} s_k(n) + b(n) = \sum_{k=1}^{p} A_k(n) e^{j\omega_k(n)} + b(n) \tag{6.9}$$

where $A_k(n)$ and $\omega_k(n)$ are, respectively, the amplitude and the instantaneous frequency of the kth component $s_k(n)$, and where $b(n)$ is an additive complex noise. The main idea is to use an SFT for each component $s_k(n)$ to estimate properly each instantaneous component frequency $\omega_k(n)$. We thus get a p-element adaptive filter bank. A problem with this scheme is the possible cross-talk between the various SFTs due to the fact that the corresponding filters are not perfect band-pass ones. This typically happens in the case where two neighboring signal components have close frequencies. A solution to this problem, proposed in Rao et al. [28], is to use all-zero filters (AZFs) in a cross-coupled fashion to suppress interferences from other frequency components in each SFT.

6.2.2.1 Structure of the Filter-Bank

We assume that the number p of components in the signal to be analyzed is known. Each filter bank element is a concatenation of two adaptive filters (see Fig. 6.5):

- an adaptive single frequency tracker (SFT): its goal is to estimate the instantaneous frequency of the kth component. The transfer function of the band-pass filter is given by (6.1) and the adaptation is performed as described in the previous section as in (6.7) and (6.8).
- an adaptive all-zero filter (AZF): the goal of this filter is to suppress the interference due to the other frequency components. This kth AZF filter is placed in front of the kth SFT. It is composed of $p-1$ complex zeros $\left\{\Upsilon_i(n) = \rho e^{j\omega_i(n)}\right\}$ whose angles are defined by the frequencies$\{\omega_i(n)\}_{i=1, i\neq k}^{p}$ in order to prevent the components at those frequencies to enter the kth SFT. Each notch has the same selectivity which is controlled by the parameter ρ. The transfer function of the kth AZF filter is given by:

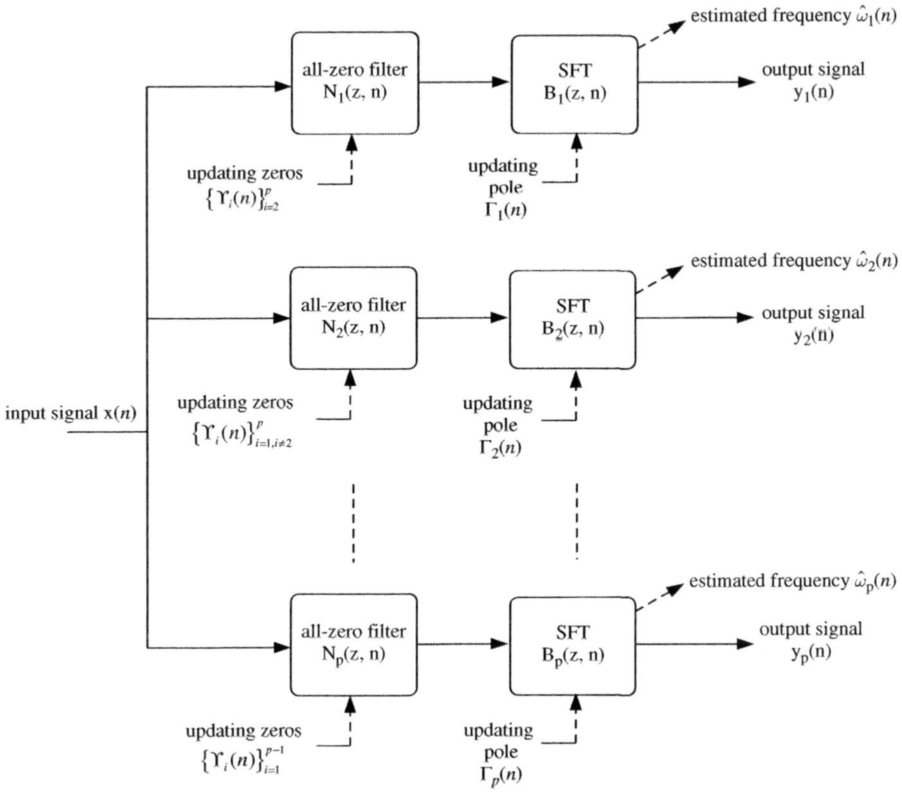

Fig. 6.5 Structure of the adaptive filter-bank MFT. Each element is composed of an SFT $B_k(z,n)$ whose value is updated according to the rule defined in (6.8). In front of the SFT an all-zero filter $N_k(z,n)$ operates in order to suppress the components processed by the other elements

$$N_k(z, n) = C_k(n) \prod_{\substack{i=1 \\ i \neq k}}^{p} \left[1 - \Upsilon_i(n) z^{-1}\right]$$

where $C_k(n)$ is a normalization coefficient to ensure a unit gain at frequency $\omega_k(n)$. In practice, the unknown true values of frequencies are replaced by their estimates provided by the other SFTs. This scheme is referred to as multiple frequency tracker (MFT) in this chapter.

The insertion of the AZFs has two harmful effects. First, each element of the filter-bank is no longer zero-phase at the BP central frequency. This causes a phase lag between the signal estimates $y_k(n)$ and true components $s_k(n)$. Second, longer filters lead to lower convergence rates. We have determined that the alternative in which the kth AZF has only one zero at the tracked frequency closest to the kth frequency reduces this effect by offering a good compromise between interference cancellation on the one hand and tracking speed and phase lag on the other.

6.2.2.2 Example of Tracking Ability

We illustrate the performance of the proposed filter-bank for a signal composed of three non-stationary oscillatory components of unit power and an additive complex, Gaussian, and independent noise for a total SNR of 20 dB. Two components have a linear frequency modulation with opposite chirp rates.[4] The third one contains an abrupt instantaneous frequency change. The three instantaneous frequency estimates obtained with the MFT are plotted in Fig. 6.6.

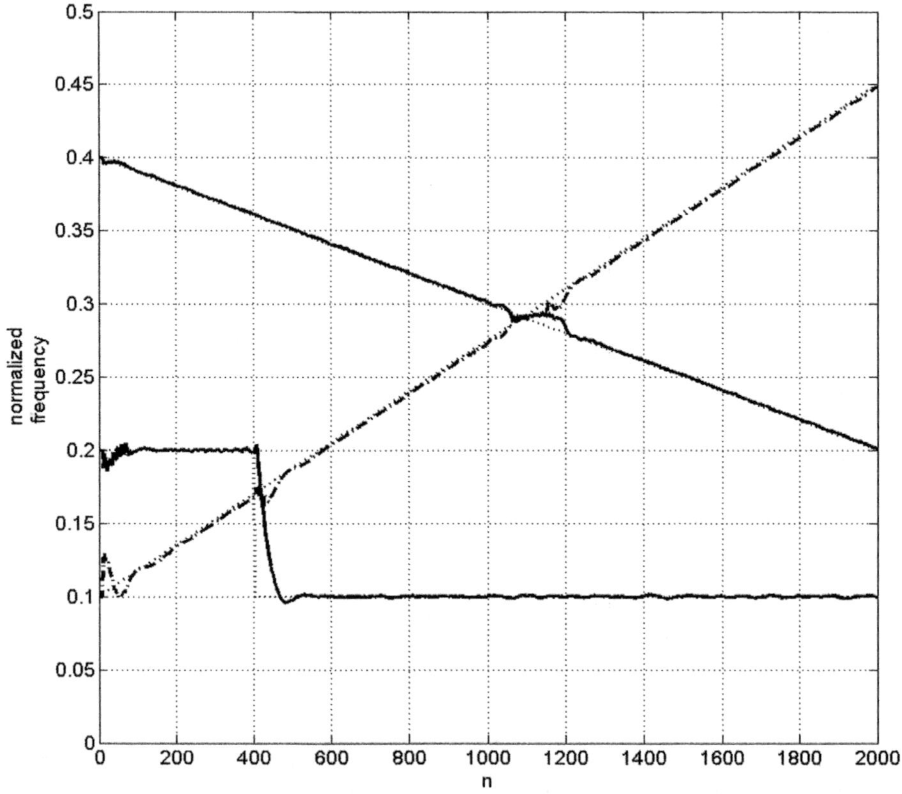

Fig. 6.6 An example of multiple tracking ability of the MFT for a three-component input signal. The *dotted lines* represent the true frequency profiles. The *solid lines* are their estimates. The update parameter values are: $\beta = 0.97$, $\rho = 0.97$, and $\delta = 0.9$

---

[4]The chirp rate is defined as the derivative of the instantaneous frequency with respect to time. For a linear frequency modulation the instantaneous frequency is an affine function. In that case the chirp rate is simply the slope coefficient.

The estimates are quite close to the true values of the instantaneous frequencies of each component. Even when the frequency trajectories of two components cross, the estimates remain coherent. To the best of our knowledge, this behavior compares favorably with that of tracking strategies proposed earlier in the literature [35].

Of course the adaptive filter bank solution assumes that the user has enough knowledge of the signal to be analyzed. If the number of components is over- or under- determined then this can lead to erroneous estimates.

6.2.3 An Extension to the Multi-Signal Case

In many practical situations the phenomena of interest are observed across multiple sensors. This is particularly the case with EEG in which neighboring lead signals are often highly correlated. Compared to the separate analysis of each EEG channel, a multi-signal approach for tracking the common oscillatory activity of a cluster of electrodes could improve largely the tracking performance, the rate of convergence and the overall robustness of the MFT. The main idea is to use the same filter bank on each signal selected among the measurement set. Then, the update of the center frequency of each band-pass filter is obtained as a weighted average of the updates that would be computed separately for each signal. A sensible choice for each weight is one minus the ratio of the instantaneous estimate of the oscillation criterion (i.e., $E[|e(n)|^2]$ for the error $e(n)$ in (6.5)) divided by the instantaneous estimate of the BP filter output variance. This means that signals for which the oscillation criterion is better satisfied influence more the frequency update. Dividing by the BP filter output variance makes the scheme scale independent.

With $y_{k,m}(n)$ the output of the kth BP filter for the mth signal, the instantaneous estimate $J_{k,m}(n)$ of the error variance is obtained recursively as:

$$J_{k,m}(n) = \delta J_{k,m}(n-1) + (1-\delta)\left|y_{k,m}(n) - \Gamma_{k,m}(n) y_{k,m}(n-1)\right|^2 \tag{6.11}$$

where we use the same forgetting factor δ, $0 << \delta < 1$ as for (6.7). Similarly the instantaneous estimate $K_{k,m}(n)$ of the BP output filter variance is given by:

$$K_{k,m}(n) = \delta K_{k,m}(n-1) + (1-\delta)\left|y_{k,m}(n)\right|^2 \tag{6.12}$$

The weight for the mth signal is:

$$W_{k,m}(n) = 1 - \frac{J_{k,m}(n)}{K_{k,m}(n)} \tag{6.13}$$

Finally the update of the kth BP central frequency is:

$$\omega_k(n+1) = \frac{\sum_{m=1}^{M} W_{k,m}(n)\omega_{k,m}(n+1)}{\sum_{m=1}^{M} W_{k,m}(n)} \tag{6.14}$$

with $\omega_{k,m}(n+1)$ is the central frequency update that would take place on the mth signal, separately. Note that the update (6.14) has to be performed on the central frequencies rather than directly on the poles $\Gamma_k(n)$ because the weighted sum of unit-modulus complex variables is not necessarily unit-modulus itself. This scheme is referred to as the multivariate MFT (MMFT) in this chapter.

6.3 Tracking EEG Oscillations

6.3.1 Tracking of a Single EEG Oscillation

Here, we illustrate the performance of the SFT on real single-trial EEG signals from a visual evoked potential (VEP) experiment studying illusory contour processing [25]. As a systematic preprocessing for all the signals of Sect. 6.3, we band-pass filtered the raw EEG traces between 20 Hz and 100 Hz, in order to observe high beta and gamma frequency bands. To initialize the SFT for a given trial, we first ran the algorithm on the trial with null initial conditions. After this first pass, we averaged the value of the measured $P(n)$ and $Q(n)$ to get the new initial values of P and Q; a relevant initial frequency could thus be computed from these values. This initialization allowed us to rapidly converge to an interesting oscillatory component in the observed signal, as shown below.

In a first example, we present the tracking performance of the SFT. In the selected single-trial EEG of Fig. 6.7, the gamma component (30–50 Hz) is particularly complex, since it presents three different periods during the trial, as illustrated by the temporal output of the SFT (bottom panel) as well as the frequency estimate of the algorithm (white line in the top panel), which was superimposed on a Wigner-Ville time-frequency representation for visualization and comparison purposes. During the first part of the trial, the oscillation has increasing magnitude and decreasing frequency; then, from 180 to 800 ms post-stimulus onset, the oscillatory component goes up to 40 Hz and remains on stable, while the magnitude is clearly decreased. Finally, from 800 ms post-stimulus onset onwards, the gamma oscillation increases again up to 50 Hz, and slightly decreases to 45 Hz at the end of the trial. While the time-frequency representation does not precisely describe the entire evolution of the gamma oscillation, especially in the second oscillatory period, we see that the SFT tracks the spectral component successfully through the three consecutive periods, therefore providing a continuous description of the oscillatory phenomenon with reliable tracking precision. Moreover, the changes in both frequency and magnitude are easily observable in the temporal output of the SFT when switching from one period to the next.

In a second example, we demonstrate the robustness of the tracking algorithm through the precise analysis of a gamma oscillation during illusory contour perception (same dataset as before), as well as its benefits for the physiological interpretation of EEG signals. In Fig. 6.8, several oscillatory responses are triggered by the presentation of an illusory contour (green line at stimulus onset), as shown on

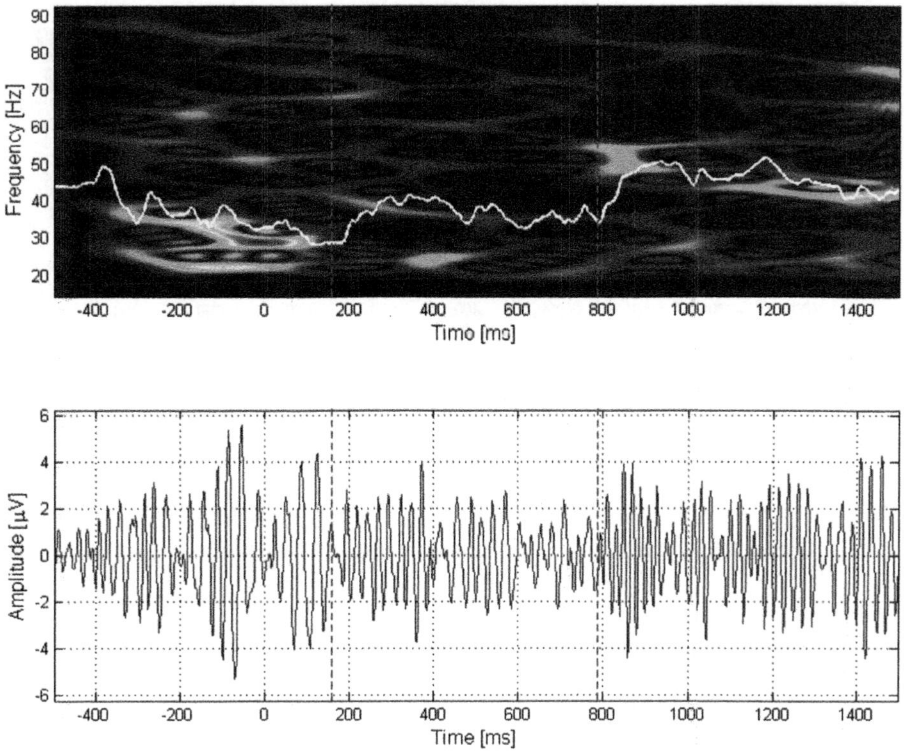

Fig. 6.7 Tracking of a varying gamma oscillation. *Top panel*: frequency tracking of the SFT algorithm (superimposed *white line*) compared to a smoothed pseudo Wigner-Ville distribution of the band-pass filtered (20–100 Hz) data from a parieto-occipital electrode (PO4). *Bottom panel*: output of the SFT for the trial shown in the *top panel*. Dashed red lines separate the different oscillatory periods

the time-frequency representation (top panel). Among them, a 50 Hz gamma band response (GBR) arises 100 ms post-stimulus onset and shows highly increased magnitude during a time interval of 500 ms. But how does this gamma oscillation behave during the remainder of the trial? Especially, how does the gamma frequency vary during periods of low magnitude, such as those indicated in Fig. 6.8 by the red squares? Such information is crucial to globally understand the neuronal response. Because time-frequency analysis focuses on the detection of spectral magnitude, it cannot reliably track a single oscillation during periods of low magnitude, particularly when other oscillatory components exhibit larger magnitudes over the same time period. A typical example of this is shown in Fig. 6.8, where high-gamma and beta oscillations are competing with the 50 Hz GBR during the whole trial, and particularly during time periods indicated by the red squares (top panel). In such situations, another criterion is needed to allow robust and continuous tracking of a single oscillation. Instead of detecting spectral power, the SFT has been designed to

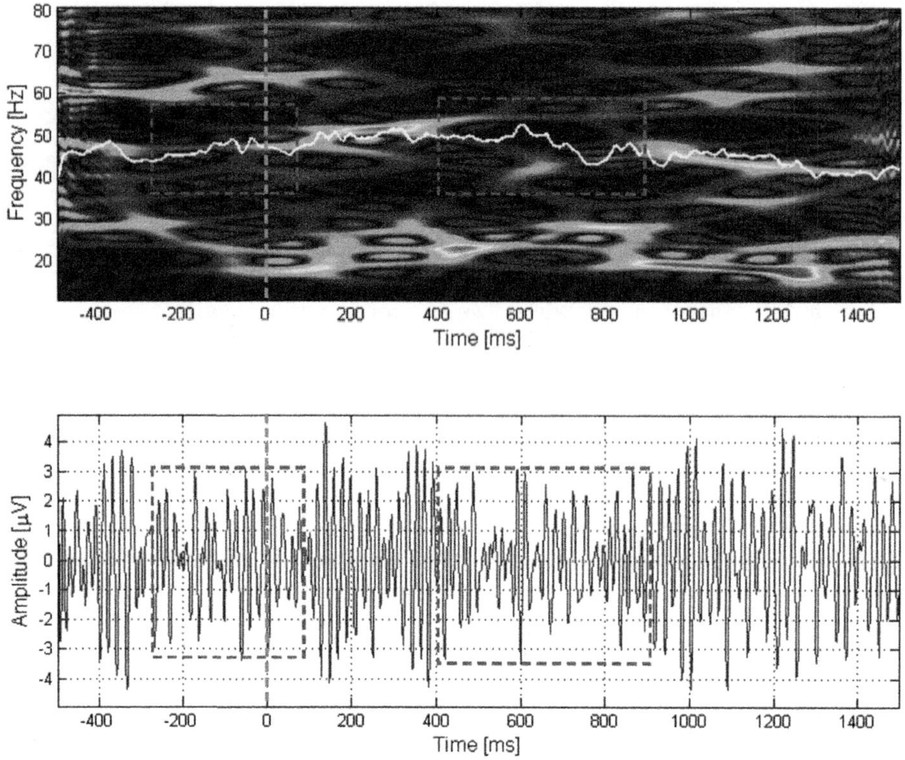

Fig. 6.8 Robust tracking of a stimulus-related gamma band response. *Top panel*: frequency tracking of the SFT algorithm (superimposed *white line*) compared to a smoothed pseudo Wigner-Ville distribution of the band-pass filtered (20–100 Hz) trial. *Bottom panel*: temporal output of the SFT for the trial depicted in the *top panel*. Red squares indicate periods of low gamma magnitude. Green lines indicate stimulus onset

track oscillatory properties independently from their magnitude, through the adaptive maximization of the complex oscillator equation given by (6.4). In doing so, the SFT is able to lock on the 50 Hz GBR during the whole trial, regardless of neighboring frequencies with transiently larger magnitudes. The SFT frequency estimation is shown in Fig. 6.8 (white line), superimposed on the time-frequency representation.

Thus, robust tracking is achieved before and after stimulus presentation, providing valuable physiological information about the general process of illusory contour processing at a single-trial level. The temporal output of the filter gives precious details about the variation in amplitude of the GBR during illusory contour processing (Fig. 6.8, bottom panel). According to the SFT, a first burst in GBR magnitude occurs around ~90 ms post-stimulus and decays slowly until 250 ms, when a second burst is triggered. The GBR comes back to a stable level only 400 ms post-stimulus. It is worth noting that these changes in magnitude of the extracted GBR are neither visible in the raw EEG signal (not shown) nor in the time-frequency representation (see Fig. 6.8). Interestingly, the timing of these transitions is exactly in line with recent event-related potential (ERP) studies from our group [25, 26]. In these studies, a multi-stage model for visual object processing under degraded conditions was proposed based on the average of numerous EEG trials: the first stage of this model, occurring over the ~65–90 ms period, is sensitive to the spatial extent of stimuli, independent of whether they form illusory contours or shapes. The second stage (~90–190 ms) is sensitive to boundary completion, while the third stage (~240–400 ms) is involved in degraded object identification and fine discrimination. Because the sequence of GBR bursts extracted by the SFT in this trial tightly matches the timing of the above-mentioned multi-stage model, we propose that the single-trial SFT method could be coupled with averaged ERP analysis to bring new elements to our model of visual object processing under degraded conditions. More generally, this method shows great promise for analyzing oscillatory events related to neuronal responses in sensory-cognitive neuroscience studies using EEG signals.

6.3.2 Tracking of Multiple Neuronal Oscillations

In this section, we present the performance of the multiple frequencies tracker (MFT) as well as its multivariate extension (MMFT). For this purpose, we use the exemplar trial presented in Sect. 6.1.2.1 that was shown to be difficult to interpret when investigated by means of a time-frequency analysis (see Fig. 6.1). Here, we show that the MFT analysis permits finer interpretation of the (neurophysiologic) process under investigation. To initialize the MFT we proceeded as follows: after the usual band-pass filtering (20–100 Hz), we added at the beginning of the trial a time segment representing the mirrored version of the first 200 ms of the trial. This additional period allowed the MFT to smoothly converge towards spectral components of interest, therefore initiating the MFT algorithm at time zero in a correct manner. Once the initialization of the different frequencies was done, we removed this additional time period. The same procedure was applied on each selected electrode signal in the case of a multivariate input. For the selected example below, we chose to track four simultaneous oscillations; the results of which are shown in Fig. 6.9.

The tracking results for a single parieto-occipital electrode PO4 (red traces) show that the MFT algorithm succeeds in tracking mixed oscillations in single-trial EEG. Compared to time-frequency analysis, the MFT provides a continuous tracking of each frequency, even when a component shows low magnitude during a given time period. This advantage over time-frequency methods allows for facilitated neurophysiologic interpretations. For example, the evoked response of the 60 Hz high-gamma oscillation can finally be described. According to the MFT, the presentation of the stimulus provokes a transient increase in frequency of the high-gamma component with duration of 300 ms. Intuitively, this increase in frequency could possibly weaken the magnitude of the oscillation during the transient period, which could explain why time-frequency methods have difficulties capturing the phenomenon.

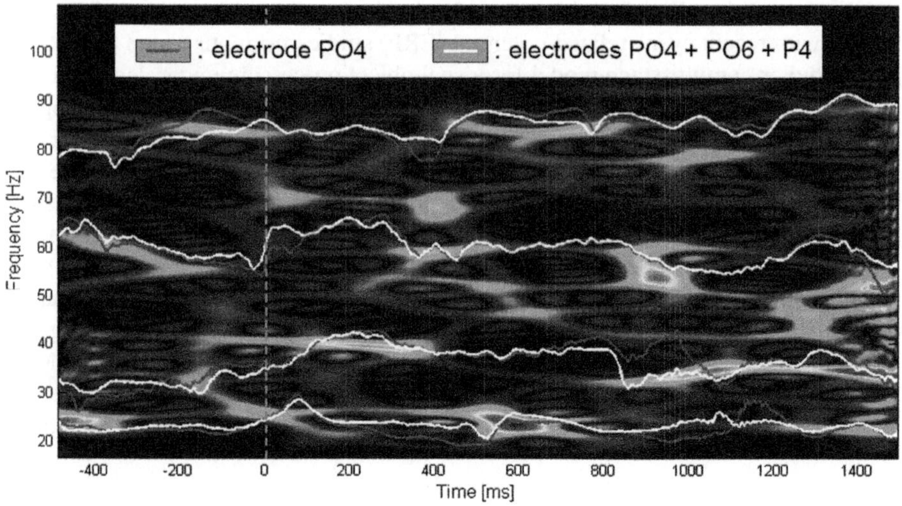

Fig. 6.9 Univariate and multivariate outcomes of the MFT algorithm compared to a smoothed pseudo Wigner-Ville distribution of the band-pass filtered (20–100 Hz) trial. The *red lines* indicate the univariate MFT estimation of electrode PO4 and the *white lines* indicate the multivariate MFT estimation of cluster PO4 – PO6 – P4

Similarly, the MFT indicates that the 40 Hz gamma oscillation decreases towards 30 Hz during the post-stimulus period. This alternative possibility is in principle also likely to occur, and the phenomenon could then be interpreted as an oscillatory relaxation after stimulus processing.

Furthermore, improvements in the quality of the frequencies estimation can be obtained through the use of the multivariate extension of the MFT (MMFT). Here, we consider a parieto-occipital cluster of electrodes (PO4, PO6 and P4) in order to maximize the amount of extracted information from this scalp region. The results for multivariate tracking (white traces) indicate that more robust estimations of oscillatory components are observed when using several correlated electrode signals. One feature is that the frequency estimations from the MMFT look smoothed compared to the univariate MFT tracking solutions. In addition, abrupt frequency modulations are also better described, such as the 60 Hz increase at stimulus onset or the rapid 40 Hz oscillation decrease towards 30 Hz during the post-stimulus period.

6.4 Discussion

In this chapter, we presented a novel method for adaptively tracking single frequencies as well as multiple oscillating components in single-trial EEG signals. This tracking algorithm is proposed as an efficient tool to complement standard spectral analyses such as time-frequency representations, and can be used alone or simultaneously with these methods in order to allow finer interpretations of the neurophysiologic phenomena under investigation.

Compared to existing methods, the proposed approach presents several interesting properties. Firstly, the method tracks oscillatory properties of the signal, as opposed to the majority of existing algorithms that tend to track spectral power more than oscillatory behaviors. Thus, it will succeed in tracking simultaneously oscillating components of varying magnitudes, or components with transiently low magnitude. This specificity is very useful in the context of EEGs, where neurophysiologic events, such as stimulus-related responses, are characterized by abrupt changes in the amplitude of oscillatory components over short, transient time periods. Another major advantage of the approach is that the algorithm does not only provide an estimation of the frequency, but also gives direct access to the temporal dynamics of the tracked component. Such temporal signals can subsequently be analyzed over time to give a precise description of the oscillatory behavior in terms of both frequency and amplitude modulations. We illustrated in Sect. 6.3.1 the potential benefits of such investigations in the context of single-trial ERPs.

Another attractive property of the method is its extension to multivariate input signals. In many situations of biomedical engineering, it is of crucial importance and relevance to add supplementary information by feeding the algorithm with signals from several channels. Thanks to this approach, a strengthened and more precise estimation of the oscillatory content of the measurements can be obtained. Once again, the field of EEG processing can clearly benefit from such an approach, because it is well known that the scalp-recorded EEG is a low-pass filtered version of the intracranial local field potentials. Neighboring electrodes are therefore strongly correlated, and the combined analysis of their spectral contents will surely strengthen the quality of the final tracking results. A convincing illustration of this approach was presented in Sect. 6.3.2.

Some drawbacks of the proposed method are also worth mentioned. First, several parameters must be fixed before running the algorithm, and bad choices of parameters can lead to unsatisfactory tracking results. Especially, a compromise has to be found between robustness of the tracker against noise and precise tracking of a quickly varying frequency. This trade-off between spectral accuracy and temporal resolution is inherent to spectral analysis methods in signal processing. In addition, another weakness of the method is the choice of the initial frequencies before the tracking algorithm starts; a poor choice of initial frequencies could also bias the algorithm towards erroneous estimates. Concerning this issue, several alternatives can be proposed in order to find the most likely initial frequencies for a given signal. For instance, the algorithm can be firstly applied on a part of the signal in order to get a first idea of its spectral content. The initial frequencies can then be fixed accordingly. This initialization proved to be efficient and relatively robust to different signals, but further developments are needed in this direction.

Another possible improvement of the method that could be very fruitful for the area of EEG analysis is the detection of arising and vanishing oscillations during a given single-trial. Studies on event related potentials (ERPs) have extensively reported induced oscillatory components triggered by an experimental stimulus that dissipate after a finite duration. Detecting these transient oscillations and tracking them over time would provide precious information about ERPs at a single-trial level, and is therefore of crucial interest in the context of EEG/ERP research. To allow such complex analysis, the method should foresee the start of an oscillation and create a supplementary channel to track the frequency once it appears. An inverse procedure should suppress the created channel once the oscillation dissipates.

The quest for understanding the mechanisms of neuronal oscillations and their role in neurophysiologic processes as well as behavioral outcomes has been a captivating research area in the last decades, and the need for new analytical methods describing oscillatory phenomena is now obvious [19]. Our approach is likely to have a beneficial impact on the field of neuroscience in both fundamental and clinical research settings. For instance, increasing evidence indicates that oscillatory phenomena have a crucial role in the onset of epileptic seizure [20]; a fine description of oscillatory patterns before and during the seizure may add new information about the oscillatory nature of epilepsy and allow a better monitoring of epileptic activity. Similarly, the importance of neuronal oscillations in sleep is now clearly established. For instance, the role of slow oscillations in memory consolidation has been recently demonstrated [1], as well as the coupling between different types of sleep-related brain waves [5]. Once again, applying the methods on such datasets will hopefully bring new insights on the oscillatory description of sleep states. Finally, the fields of cognitive neuroscience and ERP analysis still need competitive methods for describing stimulus-induced oscillatory processes at a single-trial level. Our approach will help bridge gaps between the complementary approaches of ERP and single-trial analysis.

References

1. Born J (2006) Sleep to remember. Neuroscientist, 12, 410–424
2. Buzsaki G (2006) Rhythms of the brain. Oxford University Press, New York
3. Canolty RT, Edwards E, Dalal SS, Soltani M, Nagarajan SS, Kirsch HE, Berger MS, Barbaro NM, Knight RT (2006) High Gamma Power is Phase-Locked to Theta Oscillations in Human Neocortex. Science, 313, 1626
4. Cho N I and Lee S U (1995) Tracking analysis of an adaptive lattice notch filter. IEEE Trans Circuits and Systems II, 42, 186–195
5. Clemens Z, Mölle M, Eross L, Barsi P, Halasz P, Born J (2007) Temporal coupling of parahippocampal ripples, sleep spindles and slow oscillations in humans. Brain, 130, 2868–2878
6. Cohen L (1995) Time-Frequency Analysis. Prenctice-Hall PTR, Englewood Cliffs, NJ
7. Dalal SS, Guggisberg AG, Edwards E, Sekihara K, Findlay AM, Canolty RT, Berger MS, Knight RT, Barbaro NM, Kirsch HE, Nagarajan SS (2008) Five-dimensional neuroimaging: Localization of the time-frequency dynamics of cortical activity. Neuroimage, 40(4), 1686–1700
8. Engel AK, Singer W (2001) Temporal binding and the neural correlates of sensory awareness. Trends Cogn Sci, 5, 16–25
9. Engel AK, Fries P, Singer W (2001) Dynamic predictions: oscillations and synchrony in top-down processing, Nature, 2(10), 704–716
10. Flandrin P (1999) Time-frequency / Time-scale analysis, Academic Press, San Diego, CA

11. Fries P (2005) A mechanism for cognitive dynamics: neuronal communication through neuronal coherence. Trends Cogn Sci, 9, 474–480
12. Fries P, Reynolds JH, Rorie AE, Desimone R (2001) Modulation of oscillatory neuronal synchronization by selective visual attention. Science, 291, 1560–1563
13. Hahn SL (1996) Hilbert Transforms in Signal Processing. Artech House, Norwood, MD
14. Haykin S (2001) Adaptive Filter Theory. Prentice Hall, Englewood Cliffs, NJ
15. Huang NE, Shen Z, Long SR, Wu ML, Shih HH, Zheng Q, Yen NC, Tung CC, Liu HH (1998) The empirical mode decomposition and Hilbert spectrum for nonlinear and nonstationary time series analysis. Proc Roy Soc London A, 454, 903–995
16. Hush D R, et al. (1986) An Adaptive IIIR Structure for Sinusoidal Enhancement, Frequency Estimation, and Detection. IEEE Trans Signal Process, 34, 1380–1390
17. Jacobs J, Kahana MJ, Ekstrom AD, Fried I (2007) Brain oscillations control timing of single-neuron activity in humans, J Neurosci, 27(14), 3839–3844
18. Ko C, Li C (1994) An adaptive IIR structure for the separation, enhancement, and tracking of multiple sinusoids Signal Processing. IEEE Trans Signal Process, 42, 2832–2834
19. Le Van Quyen M, Bragin A (2007) Analysis of dynamic brain oscillations: methodological advances. Trends Neurosci, 30(7), 365–373
20. Le Van Quyen M, Soss J, Navarro V, Robertson R, Chavez M, Baulac M, Martinerie J (2005) Preictal state identification by synchronization changes in long-term intracranial EEG recordings. Clin Neurophysiol, 116, 559–568
21. Liang H, Bressler SL, Buffalo EA, Desimone R, Fries P (2005) Empirical mode decomposition of field potentials from macaque V4 in visual spatial attention. Biol Cybern, 92, 380–392
22. Liao H (2005) Two discrete oscillator based adaptive notch filters (OSC ANFs) for noisy sinusoids Signal Processing, IEEE Trans Signal Process, 53(2), 528–538
23. Linkenkaer-Hansen K, Nikulin VV, Palva S, Ilmoniemi RJ, Palva JM (2004) Prestimulus oscillations enhance psychophysical performance in humans. J Neurosci, 24(45), 10186–10190
24. Mallat S (1999) A Wavelet Tour of Signal Process. Academic Press, San Diego, CA
25. Murray MM, Wylie GR, Higgins BA, Javitt DC, Schroeder CE, Foxe JJ (2002) The spatiotemporal dynamics of illusory contour processing: combined high-density electrical mapping, source analysis, and functional magnetic resonance imaging. J Neurosci 22, 5055–73
26. Murray MM, Imber ML, Javitt DC, Foxe JJ. (2006) Boundary completion is automatic and dissociable from shape discrimination. J Neurosci 26, 12043–12054
27. Prudat Y, Vesin J-M (2007) Two-signal extension of an adaptive notch filter for frequency tracking, 15th European Signal Processing Conference EUSIPCO Poland, 198–202
28. Rao A, Kumaresan R (2000) On decomposing speech into modulated components Speech and IEEE Trans Audio Process, 8, 240–254
29. Salinas E, Sejnowski TJ (2001) Correlated neuronal activity and the flow of neuronal information. Nat Rev Neurosci, 2, 539–550
30. Schall JD (2001) Neural basis of deciding, choosing and acting. Nat Rev Neurosci, 2, 33–42
31. Schoffelen JM, Oostenveld R, Fries P (2005) Neuronal coherence as a mechanism of effective corticospinal interaction. Science, 308, 111–113
32. Singer W, Gray CM (1995) Visual feature integration and the temporal correlation hypothesis. Annu Rev Neurosci, 18, 555–586
33. Sweeney-Reed CM, Nasuto SJ (2007) A novel approach to the detection of synchronization in EEG based on empirical mode decomposition. J Comput Neurosci, 23, 79–111
34. Tallon-Baudry C, Bertrand O (1999) Oscillatory gamma activity in humans and its role in object representation. Trends Cogn Sci, 3, 151–162
35. Tichavsky P, Nehorai (1997) A Comparative study of four adaptive frequency trackers Signal Processing. IEEE Trans Signal Process, 45(6), 1473–1484
36. Varela FJ, Thompson E, Rosch E (1991) The Embodied Mind. MIT press, Cambridge, Massachusetts

37. Varela F, Lachaux JP, Rodriguez E, Martinerie J (2001) The brainweb: phase synchronization and large-scale integration. Nat Rev Neurosci, 2, 229–239
38. Womelsdorf T, Fries P (2007) The role of neuronal synchronization in selective attention. Curr Opin Neurobiol, 17, 154–160
39. Womelsdorf T, Fries P, Mitra PP, Desimone R (2006) Gamma-band synchronization in visual cortex predicts speed of change detection. Nature, 439(7077), 733–736

Chapter 7
From EEG Signals to Brain Connectivity: Methods and Applications in Epilepsy

Lotfi Senhadji, Karim Ansari-Asl and Fabrice Wendling

Abstract During the past decades, considerable effort has been devoted to the development of signal processing techniques aimed at quantifying the temporal evolution of the cross-correlation (in a wide sense) between signals recorded from spatially-distributed regions in order to characterize brain functional connectivity during normal or pathological (as in epilepsy) conditions. Besides linear methods introduced in the field of EEG analysis fifty years ago, a number of studies have been dedicated to the development of nonlinear methods, mostly because of the nonlinear nature of mechanisms at the origin of EEG signals. Recent studies showed the potential value of methods commonly used in nonlinear physics (see Chap. 15). Three families of methods (linear and nonlinear regression, phase synchronization, and generalized synchronization) are reviewed. Their performances are evaluated on the basis of a simulation model in which a coupling parameter can be tuned between populations of neurons generating bivariate EEG time-series. This evaluation is performed according to quantitative criteria. The main findings of this evaluation are the following. First, some of the methods are insensitive to the coupling parameter. Second, results were found to be dependent on signal properties. In particular, the broadening of the frequency band is a parameter that strongly influences the performances. Third, and generally speaking, there is no 'universal' method for measuring statistical couplings among signals. Indeed, none of the studied methods performs better than the other ones for the two studied situations (background and epileptic activity). Finally, linear and nonlinear regression methods were found to be sensitive to the coupling parameter in all situations and showed either average or good performances. This latter point leads the authors to conclude that these "robust" methods should be applied before using more sophisticated methods.

L. Senhadji (✉)
INSERM, U642, Rennes, F-35000, France; Université de Rennes 1, LTSI, Campus de Beaulieu, Université de Rennes 1, 263 Avenue du General Leclerc – CS 74205 – 35042 Rennes Cedex, France
e-mail: lotfi.senhadji@univ-rennes1.fr

A. Naït-Ali (ed.), *Advanced Biosignal Processing*,
DOI 10.1007/978-3-540-89506-0_7, © Springer-Verlag Berlin Heidelberg 2009

7.1 Introduction

In the field of brain research, the past decades have witnessed a considerable increase of interest for methods aimed at estimating the functional connectivity between spatially-distributed regions, both under normal (cognitive research) or pathological (clinical research) conditions. The reason for this increased interest is related to the commonly-admitted assumption according to which most of the brain functions are based on interactions between neuronal assemblies distributed within and across different cerebral regions.

Indeed, it has been shown that specific networks activate in response to a particular cognitive task, although underlying processes such as the way coordination between distant areas is achieved are not resolved yet [44]. In the context of neurological disorders like epilepsy, it has also been shown that abnormal activity may occur in networks extending over rather large regions. For instance, in certain forms of partial epilepsy, both the temporal and the frontal lobe may be involved at the onset of seizures.

The identification of such networks from electrophysiological (scalp EEG, MEG or intracerebral EEG) and imaging (fMRI) data available in human subjects is still considered as a difficult and unsolved problem. Difficulties arise from (i) the characteristics of recording techniques that bring a more or less direct measurement of the activity in neuronal networks, (ii) the lack of knowledge about the links between the observations and the neuronal mechanisms involved during coordinated interactions between distant areas and (iii) the plethora of available methods for estimating functional connectivity, all based on different assumptions about the underlying model of relationship between analyzed signals.

This chapter deals with signals recorded in epileptic patients. Generally speaking, epilepsy is a complex neurological disease that is characterized by recurring seizures. It affects 50 million people worldwide. In 30% of patients, the frequency of seizures cannot be significantly reduced by anti-epileptic drugs. In the case where seizures are generated in a relatively circumscribed region of the brain, epilepsy is said to be partial and a surgical procedure can be indicated based on a comprehensive pre-surgical evaluation that includes long-term monitoring of scalp and intracerebral EEG signals (generally coupled with video monitoring).

The main objectives of this chapter are: (i) to review most of the methods aimed at characterizing brain connectivity, (ii) to provide the reader with concrete examples of application in the context of drug-refractory epilepsies, and (iii) to highlight some key points emerging from the comparison of the various methods, not only on real intracerebral EEG data but also on signals simulated from models in which the underlying connectivity patterns are known a priori (ground truth).

This chapter is organized as follows. Section 7.2 provides a review of different methods used in the field of EEG analysis. Methods are organized as families according to their common features. This section also introduces two important problems: the non stationarity of analyzed signals and the necessity of using simulation models for comparing methods and for interpreting results. In Sect. 7.3, a protocol for comparing methods is presented based on a neurophysiologically-relevant model of coupled neuronal populations able to simulate local field potentials (analogous to intra-cerebral EEG signals) for different values of the degree of coupling between populations. Section 7.4 provides results obtained from simulated signals and shows the behavior of some selected methods on real EEG data. Methods and results are then discussed before the conclusion which summarizes the main points developed in the chapter.

7.2 State of the Art

During the past decades, numerous techniques have been introduced for measuring the temporal evolution of the cross-correlation (in a wide sense) between signals recorded from spatially-distributed brain regions. The first proposed methods [4] were developed in the late 50's, based on the cross-correlation function in the time domain and on coherence function [7, 31] in the frequency domain, after fast Fourier transform (FFT) algorithms were introduced [10]. Brazier [8] was the first to study the propagation of inter-ictal events from intracerebral EEG data obtained in human. Gotman et al. [16] also used the averaged coherence on signals acquired from both hemispheres in order to study the evolution of inter-hemispheric interactions on the whole duration of partial seizures. In [13], the coherence was used to reveal possible synchronization mechanisms occurring at the onset of seizures as well as the existence of activities that could propagate over short-range or long-range connection fibers [43]. A frequently addressed issue was also the estimation of time delays (see also Sect. 14.2, Chap. 14) from coherence values [15, 21] as this quantity can provide insights into causality relationships among signals recorded from distant structures.

Some studies also reported the use of time-varying linear models in the estimation of the coherence function (autoregressive models). These parametric methods were used to measure the degree of synchronization of interictal and ictal EEG signals and to characterize the relationship between brain oscillations in the time and/or frequency domain [14, 18]. Besides these methods other families of approaches where developed to estimate directional properties of the relationship between signals while taking into account possible influences of external sources [17].

The aforementioned methods are said to be linear in the sense that the estimator they use or the model they assume for signals can only capture the linear properties of the relationship between analyzed time series. However, most of the mechanisms at the origin of EEG signals are nonlinear. Therefore many studies were also dedicated to the development of nonlinear methods [34]. A first family of methods based on mutual information [25] or on nonlinear regression [33, 47] was introduced in the field of EEG about twenty years ago. A second family developed later on, based on works related to the study of nonlinear dynamical systems and chaos [19, 23], see also Chap. 15. This second family can be divided into two groups: (i) phase synchronization (PS) methods [6, 37] which first estimate the instantaneous phase of each signal and then compute a quantity based on co-variation of extracted phases to determine the degree of relationship. (ii) generalized synchronization (GS) methods [3, 41] which also consist of two steps; in the first one, state space trajectories are reconstructed from scalar time series signals and in the second one, an index of similarity is computed to quantify the similarity between these trajectories.

As shown by this brief literature review, the number of developed methods and variants is large. In this chapter, we will only present the principles and the equations of the most widely used methods for characterizing interactions between neuronal systems. Presented methods can also be organized into the three following families as they share some commonalties: (1) linear and nonlinear regression: Pearson correlation coefficient (R^2), coherence function (CF) and nonlinear regression (h^2); (2) phase synchronization: Hilbert phase entropy (HE), Hilbert mean phase coherence (HR), wavelet phase entropy (WE) and wavelet mean phase coherence (WR); (3) generalized synchronization: three similarity indexes (S, H, N) and synchronization likelihood (SL). Main theoretical aspects are reviewed in the following.

7.2.1 Linear and Nonlinear Regression Based Methods

For two time series $x(t)$ and $y(t)$, the Pearson correlation coefficient is defined in the time domain as follows [2]

$$R^2 = \max_{\tau} \frac{\mathrm{cov}^2\left(x(t), y(t+\tau)\right)}{\mathrm{var}\left(x(t)\right) \cdot \mathrm{var}\left(y(t+\tau)\right)} \tag{7.1}$$

where *var*, *cov*, and τ denote respectively variance, covariance, and time shift between the two time series.

The magnitude-squared coherence function (CF) can be formulated as [5]:

$$\left|\rho_{xy}(f)\right|^2 = \frac{\left|S_{xy}(f)\right|^2}{S_{xx}(f) \cdot S_{yy}(f)} \tag{7.2}$$

where $S_{xx}(f)$ and $S_{yy}(f)$ respectively denote the power spectral densities of $x(t)$ and $y(t)$, and where $S_{xy}(f)$ denotes their cross-spectral density. It is the counterpart of the R^2 coefficient in the frequency domain. It can be interpreted as the squared modulus of a frequency-dependent complex correlation coefficient.

Among nonlinear regression analysis methods, we chose a method introduced in the field of EEG analysis by Lopes da Silva and colleagues [24]. It was more recently evaluated in a model of coupled neuronal populations as well as in coupled oscillators [1, 46]. Main theoretical aspects regarding this approach are revisited in [20]. In brief, this method provides a nonlinear correlation coefficient referred to as h^2, based on the fitting of a nonlinear curve $g(\cdot)$ which approximates the statistical relationship from $x(t)$ to$y(t)$:

$$h^2_{xy} = \max_{\tau} \left(1 - \frac{\mathrm{var}\left(y(t+\tau)/x(t)\right)}{\mathrm{var}\left(y(t+\tau)\right)}\right) \tag{7.3}$$

where

$$\operatorname{var}\left(y\left(t+\tau\right)/x\left(t\right)\right)\stackrel{\Delta}{=}\arg\min_{g}\left(E\left[y\left(t+\tau\right)-g\left(x\left(t\right)\right)\right]^{2}\right)$$

In practice, function $g(\cdot)$ can be obtained from the piece-wise linear approximation between the samples of the two time series $x(t)$ and $y(t)$ [32].

7.2.2 Phase Synchronization Based Methods

Phase synchronization estimation consists of two steps [37]. The first step corresponds to the extraction of the instantaneous phase from each signal and the second step is the quantification of the degree of synchronization between estimated instantaneous phases using an appropriate index (see also, Chap. 16). Phase extraction can be performed according to different techniques. Two of them are described in this paragraph: the Hilbert transform and the wavelet transform.

Using the Hilbert transform, analytical signal associated to a real time series $x(t)$ is derived:

$$Z_x(t) = x(t) + iH\left[x(t)\right] = A_x^H(t)\, e^{i\phi_x^H(t)} \tag{7.4}$$

Where H, ϕ_x^H, and $A_x^H(t)$ are respectively the Hilbert transform, the phase, and the amplitude of $x(t)$.

Complex continuous wavelet transform can also be used to estimate the phase of signal [12, 22, 39]:

$$W_x(t) = (\psi * x)(t) = \int \psi\left(t'\right) x\left(t - t'\right) dt' = A_x^W(t) \cdot e^{i\phi_x^W(t)} \tag{7.5}$$

where ψ, ϕ_x^W, and $A_x^W(t)$ are respectively a wavelet function (e.g., Morlet used here), the phase, and the amplitude of $x(t)$. Once phase extraction is performed on the two signals under analysis, several synchronization indexes can be used to quantify the phase relationship. In this chapter, we present two of them, both based on the shape of the probability density function (pdf) of the modulo 2π phase difference ($\phi = \left(\phi_x - \phi_y\right) \bmod 2\pi$). The first index is stemmed from the Shannon entropy and defined as follows [28].

$$\rho = \frac{H_{\max} - H}{H_{\max}} \tag{7.6}$$

Where $H = -\sum_{i=1}^{M} p_i \ln p_i$

where M is the number of bins used to obtain the pdf, p_i is the probability of finding the phase difference ϕ within the *ith* bin, and $H_{\max}$ is given by $\ln M$. The second index is named "mean phase coherence". It corresponds to $\left|\mathbb{E}\left[e^{i\phi}\right]\right|$. As described in [27], it can be estimated by: $R = \left|\frac{1}{N}\sum_{t=0}^{N-1} e^{i\phi(t)}\right|$ where N is the length of time series.

Therefore, if we take into account the two ways to estimate the instantaneous phase and the two indices to quantify the phase relationship, we end with four different measures of interdependencies, denoted as follows: Hilbert entropy (HE), Hilbert mean phase coherence (HR), wavelet entropy (WE), and wavelet mean phase coherence (WR).

7.2.3 Generalized Synchronization Based Methods

Generalized synchronization based methods aim at investigating the interaction between two nonlinear dynamical systems without any knowledge about governing equations. They generally proceed according to two steps. First, a state space trajectory is reconstructed from each scalar time series using a time delay embedding method [42]. For each discrete time n a delay vector corresponding to a point in the state space reconstructed from x is defined as:

$$X_n = \left(x_n, x_{n+\tau}, \ldots, x_{n+(m-1)\tau}\right) \quad n = 1, \ldots, N \tag{7.7}$$

where m is the embedding dimension and τ denotes time lag. The state space trajectory of y is reconstructed in the same way. Second, a synchronization degree is determined using a suitable measure. Four measures, all based on conditional neighborhood, are presented in this section. The general principle is to quantify the proximity, in the second state space, of the points whose temporal indices corresponds to neighbor points in the first state space. Three of these measures S, H, and N [3], which are also sensitive to the direction of interaction, originate from this principle. They are based on an Euclidean distance:

$$S^{(k)}(X|Y) = \frac{1}{N}\sum_{n=1}^{N} \frac{R_n^{(k)}(X)}{R_n^{(k)}(X|Y)} \tag{7.8}$$

$$H^{(k)}(X|Y) = \frac{1}{N}\sum_{n=1}^{N} \log \frac{R_n^{(N-1)}(X)}{R_n^{(k)}(X|Y)} \tag{7.9}$$

$$N^{(k)}(X|Y) = \frac{1}{N}\sum_{n=1}^{N} \frac{R_n^{(N-1)}(X) - R_n^{(k)}(X|Y)}{R_n^{(N-1)}(X)} \tag{7.10}$$

where $R_n^{(k)}(X)$ is expressed as:

$$R_n^{(k)}(X) = \frac{1}{k}\sum_{j=1}^{k} \left(X_n - X_{r_{n,j}}\right)^2 \tag{7.11}$$

and

$$R_n^{(k)}(X|Y) = \frac{1}{k}\sum_{j=1}^{k}\left(X_n - X_{s_{n,j}}\right)^2 \tag{7.12}$$

where $r_{n,j}, j = 1,\ldots,k$ and $s_{n,j}, j = 1,\ldots,k$ respectively stand for the time indices of the k nearest neighbors of X_n and Y_n.

It is noteworthy that the fourth measure, referred to as the synchronization likelihood (SL) [41], is a measure of multivariate synchronization. Here, for simplicity, we will only focus on the bivariate case. The estimated probability that embedded vectors X_n are closer to each other than a distance ε is:

$$P_{x,n}^{\varepsilon} = \frac{1}{2(w_2 - w_1)}\sum_{\substack{j=1 \\ w_1<|n-j|<w_2}}^{N} \theta\left(\varepsilon - \left|X_n - X_j\right|\right) \tag{7.13}$$

where $|\cdot|$ is the Euclidean distance; θ stands for the Heaviside step function, w_1 is the Theiler correction and w_2 determines the length of sliding window. Letting $P_{x,n}^{\varepsilon} = P_{y,n}^{\varepsilon} = P_{ref}$ be a small arbitrary probability, the above equation for X_n and its analogous for Y_n, gives the critical distances $\varepsilon_{x,n}$ and $\varepsilon_{y,n}$ from which we can determine if simultaneously X_n is close to X_j and Y_n is close to Y_j, i.e., $H_{n,j} = 2$ in the equation below

$$H_{n,j} = \theta\left(\varepsilon_{x,n} - \left|X_n - X_j\right|\right) + \theta\left(\varepsilon_{y,n} - \left|Y_n - Y_j\right|\right) \tag{7.14}$$

Synchronization likelihood at time n can be obtained by averaging over all values of j

$$S_n = \frac{1}{2P_{ref}(w_2 - w_1)}\sum_{\substack{j=1 \\ w_1<|n-j|<w_2}}^{N} (H_{n,j} - 1) \tag{7.15}$$

All aforementioned measures but H, are normalized between 0 and 1; the 0 value means that the two signals are completely independent. On the opposite, the 1 value means that the two signals are completely synchronized.

Finally, in order to deal with the evolution, in time, of brain connectivity the measures described above can be estimated over a sliding window.

7.2.4 Frequency-Dependence of Brain Connectivity Measures

Nonlinear "time-dependent" methods have the capability to account for the nonlinearity of relationship. However, they are generally independent from frequency, a key parameter in EEG analysis that can be related to the oscillatory behavior of recorded neural populations. On the opposite, linear time-dependent methods can not be used to analyze the nonlinear property of relationship but they can characterize its dependence on frequency. However, as underlined by Zaveri et al. [49] who performed a review on the use of coherence in the field of EEG, proposed estimators generally have strong bias and variance which make the interpretation of real data intricate. To alleviate these difficulties, frequency bands may be defined. For instance, classical delta, theta, alpha, beta and gamma EEG bands can be used to average the coherence function [36] or to compute the cross-correlation of filtered signals [29, 45]. Again, this is not entirely satisfactory since the choice of frequency bands becomes critical in this case (relevant phenomena may overlap two bands). Therefore there is a need to derive high resolution time-frequency estimators of signal interdependencies. Readers may refer to Ansari et al. [1] for more details about this topic that will not be dealt with in this chapter.

7.2.5 Performance Evaluation of Brain Connectivity Measures

Given the number and the variety of methods introduced for characterizing brain signal interdependencies and considering the diversity of situations in which these methods are applied, there is a need for identifying objectively, among available methods, those which offer the best performances.

Recently, some efforts have been made for comparing methods but mainly qualitatively [11, 35] and for particular applications [26, 30]. Therefore, there is a need for objective comparison of methods based on well controlled scenarios in which *a priori* knowledge about the relationship between brain structures is available. In this context, physiologically-plausible models of coupled neuronal populations can be used as they allow for the generation of multidimensional EEG signals for different values of the coupling parameter between populations. Such a model is presented in Sect. 3. Moreover, quantitative criteria must also be introduced in order to objectively compare candidate methods. In Sect. 4, three criteria will be considered: (*i*) the mean square error (MSE) under null hypothesis (independence between the two analyzed signals); (*ii*) the mean variance (MV) computed over all values of the coupling parameter in the EEG model; (*iii*) a criterion related to the method sensitivity.

7.3 Model-Based Comparison of Methods Aimed at Characterizing the Connectivity Between Brain Structures

7.3.1 General Model of Interdependence Between Two Time-Series

In this section, the general features of the model introduced in the perspective of methods comparison are described. The model is a more or less simplified version of a general finite dimensional state-space model with three inputs and two

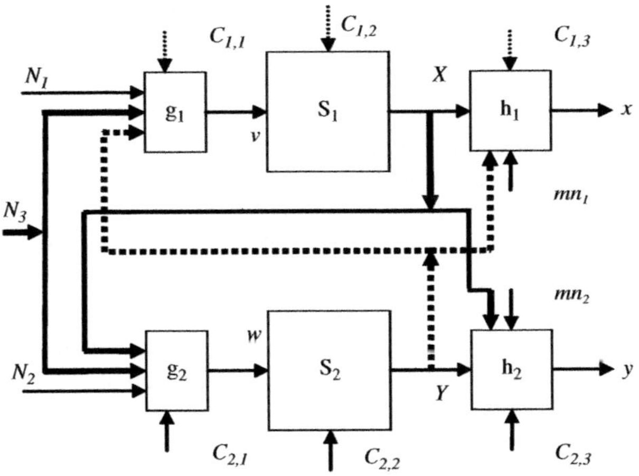

Fig. 7.1 General finite dimensional state-space model (composed of two coupled sub-systems S_1 and S_2) with three inputs N_1, N_2, N_3 and two outputs x, y. Models considered in this study correspond to simplified versions of this model. See text for details

outputs [1]. This general model denoted by $M^C_{X,Y}$ is decomposed in two sub-systems S_1 and S_2 as illustrated in Fig. 7.1. To describe state evolution (on discrete time or on continuous time) of the global system two finite dimensional marginal state vectors, respectively denoted X and Y, must be introduced. In an EEG measurement perspective, X and Y macroscopically represent dynamical states of two functionally interdependent neuronal subpopulations, respectively. Each subsystem is specified by a state evolution equation:

$$\begin{cases} X(t+\tau) = F^C_\tau(X(t); v(\theta), t \leq \theta \leq t+\tau), X(t) \in \mathbb{R}^m \\ Y(t+\tau) = G^C_\tau(Y(t); w(\theta), t \leq \theta \leq t+\tau), Y(t) \in \mathbb{R}^n \end{cases} \tag{7.16}$$

where matrix $C = (C_{i,j})$ is a matrix of positive numbers interpreted in the sequel as a coupling parameter which weights the effect of "non-autonomous" terms v and w respectively on states X and Y (Fig. 7.1).

The inputs N_1, N_2, and N_3 are mutually independent, zero-memory, zero-mean and unit-variance stochastic processes (white noises) which can be interpreted, in a physiological perspective, as influences from distant neuronal populations. Input N_3 corresponds to a possible shared afference. The scalar outputs x and y, in the same perspective, correspond to two EEG channels. If it exists, the dynamical 'coupling' between states X and Y is represented through a functional dependence of v on Y and on the shared input N_3 and through the dependence of w on X and N_3:

$$\begin{cases} v(t) = g_1(C_{1,1}, N_1(t), N_3(t), Y(t)) \\ w(t) = g_2(C_{2,1}, N_2(t), N_3(t), X(t)) \end{cases} . \tag{7.17}$$

The models for the two output scalar signals are:

$$\begin{cases} x(t) = h_1(X(t), Y(t), mn_1(t)) \\ y(t) = h_2(X(t), Y(t), mn_2(t)) \end{cases} \tag{7.18}$$

g_1, g_2, h_1 and h_2 are deterministic functions. The measurement noises, if present, are modeled by two independent random processes $mn_1(t)$, $mn_2(t)$.

If v does not depend on $Y(.)$ and w does not depend on $X(.)$ and if furthermore $N_3 = 0$, then the two subsystems S1 and S2 are disconnected. In this case and when inputs N_1 and N_2 are present, outputs x and y are statically independent if h_1 (resp h_2) is not a function of Y (resp X). Then equations become $x(t) = h_1(X(t))$, $y(t) = h_2(Y(t))$, in absence of measurement noise.

The dashed lines in Fig. 7.1 denote the influences of S_2 on signal x. It corresponds to the most general bidirectional situation which is beyond the scope of this chapter that will only deal with the causal influence directed from S_1 to S_2. In this context, matrix C is the parameter which tunes the dependence of y on X. When C is null no dependence exists. Dependence between the two systems is expected to increase with C coefficients. For large values of these coefficients and when $N_2 = 0$ and $mn_2 = 0$, output y becomes a deterministic function of state X and of N_3.

7.3.2 Physiologically-Relevant Model of Depth-EEG Signals

In order to comprehensively simulate a wide range dynamics encountered in real depth-EEG signals (intracerebral electrodes), as recorded in patients with drug-resistant epilepsy during pre-surgical evaluation, we now introduce a physiologically-relevant computational model of EEG generation. As illustrated in Fig. 7.2, this model corresponds to a particular case of the general model with $C_{2,1} = c$, $v = N_1$, $w = g_2(X, c, N_2) = N_2 + c + GX$, $N_3 = 0$, $x = h_1(X) = HX$ and $y = h_2(Y) = HY$, where, GX, HX and HY are linear forms of the state vectors. The objective is to describe the field activity of two distant – and possibly coupled\INTtie;– neuronal populations whose dynamical states are represented by X and Y, respectively. Each population generates a local field potential ($x(t)$ and $y(t)$) that can be seen as a depth-EEG signal if one does not consider the source-electrode transfer function. Readers may refer to [48] for generalization to more

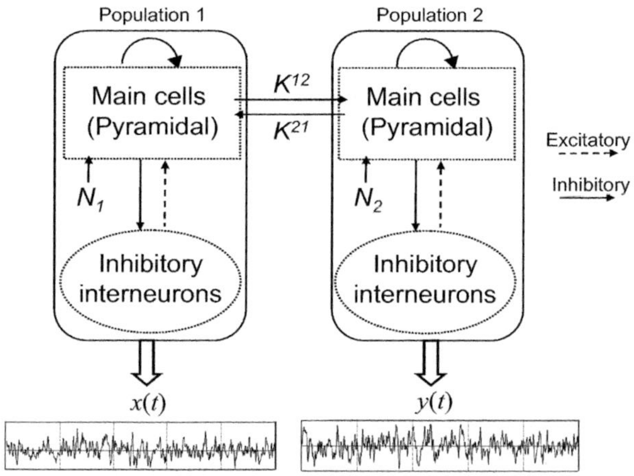

Fig. 7.2 Model of coupled neuronal populations used to generate bivariate time-series EEG data

than 2 populations. In the model, each population contains two subpopulations of neurons that mutually interact via excitatory or inhibitory feedback: main pyramidal cells and local interneurons.

The excitatory influence from neighboring is modeled by a Gaussian input noise ($N_1(t)$ or $N_2(t)$) that globally represents the average density of afferent action potentials on population 1 and population 2. Since pyramidal cells are excitatory neurons that project their axons to other areas of the brain, the model accounts for this organization by using the average pulse density of action potentials from the main cells of population 1 as an excitatory input to population 2. In addition, this connection from population 1 to population 2 is characterized by parameter K^{12} which represents the degree of coupling associated with this connection. Other parameters include excitatory and inhibitory gains in feedback loops as well as average number of synaptic contacts between subpopulations. Appropriate setting of parameters K^{12} and K^{21} allows for building systems where the two neuronal populations are unidirectionally or bidirectionally coupled. In the following, we will only consider the case where the two populations of neurons are unidirectionally coupled ($K^{12} = c$ is varied and K^{21} stays equal to 0). This model was used to generate two kinds of signals: background (BKG) and spiking (SPK) EEG activity. A comparison between simulated and real intracerebral EEG data during background and ictal activity is provided in Fig. 7.3. For both cases, bivariate data were simulated. The normalized coupling

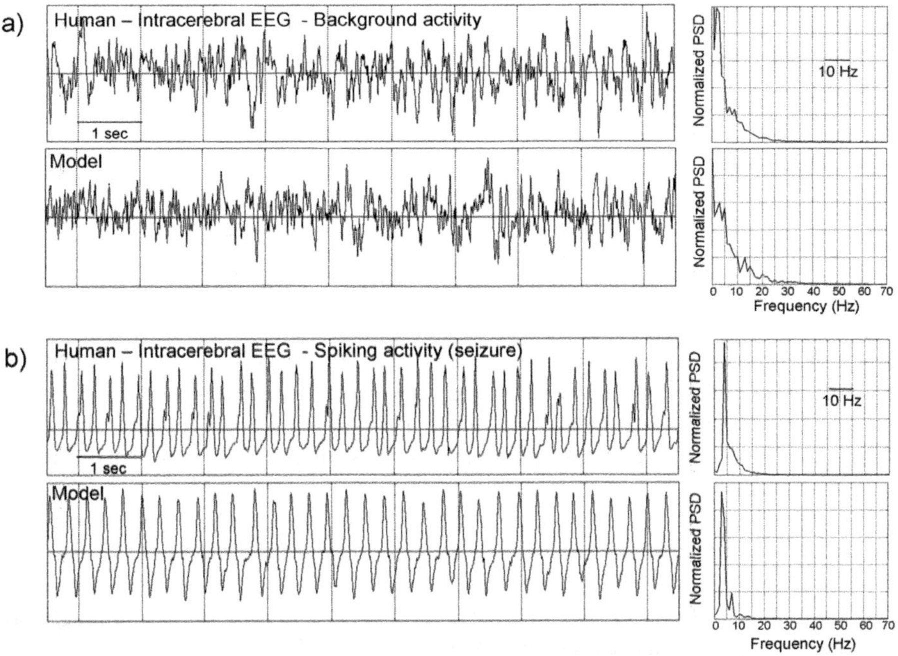

Fig. 7.3 Comparison between real (intracerebral EEG recorded in a patient during pre-surgical evaluation) and simulated activity produced by the model. (**a**) normal background activity, (**b**) typical rhythmic spiking activity observed during seizures. PSD: Power Spectral Density

parameter c was varied from 0 (independence between signals) to 1 (dependence between signal, similar temporal dynamics).

7.4 Comparison Criteria

In this section, we define three criteria. These are used to quantitatively evaluate the behavior of the methods presented in Sect. 7.2 on time-series simulated from the model presented in Sect. 7.3 for different values of the coupling parameter $K^{12} = c$. The two first criteria are classical:

- the *mean square error* (MSE) under null hypothesis (i.e., independence between two signals) can be interpreted as a quadratic bias, defined by $E\left\{\left(\hat{\theta}_0 - \theta_0\right)^2\right\}$ where E is the mathematical expectation, $\theta_0 = 0$ and $\hat{\theta}_0$ is the estimation of θ_0,
- the *mean variance* (MV) computed over all values c_i, $i = 1, 2, \ldots, I$ of the degree of coupling and defined as $\frac{1}{I}\sum_{i=1}^{I} E\left\{\left(\hat{\theta}_i - E\left(\hat{\theta}_i\right)\right)^2\right\}$ where I is number of coupling degree points and $\hat{\theta}_i$ is the estimated relationship for the coupling degree c_i.

Methods with lower values of MSE and MV can be considered to have better performances.

The third criterion is introduced in order to quantify the sensitivity of the method with respect to changes in the coupling degree:

- the *median of local relative sensitivity* (MLRS) given by:

$$MLRS = Median\left(S_i/\bar{\sigma}_i\right), S_i = \frac{\hat{\theta}_{i+1} - \hat{\theta}_i}{c_{i+1} - c_i}, \bar{\sigma}_i = \sqrt{\frac{\hat{\sigma}_i^2 + \hat{\sigma}_{i+1}^2}{2}} \tag{7.19}$$

where S_i is the increase rate of the estimated relationship and $\bar{\sigma}_i$ is the square root of the average of estimated variances associated to two adjacent values of the coupling degree. One can also use the median of the distribution of local relative sensitivity instead of its mean because the fluctuation in its estimation may make this distribution very skewed.

Conversely to MSE and MV, higher MLRS values indicate better performances.

Finally, for all values of parameter c (degree of coupling), Monte Carlo simulations must be conducted in order in order to assess statistical properties of interdependence measures provided by methods described in Sect. 7.2 and to comparatively evaluate their performances. For parameter τused in GS methods, first the mutual information as a function of positive time lag is plotted and then, as described in [40], time lag τ was chosen as the abscissa value corresponding to the first minimum this curve. The embedded dimension *m*, in this family of methods, was determined from the Cao method [9].

7.5 Results

In the neuronal population model, signals were generated to reproduce normal background EEG activity (BKG) or spiking activity (SPK) as observed during epileptic seizures, as respectively shown in Figs. 7.4a and 7.5a. Properties of these signals are very close to those reported in a previous attempt for comparing relationship estimators [35]. In this study, the relationship between the two modeled populations of neurons was unidirectional. As shown in [11] in the case of background activity using surrogate data techniques, the relationship between signals in this model are mainly linear. Thus we expected all methods to exhibit similar behavior in this case. Results are displayed in Figs. 7.4(b,c) and 7.5(b,c) which gives the mean value and the variance of the quantity estimated from each method a function of the coupling degree in the model. Results show that increasing the degree of coupling between neuronal populations does not lead to significant increase of computed quantities.

In the BKG situation (Fig. 7.4) CF and all the PS methods but HR do not detect any relationship; other methods detect a weak relationship. For the spiking activity (Fig. 7.5), most of the methods detect the increase of the coupling parameter in the model. As an interesting result, we observed that WE and CF were almost blind to the established relation. Similarly, HE and WR only displayed small increase

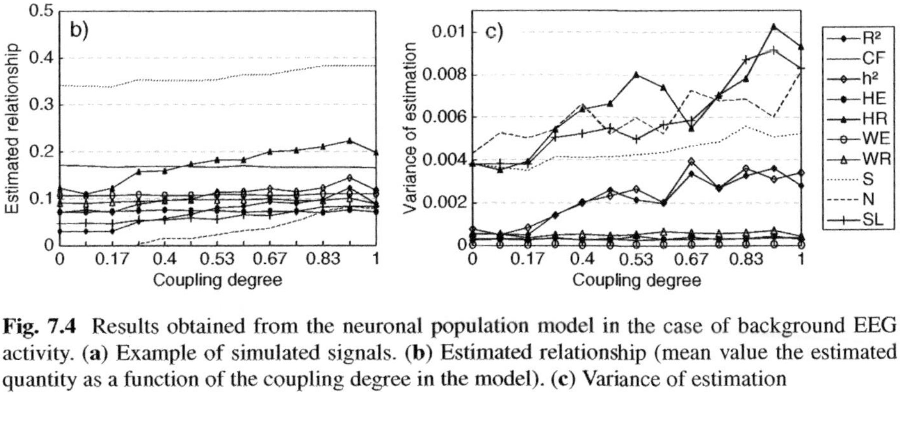

Fig. 7.4 Results obtained from the neuronal population model in the case of background EEG activity. (**a**) Example of simulated signals. (**b**) Estimated relationship (mean value the estimated quantity as a function of the coupling degree in the model). (**c**) Variance of estimation

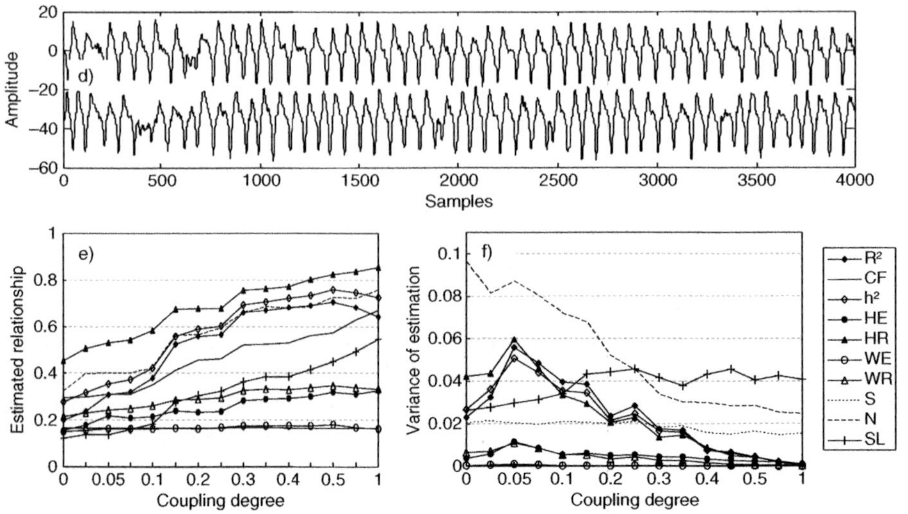

Fig. 7.5 Results obtained from the neuronal population model in the case of epileptic EEG activity (sustained spiking activity as observed during ictal periods). (**d**) Example of simulated signals. (**e**) Estimated relationship (mean value the estimated quantity as a function of the coupling degree in the model). (**f**) Variance of estimation

with increasing of degree of coupling but their variance was low. R^2, h^2, S and HR methods exhibited good sensitivity. However, MSE under null hypothesis was found to be high for HR.

Results presented in Figs. 7.4 and 7.5 are summarized in Tables 7.1–7.3 which give the MSE under null hypothesis, the MV and the MRLS for all methods. For each of the two studied situations, the best method is highlighted with grey color.

Table 7.1 Mean square error (MSE) values and standard deviations for studied methods and models. "*" denotes methods that are nearly insensitive to changes in the coupling degree and for which this criterion is not applicable. For both situations (normal background and epileptic activity), the "best" method, according to this criterion, is highlighted in grey color

	BKG	SPK
R^2	**1.54 +/− 0.1**	63.2 +/− 3.1
CF	*	*
h^2	6.0 +/− 0.1	103.8 +/− 3.3
HE	*	**28.8 +/− 0.5**
HR	19.0 +/− 0.6	249.3 +/− 5.3
WE	*	*
WR	*	53.4 +/− 0.9
S	120.0 +/− 0.1	107.5 +/− 2.5
H	4.2 +/− 11.6	651.1 +/− 45.7
N	4.9 +/− 1.9	201.5 +/− 15.3
SL	6.2 +/− 0.9	41.3 +/− 5.6

Table 7.2 Mean variance (MV) values and standard deviations. "∗" denotes methods that are nearly insensitive to changes in the coupling degree and for which this criterion is not applicable. For both situations (normal background and epileptic activity), the "best" method, according to this criterion, is highlighted in grey color

	BKG	SPK
R^2	**21.3 +/− 0.3**	215.8 +/− 2.5
CF	∗	∗
h^2	22.6 +/− 0.3	205.0 +/− 2.4
HE	∗	45.3 +/− 0.5
HR	65.5 +/− 0.8	217.5 +/− 2.8
WE	∗	∗
WR	∗	**38.4 +/− 0.5**
S	44.1 +/− 0.9	183.7 +/− 2.0
H	70.6 +/−15.3	2942.3 +/− 25.8
N	60.1 +/− 5.7	501.1 +/− 5.1
SL	59.2 +/− 1.0	383.8 +/− 3.5

Table 7.3 Median of local relative sensitivity (MRLS) values. "∗" denotes methods that are nearly insensitive to changes in the coupling degree and for which this criterion is not applicable. For both situations (normal background and epileptic activity), the "more sensitive" method is highlighted in grey color

	BKG	SPK
R^2	**1.2**	1.3
CF	∗	∗
h^2	1.1	**1.8**
HE	∗	1.2
HR	0.7	1.2
WE	∗	∗
WR	∗	1.2
S	0.05	0.9
H	0.9	0.7
N	0.9	0.4
SL	0.007	1.3

Methods that were found to be insensitive with respect to changes in the coupling degree are denoted by symbol "∗". From this table, we deduced that for the neuronal population model, in the background activity situation, R^2 and h^2 methods detected the presence of a relationship and performed better than other methods; this tendency was also confirmed in the spiking activity situation. However, it was difficult to determine the overall best method in this second case since criteria did not lead to consensual results.

In order to globally compare the three groups of methods, we averaged results obtained for each criterion. Results are synthesized in Fig. 7.6. For the neuronal population model M, regression methods outperform others in the case of normal background EEG activity. For spiking epileptic-like activity, these methods, in addition to PS methods have also higher performances than GS methods.

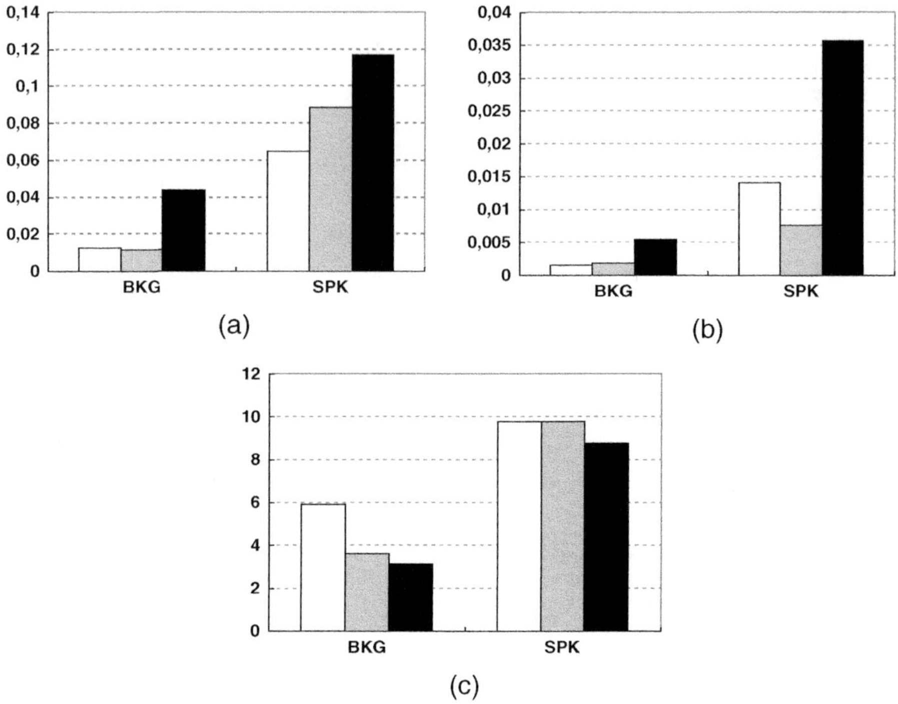

Fig. 7.6 Mean values of (**a**) MSE, (**b**) MV, and (**c**) MRLS for the three categories of methods (white: regression methods, grey: PS methods, and black: GS methods). Note that, inversely to MSE and MV, higher MLRS values indicate better performances

7.6 Discussion

Numerous methods have been introduced to tackle the difficult problem of characterizing the statistical relationship between brain signals without *a priori* knowledge about the nature of this relationship. This question is of great interest for understanding brain functioning in normal or pathological conditions. Therefore, these methods play a key role as they are supposed to give important information regarding brain connectivity from electrophysiological recordings. In this chapter, we presented three families of methods (regression, phase synchrony and general synchronization) and we compared their performances on the basis of simulations produced by a neurophysiologically-relevant model. In this regard, this approach differs from that of Schiff et al. [38] who evaluated one method to characterize dynamical interdependence (based on mutual nonlinear prediction) on both simulated (coupled identical and non identical chaotic systems) and real (activity of motoneurons within a spinal cord motoneuron pool) data. It also differs from other evaluation studies which mainly focused on qualitative comparisons [11, 35] and for specific applications [26 30]. In the particular field of EEG analysis, the model of coupled neuronal populations is of particular relevance since it generates realistic temporal dynamics. In this model, for background activity (that can be considered as a broadband random signal), we found that coherence and phase synchrony methods (except HR) were not sensitive to the increase of the coupling parameter whereas regression methods (linear and nonlinear) exhibited better sensitivity. This result may be explained by the fact that the interdependence between simulated signals is not entirely determined by a phase relationship. This point is crucial since it illustrates the fact that the choice of the method used to characterize the relationship between signals is critical and may lead to possible incorrect interpretation of results obtained on EEG data.

In addition, as background activity can be recorded in epileptic patients during interictal periods, our results also relate to those recently published by Morman et al. [26] in the context of seizure prediction. For thirty different measures obtained from univariate and bivariate approaches, authors evaluated their ability to distinguish between the interictal period and the pre-seizure period (sensitivity and specificity of all measures were compared using receiver-operating-characteristics). In both types of approach (and consequently for bivariate methods similar to those implemented in the present study) they also found that linear methods performed equally good or even better than nonlinear methods.

In this report, results about the characterization of the direction of coupling were not dealt with. This difficult issue has already been addressed in various reports. For instance, Quian Quiroga et al. [35] quantitatively tested two interdependence measures on coupled nonlinear oscillators for their ability to determine whether one the two systems drives the other.

To sum up, the main findings of this study are the following: (*i*) some of the compared methods are insensitive to the coupling parameter in the model; (*ii*) results are dependent on signal properties (broad band versus narrow band); (*iii*) generally speaking, there is no universal method to deal with signal coupling, i.e., none of the studied methods performed better than the other ones in the two studied situations.

Nevertheless, we notice that simple methods like R^2 and h^2 methods showed to be sensitive to the coupling parameter in the model with average or good performances. Therefore, it might be reasonable to first apply these "robust" regression methods in order to characterize brain connectivity before using more sophisticated methods that require specific assumptions about the underlying model of relationship.

References

1. Ansari-Asl, K., Senhadji, L., Bellanger, J. J., and Wendling, F. (2006). "Quantitative evaluation of linear and nonlinear methods characterizing interdependencies between brain signals." *Phys Rev E Stat Nonlin Soft Matter Phys*, 74(3 Pt 1), 031916.
2. Ansari-Asl, K., Wendling, F., Bellanger, J. J., and Senhadji, L. (2004). "Comparison of two estimators of time-frequency interdependencies between nonstationary signals: application to epileptic EEG." *Engineering in Medicine and Biology Society, 2004. EMBC 2004. Conference Proceedings. 26th Annual International Conference of the*, San Francisco, 263–266.
3. Arnhold, J., Grassberger, P., Lehnertz, K., and Elger, C. E. (1999). "A robust method for detecting interdependences: application to intracranially recorded EEG." *Physica D: Nonlinear Phenomena*, 134(4), 419–430.

4. Barlow, J. S., and Brazier, M. A. (1954). "A note on a correlator for electroencephalographic work." *Electroencephalogr Clin Neurophysiol*, 6(2), 321–325.
5. Bendat, J., and Piersol, A. (1971). *Random data: analysis and measurement procedures*, Willey-Interscience.
6. Bhattacharya, J. (2001). "Reduced degree of long-range phase synchrony in pathological human brain." *Acta Neurobiol Exp (Wars)*, 61(4), 309–318.
7. Brazier, M. A. (1968). "Studies of the EEG activity of limbic structures in man." *Electroencephalogr Clin Neurophysiol*, 25(4), 309–318.
8. Brazier, M. A. (1972). "Spread of seizure discharges in epilepsy: anatomical and electrophysiological considerations." *Exp Neurol*, 36(2), 263–272.
9. Cao, L. (1997). "Practical method for determining the minimum embedding dimension of a scalar time series." *Physica D: Nonlinear Phenomena*, 110(1–2), 43–50.
10. Cooley, J. W., and Tukey, J. W. (1965). "An Algorithm for the Machine Calculation of Complex Fourier Series." *Math Comput*, 19, 297–301.
11. David, O., Cosmelli, D., and Friston, K. J. (2004). "Evaluation of different measures of functional connectivity using a neural mass model." *Neuroimage*, 21(2), 659–73.
12. Delprat, N., Escudie, B., Guillemain, P., Kronland-Martinet, R., Tchamitchian, P., and Torresani, B. (1992). "Asymptotic wavelet and Gabor analysis: extraction of instantaneous frequencies." *Inf Theory, IEEE Trans*, 38(2), 644–664.
13. Duckrow, R. B., and Spencer, S. S. (1992). "Regional coherence and the transfer of ictal activity during seizure onset in the medial temporal lobe." *Electroencephalogr Clin Neurophysiol*, 82(6), 415–422.
14. Franaszczuk, P. J., and Bergey, G. K. (1999). "An autoregressive method for the measurement of synchronization of interictal and ictal EEG signals." *Biol Cybern*, 81(1), 3–9.
15. Gotman, J. (1983). "Measurement of small time differences between EEG channels: method and application to epileptic seizure propagation." *Electroencephalogr Clin Neurophysiol*, 56(5), 501–14.
16. Gotman, J. (1987). "Interhemispheric interactions in seizures of focal onset: data from human intracranial recordings." *Electroenceph Clin Neurophysiol*, 67, 120–133.
17. Gourevitch, B., Bouquin-Jeannes, R. L., and Faucon, G. (2006). "Linear and nonlinear causality between signals: methods, examples and neurophysiological applications." *Biol Cybern*, 95(4), 349–369.
18. Haykin, S., Racine, R. J., Xu, Y., and Chapman, C. A. (1996). "Monitoring neural oscillation and signal transmission between cortical regions using time-frequency analysis of electroencephalographic activity." *Proc IEEE*, 84, 1295–1301.
19. Iasemidis, L. D. (2003). "Epileptic seizure prediction and control." *IEEE Trans Biomed Eng*, 50(5), 549–558.
20. Kalitzin, S. N., Parra, J., Velis, D. N., and Lopes da Silva, F. H. (2007). "Quantification of unidirectional nonlinear associations between multidimensional signals." *IEEE Trans Biomed Eng*, 54(3), 454–461.
21. Ktonas, P. Y., and Mallart, R. (1991). "Estimation of time delay between EEG signals for epileptic focus localization: statistical error considerations." *Electroencephalogr Clin Neurophysiol*, 78(2), 105–110.
22. Le Van Quyen, M., Foucher, J., Lachaux, J., Rodriguez, E., Lutz, A., Martinerie, J., and Varela, F. J. (2001). "Comparison of Hilbert transform and wavelet methods for the analysis of neuronal synchrony." *J Neurosci Methods*, 111(2), 83–98.
23. Lehnertz, K. (1999). "Non-linear time series analysis of intracranial EEG recordings in patients with epilepsy – an overview." *Int J Psychophysiol*, 34(1), 45–52.
24. Lopes da Silva, F., Pijn, J. P., and Boeijinga, P. (1989). "Interdependence of EEG signals: linear vs. nonlinear associations and the significance of time delays and phase shifts." *Brain Topogr*, 2(1–2), 9–18.
25. Mars, N., and Lopes da Silva, F. (1983). "Propagation of seizure activity in kindled dogs." *Electroencephalography and Clinical Neurophysiology*, 56, 194–209.

26. Mormann, F., Kreuz, T., Rieke, C., Andrzejak, R. G., Kraskov, A., David, P., Elger, C. E., and Lehnertz, K. (2005). "On the predictability of epileptic seizures." *Clin Neurophysiol*, 116(3), 569–587.
27. Mormann, F., Lehnertz, K., David, P., and Elger, C. E. (2000). "Mean phase coherence as a measure for phase synchronization and its application to the EEG of epilepsy patients." *Physica. D*, 144, 358–369.
28. Munari, C., Tassi, L., Kahane, P., Francione, S., DiLeo, M., and Quarato, P. (1994). "Analysis of clinical symptomatology during stereo-EEG recorded mesiotemporal lobe seizures." Epileptic seizures and syndromes, W. P, ed., John Libbey & Co, London.
29. Nikolaev, A. R., Ivanitsky, G. A., Ivanitsky, A. M., Posner, M. I., and Abdullaev, Y. G. (2001). "Correlation of brain rhythms between frontal and left temporal (Wernicke's) cortical areas during verbal thinking." *Neurosci Lett*, 298(2), 107–110.
30. Pereda, E., DelaCruz, D. M., DeVera, L., and Gonzalez, J. J. (2005). "Comparing Generalized and Phase Synchronization in Cardiovascular and Cardiorespiratory Signals." *Biomed Eng, IEEE Trans*, 52(4), 578–583.
31. Pfurtscheller, G., and Andrew, C. (1999). "Event-Related changes of band power and coherence: methodology and interpretation." *J Clin Neurophysiol: Official Publication American Electroencephalographic Society*, 16(6), 512–519.
32. Pijn, J. P. (1990). "Quantitative evaluation of EEG signals in epilepsy, nonlinear associations, time delays and nonlinear dynamics," University of Amsterdam, Amsterdam.
33. Pijn, J. P., and Lopes da silva, F. H. (1993). "Propagation of electrical activity: nonlinear associations and time delays between EEG signals." *in Basic Mechanisms of the Eeg, Brain Dynamics, S. Zschocke and E. J. Speckmann, Eds. Boston: Birkhauser*, 41–61.
34. Pikovsky, A., Rosenblum, M., and Kurths, J. (2001). "Synchronization: a universal concept in nonlinear sciences." *Cambridge: Cambridge University Press.*
35. Quian Quiroga, R., Kraskov, A., Kreuz, T., and Grassberger, P. (2002). "Performance of different synchronization measures in real data: a case study on electroencephalographic signals." *Phys Rev E Stat Nonlin Soft Matter Phys*, 65(4 Pt 1), 041903.
36. Razoumnikova, O. M. (2000). "Functional organization of different brain areas during convergent and divergent thinking: an EEG investigation." *Brain Res Cogn Brain Res*, 10(1–2), 11–18.
37. Rosenblum, M., Pikovsky, A., and Kurths, J. (2004). "Synchronization approach to analysis of biological signals." *Fluctuation Noise Lett*, 4, L53–L62.
38. Schiff, S. J., So, P., Chang, T., Burke, R. E., and Sauer, T. (1996). "Detecting dynamical interdependence and generalized synchrony through mutual prediction in a neural ensemble." *Phys Rev E Stat Phys Plasmas, Fluids, and Relat Interdisci Topics*, 54(6), 6708–6724.
39. Senhadji, L., Thoraval, L., and Carrault, G. (1996). "Continuous wavelet transform: ECG recognition based on phase and modulus representations and hidden Markov models." Wavelets in medicine and biology, A. Aldroubi and M. Unser, eds., CRC Press, NewYork, 439–463.
40. Sills, G. J., Leach, J. P., Kilpatrick, W. S., Fraser, C. M., Thompson, G. G., and Brodie, M. J. (2000). "Concentration-effect studies with topiramate on selected enzymes and intermediates of the GABA shunt." *Epilepsia*, 41 Suppl 1, S30–S34.
41. Stam, C. J., and van Dijk, B. W. (2002). "Synchronization likelihood: an unbiased measure of generalized synchronization in multivariate data sets." *Physica D: Nonlinear Phenomena*, 163(3–4), 236–251.
42. Takens, F. (1981). "Lecture Nontes in Mathematics." *Springer*, 898, 366.
43. Thatcher, R., Krause, P., and Hrybyk, M. (1986). "Cortico-cortical associations and EEG coherence: a two-compartmental model." *Electroenceph Clin Neurophysiol*, 64(2), 123–143.
44. Uhlhaas, P. J., and Singer, W. (2006). "Neural synchrony in brain disorders: relevance for cognitive dysfunctions and pathophysiology." *Neuron*, 52(1), 155–168.

45. Wendling, F., Bartolomei, F., Bellanger, J. J., Bourien, J., and Chauvel, P. (2003). "Epileptic fast intracerebral EEG activity: evidence for spatial decorrelation at seizure onset." *Brain*, 126(Pt 6), 1449–1459.
46. Wendling, F., Bartolomei, F., Bellanger, J. J., and Chauvel, P. (2001a). "Interpretation of interdependencies in epileptic signals using a macroscopic physiological model of the EEG." *Clinical Neurophysiology*, 112(7), 1201–1218.
47. Wendling, F., Bartolomei, F., Bellanger, J. J., and Chauvel, P. (2001b). "Interpretation of interdependencies in epileptic signals using a macroscopic physiological model of the EEG." *Clin Neurophysiol*, 112(7), 1201–1218.
48. Wendling, F., Bellanger, J. J., Bartolomei, F., and Chauvel, P. (2000). "Relevance of nonlinear lumped-parameter models in the analysis of depth-EEG epileptic signals." *Biol Cybern*, 83(4), 367–78.
49. Zaveri, H. P., Williams, W. J., Sackellares, J. C., Beydoun, A., Duckrow, R. B., and Spencer, S. S. (1999). "Measuring the coherence of intracranial electroencephalograms." *Clin Neurophysiol*, 110(10), 1717–1725.

Chapter 8
Neural Network Approaches for EEG Classification

Amitava Chatterjee, Amine Naït-Ali and Patrick Siarry

Abstract This chapter is aimed at providing a state-of-the-art review of the prominent neural network based approaches that can be employed for EEG classification. The chapter consists of five major sections. Following a short introduction, the next two sections are devoted to the discussions about different feature extraction algorithms and ANN based classifiers employed for EEG signals. Several representative schemes are mentioned in a nutshell, which show how diverse schemes can be employed, with good effect, to solve a similar type of problem. Here it should be kept in mind that this in no way indicates that the works mentioned in the reference within a genre either reflect the most suitable ones available within this sub-category or present the exhaustive list of references. For example we have put several references of the schemes employing Discrete Wavelet Transform (DWT) based feature extraction scheme, in the context of EEG classifiers. But this does neither indicate that they present a complete list of works utilizing DWT nor do they indicate that similar types of other works carried out, utilizing DWT, are less useful. We apologize for the fact that we may have not been able to accommodate several such suitable algorithms, within one or more sub-categories, within the boundary of this discussion.

8.1 Introduction

The electroencephalogram (ECG) signals provide rich information about the electrical activity of human brain, specially characterizing the complex human brain dynamics. Hence EEG recordings are quite popular in analysis of brain activity and determining the state of a human being. For a very long period of time, the usual methods were based on visual inspection of the recordings of these EEG signals and conclusions were drawn on the basis of expert opinions of some experienced EEGers. However, on many occasions, clinical diagnosis requires analysis of

---

A. Chatterjee (✉)
Electrical Engineering Department, Jadavpur University, Kolkata, West Bengal, India.
PIN - 700 032
e-mail: cha_ami@yahoo.co.in

A. Naït-Ali (ed.), *Advanced Biosignal Processing*,
DOI 10.1007/978-3-540-89506-0_8, © Springer-Verlag Berlin Heidelberg 2009

EEG signals recorded over a long period of time, typically over a period of more than 24 h or 48 h of time. Also, for the same set of EEG recordings utilized, the inferences drawn by different expert EEGers may vary, as their knowledge gathered over a period of past years can be quite fuzzy in nature. These tedious, time-consuming and often not so accurate methods of EEG analysis are recently gradually getting replaced by automated diagnostic systems, which are gaining popularity in patient monitoring, brain computer interfacing (BCI) and in identifying/classifying whether a person is suffering from some disease or not and sometimes even can assist quite satisfactorily in diagnosing the type of disease the patient is suffering from.

One of the useful domains of EEG analysis in automated diagnostic systems is epileptic detection, as it is primarily related to the electrical activity of the brain [2, 3]. Epilepsy is a neurological disorder which is often characterized by excessive discharges of brain cells, leading to sharp recurrent and transient disturbances of mental functions and/or excessive uncontrolled movements of several body parts [9]. Although the presence of such spikes in EEG recordings is a definite indication of presence of epilepsy, there is possibility of the presence of similar spikes in EEG waveforms due to disorders produced by similar types of seizure-like activities. In accordance with the definition given by The Committee on Terminology of the International Federation of Societies for Electroencephalography and Clinical Neurophysiology, an EEG epileptiform transient (ET) is a transient waveform having a pointed peak of duration of 20–70 ms and it can be clearly distinguished from background activity [2, 4]. Such an ET may either occur alone or it may be followed by a slow wave, of duration 150–350 ms, which are together termed as "spike and slow wave complex" [2, 9]. Hence, distinguishing and correctly identifying such spikes and seizures, originating from epileptic activities, in EEG recordings is of paramount importance. Automatic detection and classification of EEG recordings have become an important thrust area of research in biomedical engineering and bio-informatics over the last ten years or so. Although several automatic diagnostic schemes have been proposed over the years, artificial neural network (ANN) based pattern recognition schemes or classifiers have gained significant prominence among them. ANNs have been successfully employed to determine complex, nonlinear, multidimensional mathematical fitting. The function approximation problems can be solved by employing either supervised learning, where the weights and biases of an ANN are learned in presence of a set of teaching data or ANNs can be employed with unsupervised learning, where the input exemplars are classified into different clusters in a multidimensional space, in absence of any teaching data. Over last fifteen years or more, artificial neural network based solutions have been successfully employed, for quite some time now, in the domains of functions approximation, pattern recognition, automated medical diagnostic systems, decision support systems, time series prediction, signal processing, image processing, etc. [6–8, 16, 18–20]. In case of EEG classification and recognition problems, several types of ANNs have been proposed till now. Most of these ANNs are usually employed in supervisory mode. Figure 8.1 shows a typical EEG classification scheme, employing artificial neural networks. This is a three stage algorithm.

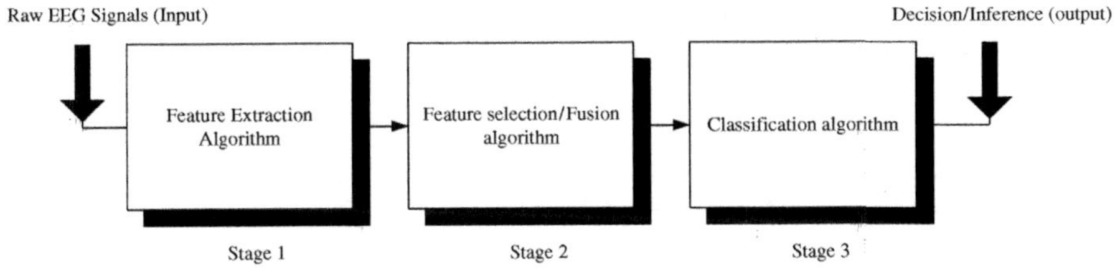

Fig. 8.1 A typical EEG-classification scheme

In stage 1, raw EEG signals are input to a feature extraction algorithm, where suitable features characterizing each signal are extracted from each signal. Each EEG signal is usually a long duration signal with huge number of samples in it. The feature extraction algorithm attempts to determine some characteristic quantities from each signal, such that they can be utilized as representatives of that signal. Naturally the size of a feature vector extracted from a raw signal should be much smaller than the original size of the signal. Sometimes the feature extraction procedure is followed by a second stage of feature selection/fusion procedure. Here the objective of the feature selection/fusion algorithm is either to select more meaningful features from a pool of features extracted from a given EEG signal or to fuse all the features extracted from a given signal, by applying another stage of transformation, to achieve a small, meaningful array of features for a given input signal. While determination of meaningful features is an extremely important requirement, another important factor is the determination of as small number of relevant features as possible to characterize a signal, so that the subsequent computational burden of the classifier to be developed can be kept reasonably light. Once such small feature arrays are formed for all raw input EEG signals, we can employ stage 3 (i.e., a classification algorithm). In this manuscript, we keep our discussions restricted to classifiers developed using ANN algorithms only. The output of the classifier algorithm will attempt to correctly diagnose/recognize whether the EEG waveform belongs to a neurologically disordered activity or to a normal human brain activity, i.e., it will attempt to perform a binary classification job and sometimes it will even attempt to pinpoint the root causes of spike/seizure like activities in EEG waveforms, (i.e., it will attempt to perform a multi-class classification job). The feature extraction algorithms may employ some techniques either in time domain, or frequency domain or they may employ time-frequency based multi-resolution analysis, which has become very popular in recent times. Similarly, the ANN based classification algorithms may also employ several types of ANN schemes, which vary according to architecture of the network and/or learning methodologies employed to learn them. Here it should be mentioned that many algorithms do not make clear demarcations about stage 1 and stage 2, discussed in Fig. 8.1. They employ these two stages as a single stage where we directly obtain suitable feature arrays for all input EEG signals. In fact, in some other algorithms, the researchers propose to employ more than one feature extraction algorithm for the same signal and then they combine suitable features extracted from each algorithm to generate a composite, hybrid feature array and then utilize such feature arrays as input to the classification algorithm.

8.2 Feature Extraction Algorithms

Some popular feature extraction algorithms employed for classification of EEG signals are given below:

- Autoregressive (AR) model [17]

This type of feature extraction algorithm is based on a well-known system identification methodology where the underlying dynamics of an EEG signal is modeled by an AR model. In this process an EEG signal is modeled as the output of a linear filter $y(n)$, fed with white noise $u(n)$. This can be given as:

$$y(n) = \sum_{k=1}^{M} w_k . y(n-k) + u(n) \tag{8.1}$$

Here M is the order of the filter and $w_1, w_2 ... w_M$ are the coefficients of the AR model. The choice of the order M is important as this determines the number of past samples that are significant in determining the present output sample. This M can be taken equal to the maximum lag that leads to significantly different result from zero autocorrelation function [17]. Once M is known, the AR parameters w_i $(i = 1, 2 ... M)$ can be employed using Yule-Walker equation [21]:

$$\begin{bmatrix} r(0) & r(1) & \cdots & r(M-1) \\ r(-1) & r(0) & \cdots & r(M-2) \\ \vdots & \vdots & \ddots & \vdots \\ r(-M+1) & r(-M+2) & \cdots & r(0) \end{bmatrix} \begin{bmatrix} w_1 \\ w_2 \\ \vdots \\ w_M \end{bmatrix} = \begin{bmatrix} r^*(1) \\ r^*(2) \\ \vdots \\ r^*(M) \end{bmatrix} \tag{8.2}$$

Where the $M \times M$ matrix represents the autocorrelation matrix R. Hence, w_i, $(i = 1, 2 ... M)$ parameters can be found if autocorrelation functions are known. Once these w_i are calculated, they can be utilized to form the feature-array for a given EEG signal and these feature vectors can be utilized as input for a ANN based classifier.

- Discrete Wavelet Transform (DWT) [2, 3, 10, 11]

The wavelet transform (WT) is known as an extended version of the Fourier transform, which can provide time-frequency representation [1]. It provides accurate localization of transient features of a signal in both time and frequency domain by working on a multi-scale basis where each signal under consideration is decomposed into a number of scales, each scale providing a particular coarseness of the signal. In DWT, a signal is analyzed in several bands, using filters of different cutoff frequencies. The high frequency components of the signal are analyzed by passing the signal through a series of high pass filters and the low frequency components are analyzed by passing the signal through a series of low pass filters. Figure 8.2 shows the procedure of DWT in a schematic form. Here, in each stage $h(.)$ gives the high pass filter of that stage, which acts as the discrete mother wavelet and $g(.)$ gives the low pass filter, which is the mirror version of the corresponding $h(.)$. In each stage j the output of $h(.)$ is down sampled by 2 to obtain the corresponding detail D_j and the output of $g(.)$ is down sampled by 2 to obtain the corresponding approximation A_j. In each stage j, the approximation A_j is further decomposed, by employing a similar procedure, to obtain D_{j+1} and A_{j+1}, and this process is continued for subsequent scales. The coarsest scale is denoted by $j = 0$ and for finer and finer scales, j is progressively increased. Each low pass filter g satisfies the quadrature mirror condition given as [11]:

$$G(z)G(z^{-1}) + G(-z)G(-z^{-1}) = 1 \tag{8.3}$$

where $G(z)$ denotes the z-transform of the filter g. The corresponding high pass filter h can be obtained as:

$$H(z) = zG\left(-z^{-1}\right) \tag{8.4}$$

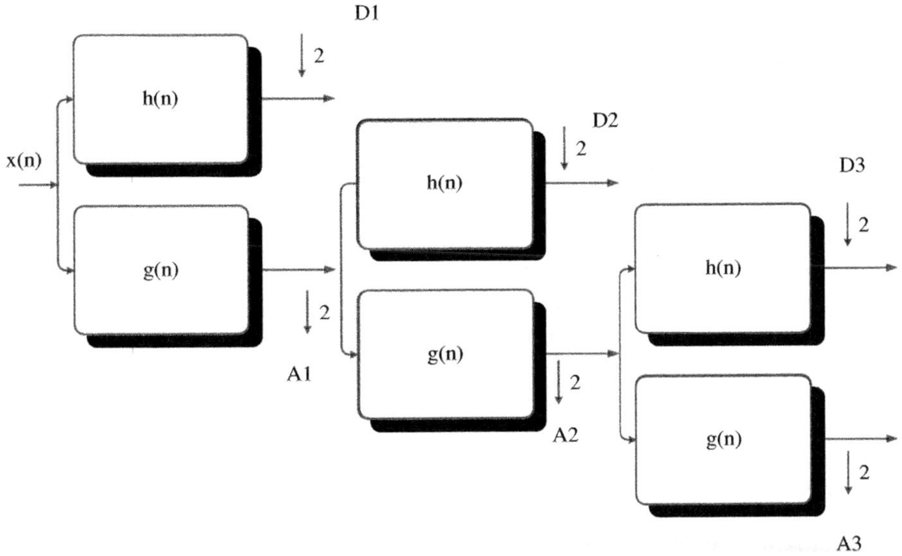

Fig. 8.2 Schematic representation of DWT employing a tree of filter banks [11]

Hence, a sequence of filters, in increasing subscript of i (increasing with increasing length) can be given as:

$$\left.\begin{aligned} G_{i+1}(z) &= G(z^{2^i})G_i(z) \\ H_{i+1}(z) &= H(z^{2^i})H_i(z) \end{aligned}\right\} \quad i = 0, \cdots, I-1 \tag{8.5}$$

Where initially $G_0(z) = 1$. The choice of the appropriate wavelet and the number of decomposition levels play a crucial rule in determining the efficiency of the wavelet coefficients extracted to be utilized as suitable features in characterizing EEG signals, for useful classification/recognition purpose in subsequent stages. In general, DWT is regarded as a useful method for detecting epileptic seizures because of its strong capability of identifying transient features in EEG signal in time-frequency domain.

– Wavelet Packet Transform (WPT) [13]

In DWT, a signal is decomposed into corresponding detail and approximation at scale j and at scale $j+1$ the corresponding approximation is further decomposed into detail and approximation. Hence the detail information obtained in each scale j is kept untouched. As opposed to this method, in wavelet packet transform (WPT), both the details and approximations are further decomposed in each scale to generate subsequent details and approximations for the next stage. This increases the flexibility of different ways in which a signal can be encoded. With increasing number of levels or stages of decomposition for a given signal, the increase in flexibility in encoding utilizing WPT, compared to DWT, grows tremendously. A lot of bases can be obtained from a single wavelet packet decomposition, which provides rich representation of the signal and provides more option of choosing the best among these bases, which suits the design objective. However, this increased flexibility is associated with corresponding increase in complexity of encoding a given signal.

– Lyapunov Exponents [11]

Lyapunov exponents are popular in determining stability of any steady-state behavior of a given system and they are particularly useful in obtaining chaotic solutions. They can be used as dynamical quantitative measures of a signal and can be given either (a) on the basis of the estimation obtained from the time evolutions of nearby points in state-space (gives only the largest Lyapunov exponent) (b) or on the basis of estimation of local Jacobi matrices (gives all Lyapunov exponents). The presence of Positive Lyapunov exponents confirms the chaotic nature of the signal under consideration.

– Approximate Entropy (ApEn) [5, 22–25]

Approximate entropy is a comparatively recently formulated statistical parameter which can be suitably utilized as a time domain feature to quantitatively express the regularity of a time series data. This can be particularly useful in extracting features to characterize the nonlinear dynamics inherent within the temporal variation of an aperiodic signal. It has come into prominence in 1990s and has been successfully employed recently in automatically detecting epileptic EEGs [5], along with in other medical diagnostic domains e.g., heart rate variability, endocrine hormone release pulsatility etc. In [25] it has been shown that, during an epileptic activity, there occurs an abrupt decrease in the value of ApEn because of the synchronous discharge of large groups of neurons and this characteristic makes it particularly useful for automated detection of epileptic EEGs. If the values of ApEn are found to be comparatively small, it indicates that there is a strong regularity in the data sequence, (i.e., the temporal EEG signal under consideration) and if the values of ApEn are comparatively large, it indicates that there is feeble regularity in the data sequence, characterized by large fluctuations [5].

8.3 ANN Based Classification Algorithms

Some popular ANN algorithms employed for classifying EEG signals on the basis of features extracted are given below:

– Multilayer Perceptron Neural Network (MLPNN) [2, 3, 9, 11–14, 17]

This is a classical supervised neural network architecture which has been most popular in a variety of application domains over the years. The MLPNN comprises an input layer, where all the input signals are connected to input nodes, followed by one or more hidden layers where several hidden layer nodes are connected to accommodate more nonlinearity which can, hopefully, help in determining an efficient multidimensional nonlinear mapping between input and output exemplars and this is followed by an output layer which produces the output of the neural network. Figure 8.3 shows the schematic form of the architecture of a typical m-input-n-output MLPNN, with a single hidden layer comprising p neurons. The general form

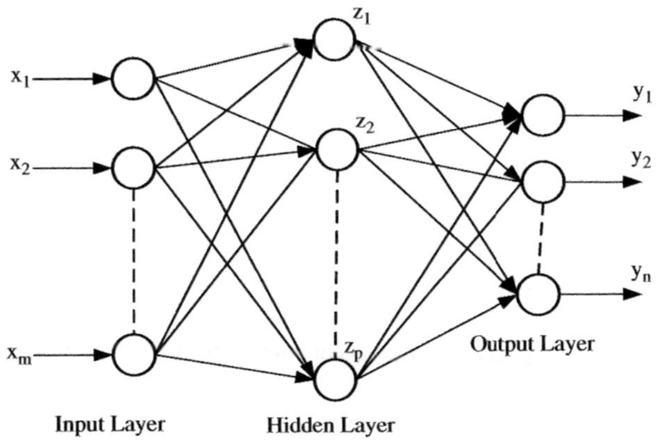

Fig. 8.3 A typical architecture of an MLPNN

of an MLPNN is a fully connected one where each node or neuron in a given layer is connected to all nodes or neurons in the previous layer through some connecting weights. Each node, in its most general form, comprises two functions: integration function and activation function. The integration function integrates or summates all the weighted inputs at the given node to produce an aggregated input for the activation function. Then the activation function applies a nonlinearity on its aggregated input by employing a continuous nonlinear function. These nonlinearities can be popularly employed in terms of log-sigmoidal or tan-sigmoidal functions, which are known as smooth functions that are differentiable everywhere. However, in case of some neurons, both the integration and activation function may not be present and, in those situations, in all probability, only the integration function will be present.

This network is trained in a supervised manner, in presence of ideal input-output exemplars, utilized in form of a training data set, which determines the suitable weights and biases of the network. In such cases, the most popular training algorithm is known as the error backpropagation algorithm, where the synaptic weights and biases are adjusted by backpropagating the error signal through different layers of the network in a chain form, with an objective of adjusting the free parameters of the network so that the actual response of the network approaches the ideal response in a statistical sense. This learning algorithm can be employed either in pattern mode, i.e., the weights and biases are adjusted every time an input exemplar is presented to the system, or in batch mode, (i.e., the weights and biases are adjusted every time all the input exemplars present in the training data set are presented once to the system). The original form of the MLPNN employing backpropagation learning is also popularly known as backpropagation neural network (BPNN). However, the original form of the backpropagation algorithms had many disadvantages, (e.g., they used to show slow convergence, as they were prone to getting trapped at local minima). Proposing improved variants of BPNNs has become an active area of research for several years and researchers have proposed popular improved variants employing adaptive learning rate, momentum etc. which can potentially reduce convergence time, can avoid problem of getting trapped in local minima, can achieve lesser computational burden, can involve reduced memory requirement etc. A detailed discussion on several aspects of MLPNN with many new variants can be obtained in [18–20, 27].

– Probabilistic Neural Network (PNN) [5, 11]

Probabilistic neural networks are proposed as a variation of radial basis function networks (RBFN) and PNN can be popularly employed for classification problems. Like MLPNN, they can be directly employed for either binary or multi-class classification problems and are also known as to provide good generalization capability. This is usually a four-layer architecture, where there are two middle layers in addition to the usual input layer and output layer. These two middle layers are called radial basis layer and competitive layer [5, 26]. The architecture of a PNN is shown in Fig. 8.4.

Fig. 8.4 The architecture of a probabilistic neural network [5]

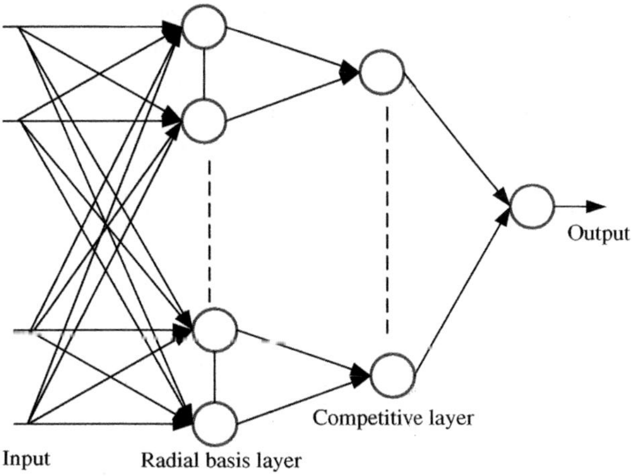

The objective of employing the radial basis layer (or pattern layer) in Fig. 8.4 is to determine the distance between the input vector x and the training input vector x_{ij} and an output vector is formed which shows the proximity between the input and the training input, in terms of distances. This can be given as [11]:

$$\varphi_{ij}(\mathrm{x}) = \frac{1}{(2\pi)^{d/2}\,\sigma^d} \exp\left[-\frac{\left(\mathrm{x}-\mathrm{x}_{ij}\right)^T\left(\mathrm{x}-\mathrm{x}_{ij}\right)}{2\sigma^2}\right] \tag{8.6}$$

where d denotes the size of the input vector x and σ is the smoothing parameter. This is input to the second middle layer called the competitive layer or the summation layer. Here the contributions for each class of inputs are summated and the output from this layer is given as a vector of probabilities. This is followed by a "compete" function, which assigns the input vector to that class for which the maximum probability value was obtained.

– Support Vector Machines [11]

Support Vector Machines (SVMs) are basically another category of universal feed-forward networks. The concept of SVM is based on the Vapnik-Chervonenkis (VC) theory of statistical learning and it is an approximate implementation of the method of structural risk minimization. A detailed discussion on SVMs is available in [18] and [27]. The SVM is originally proposed for binary pattern classification problems which has been later extended to solve multi-class pattern classification problems by employing (e.g., one-against-the rest or one-per-class approach). This essentially means each multi-class classifier SVM employs several component SVMs, each of which attempts to perform a binary classification job in order to identify a specified class. The objective of training an SVM is to search for an optimal separating hyperplane (OSH) that can provide superior generalization, particularly when the dimension of input data is large and the number of observations available for developing or training the model is limited. This OSH between two classes is determined with an objective of maximizing the margin between the two classes. For a binary classification problem at hand, let the data set available be composed of N exemplars $\{x_i, c_i\}$, $i = 1,2,\ldots,N$ where each input feature x_i is m-dimensional and the classification outputs c_i are given as $c_i \in \{+1, -1\}$. When these input exemplars are linearly separable patterns, the OSH, determined in m dimensional space, is given according to the relation

$$w^T x + b = 0 \tag{8.7}$$

Here x is the input vector, w is an adjustable weight vector and b is the hyperplane bias. Hence, for linearly separable patterns, we can write [18]:

$$\left.\begin{aligned} w^T x_i + b &\geq 0 \quad \text{for } c_i = +1 \\ w^T x_i + b &< 0 \quad \text{for } c_i = -1 \end{aligned}\right\} \tag{8.8}$$

The training objective of the SVM is to find the optimum weight vector w_0 and the optimal bias b_0 such that the separation between the corresponding hyperplane and the closest data point, called the margin of separation, is maximized. It can be shown that (w_0, b_0) must satisfy the constraint given as [18]:

$$\left.\begin{aligned} w_0^T x_i + b_0 &\geq 1 \quad \text{for } c_i = +1 \\ w_0^T x_i + b_0 &\leq -1 \quad \text{for } c_i = -1 \end{aligned}\right\} \tag{8.9}$$

Those data points, which satisfy the equality conditions given in the two row-wise equations of relation (9), are called support vectors. The determination of the OSH is solved as a constrained quadratic optimization problem.

The algorithm gets a little complicated for linearly non-separable patterns. Here the algorithm is so modified that we design to generate a soft margin so that some misclassification is permitted. In this method, the design problem is solved by introducing N number of non-negative scalar variables, called slack variables. Then this learning machine attempts to determine a separating hyperplane so that the average error due to misclassification is minimum. A very popular approach in designing SVMs for classifying linearly non-separable patterns is to employ inner-product kernel functions. These kernel functions are used to nonlinearly transform each input vector to a high dimensional feature space. Ideally, this nonlinear mapping performed with the help of the inner-product kernels is so that, although the original input vectors are not linearly separable patterns, the corresponding transformed vectors in the high-dimensional feature space can be separated by constructing a linear decision surface or hyperplane in that feature space. Hence the construction of the inner-product kernels looks very similar to the implementation of the hidden layer neurons in MLPNN. So the architecture of an SVM can be viewed as a three-layer one where the first layer comprises the input vector, the second layer comprises the hidden layer of inner-product kernels (that generate outputs in high-dimensional feature space) and the third layer produces the decision or the class as its output. Some of the very popular inner-product kernels in SVMs employ polynomial learning machines, radial basis function networks and two-layer perceptrons, given respectively in relations (10), (11) and (12) shown below [18]:

$$K(x, y) = \left(x^T y + 1\right)^p \tag{8.10}$$

$$K(x, y) = e^{-\|x-y\|^2/2\sigma^2} \tag{8.11}$$

$$K(x, y) = \tanh\left(\beta_0 x^T y + \beta_1\right) \tag{8.12}$$

– Elman Recurrent Neural Network [5]

Elman recurrent neural network (ERNN) is a special type of backpropagation neural network which employs a combination of feedback and feedforward connections, that can exhibit memory and can be very useful for classification purposes as they can efficiently learn both temporal and spatial patterns. Elman RNN is a popular variant in a general class of recurrent neural networks (RNNs), where the feedback is employed from the output of each hidden layer neuron to the input of the corresponding hidden layer neuron. The architecture of the ERNN is essentially as that of the multilayer perceptron (MLP) where tan-sigmoidal and log-sigmoidal neurons are usually employed. Figure 8.5 shows the architecture of an ERNN.

Let the input and output of a hidden layer node be y_k and z_k at the *kth* iteration. Then y_k is obtained as a summation of weighted influences of all input nodes i.e., each value at an input layer node multiplied by the weight between that input layer node and the hidden layer node under consideration, weighted influences of all feedback outputs from hidden layer nodes (i.e., each z_{k-1} multiplied by the weight

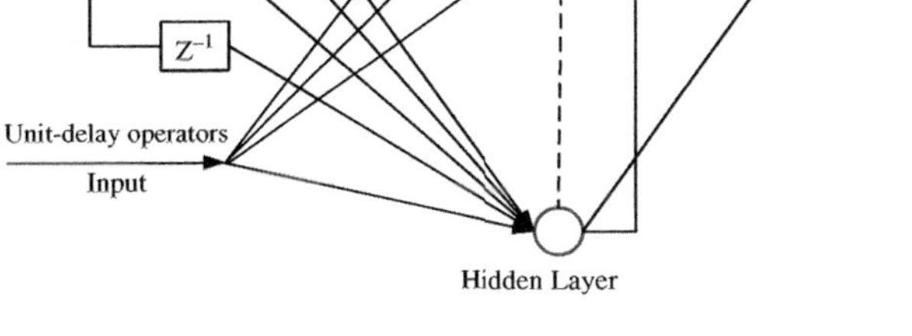

Fig. 8.5 The architecture of an Elman recurrent neural network [5]

between that delayed output of the hidden layer node and the present hidden layer node whose input calculation is under consideration) and bias. Once this y_k is calculated, one can proceed to obtain the present output from this hidden layer z_k by employing log-sigmoidal or tan-sigmoidal nonlinearity and then one can proceed in an usual manner to compute the output from the output layer, as is carried out in an MLPNN.

– Log-linearized Gaussian Mixture Neural Network [15, 28]

The log-linearized Gaussian mixture neural network (LLGMNN), which has been successfully employed in classifying EEG signals, is essentially a PNN network which determines the posteriori probability of an input feature vector belonging to each class. Figure 8.6 shows the structure of a typical LLGMNN.

At first, this three layer feedforward type neural network transforms each input vector $\mathbf{x}$ of size d to a transformed vector X of size $(1 + d(d + 3) + 2)$, which represents the probability density function corresponding to each component of the Gaussian mixture model (GMM). This transformed vector X is then input to the next layer, which determines the posteriori probability of each component of the GMM. This layer is followed by another layer where the total number of nodes corresponds to the total number of classes in which the input data is classified. The output of this layer corresponds to the posteriori probability of each candidate class. The network is trained to obtain the weight coefficients in the second layer so that successful classification can be carried out. The detailed layer-wise governing equations are available in [15].

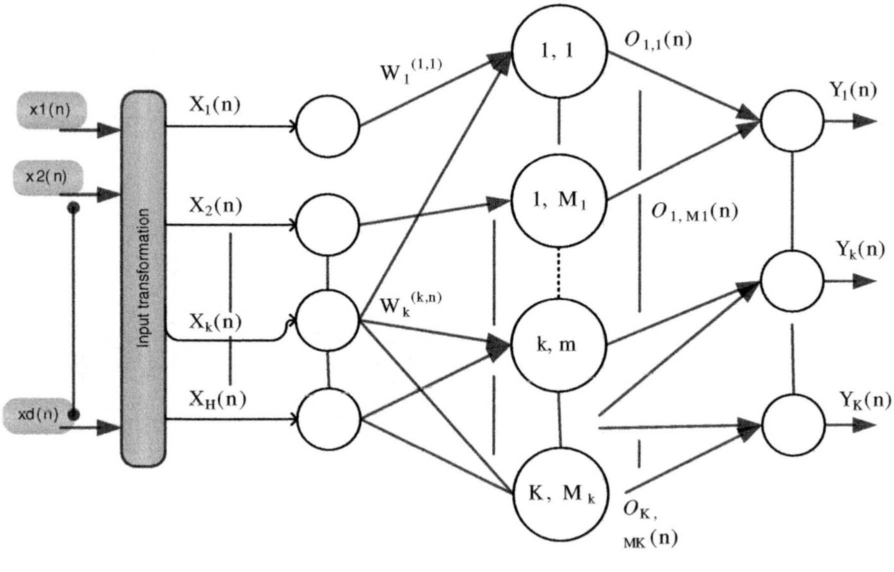

Fig. 8.6 The architecture of the LLGMNN

– Neuro-fuzzy systems [10, 33]

Neuro-fuzzy systems (NFS) are developed with the objective of capturing both the strong points of two pillars of techniques employed in computational intelligence: (i) to incorporate the linguistic interpretation of fuzzy systems (which is easier to understand) and (ii) to incorporate generalized data-driven multidimensional function approximation capability of neural networks. One of the earliest neuro-fuzzy systems proposed, which is still quite popular and extensively used, is the adaptive neuro-fuzzy inference systems (ANFIS) [34]. An NFS essentially embodies different sub-modules of a fuzzy system in different layers of an ANN, typically, e.g., one can have the input layer of the ANN followed by a fuzzification layer, where different input membership functions (MFs) are placed in different nodes. This may be followed by a rule layer which implements the fuzzy rule base, where there may be m nodes corresponding to m number of fuzzy rules. The output of the rule base layer is usually fed to the defuzzification layer which determines the output of the NFS. For Mamdani-type of inferencing the defuzzification layer may incorporate nodes for several popular defuzzification methods, e.g., center-of-gravity method, center-of-sums method, height method etc. For Sugeno-type of inferencing the most common defuzzification method is the weighted average method. When such a fuzzy system is implemented within the paradigm of an NFS, the training procedure can be implemented to adaptively determine the free parameters and/or the structure of the NFS itself. In the original ANFIS, the structure was kept unchanged and the free parameters of the NFS were adapted. Figure 8.7 shows the original structure of the ANFIS.

Here in layer 1, fuzzification is carried out, with each node signifying the parameters characterizing an input MF. Parameters in this layer form the premise parameters, which are adapted during the training procedure to adapt each input MF. Layers 2 and 3 are implemented with fixed nodes where the firing strength of each rule

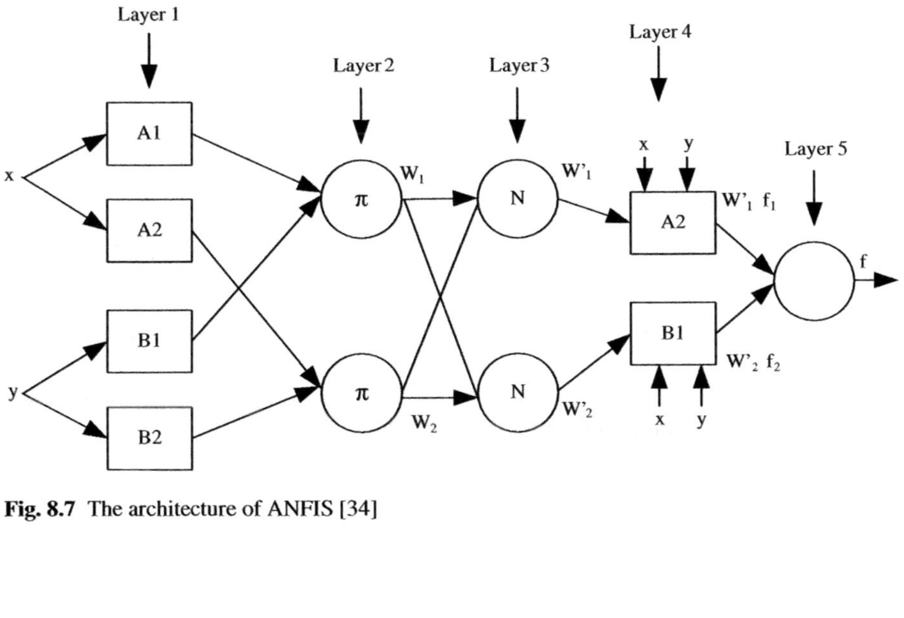

Fig. 8.7 The architecture of ANFIS [34]

and the normalized firing strength of each rule are computed, respectively. Layer 4 implements the actual fuzzy rule base and comprises adaptive nodes where the consequent parameters of each rule can be adapted to determine a suitable input-output mapping for the training data under consideration. Layer 5 is a fixed layer, which performs defuzzification to determine the final output. This structure was proposed for Sugeno-model employing a hybrid learning algorithm. In the forward pass, node outputs until layer 4 are calculated and the consequent parameters are identified using least-square method. In the backward pass, the premise parameters are adjusted using gradient-descent method. Since its development, ANFIS has been extensively used and many variants have been proposed to suit different real problems at hand. A direct application of ANFIS for epileptic seizure detection is given in [33].

Another application for an identical problem, proposed in [10], shows another useful architecture that utilizes neural networks and fuzzy systems together, called dynamic fuzzy neural networks (DFNN). DFNNs consist of "feurons", where each feuron is basically a dynamic neuron with fuzzy activation functions. A DFNN can consist of several such feurons, depending on the degree of nonlinearity of the mapping/classification problem in hand. The fuzzy activitation function is typically modeled using product inference engine, Gaussian membership function, singleton based fuzzification and center average based defuzzification. Such a DFNN is very useful in those situations where we encounter dynamic variation in data as it employs dynamic feuron units. A detailed discussion about DFNNs can be obtained in [35, 36].

8.4 EEG Data Sets

The evaluation of several proposed EEG classification or recognition schemes was carried out on the basis of some signals experimentally acquired from human volunteers/samples. In some cases, the researchers themselves acquired these signals from human volunteers, by setting up their own experimental set up. In many other cases, the researchers demonstrated the efficacy of their proposed algorithms on the basis of popular benchmark EEG signals, freely available for downloading in the internet. In case of several research works regarding epilepsy recognition, the benchmark EEG signal database is available from the Department of Epileptology, University of Bonn, [29, 30]. This complete database contains five sets of 100 single-channel EEG signals, obtained with a sampling rate of 173.61 Hz and digitized with 12 bit resolution. The same amplifier system was employed for acquiring all EEG signals from human volunteers. The EEG signals were extracted for those portions which were devoid of any artifact e.g., due to eye movements or muscle activities [31]. The spectral bandwidth of the acquisition system was 0.5–85 Hz and the first step for processing of such EEG signals requires low pass filtering with a cut-off frequency of 40 Hz. Out of five sets of data, two sets were obtained from five healthy volunteers, relaxing in awake condition, with their eyes open and eyes closed and the EEG surface recordings were acquired using standard 10–20 international system of electrode placement. The three other sets originated from EEG archive of presurgical diagnosis with recordings from five patients selected. Two of these three data sets contained EEG signals from patients during seizure-free interval and the third set comprised EEG signals during seizure activity. Researchers sometimes developed schemes for binary classification i.e., to segregate between healthy volunteers and patients suffering from epilepsy or more complicated multi-class schemes, (e.g., for segregating all five classes simultaneously). Another comprehensive collection of experimentally recorded continuous EEG signals is available from the Freiburg Centre for Data Analysis and Modeling [32]. Some EEG signal recordings are also available in PhysioBank, a popular freely available physiologic signal archive for biomedical research [37].

8.5 Performance Measures

A very common method of quantitatively presenting the classification results is in form of confusion matrices, where one can show how many of the original signals belonging to a particular class was actually recognized by the classification system output as signals belonging to that class and how many such signals were misclassified into other classes. It is common either to present these results in form of actual counts or in form of percentage classification/recognition accuracy. For example let us consider a five-class classification scheme where the objective is to classify the signals into classes ‘a’, ‘b’, ‘c’, ‘d’ and ‘e’. The confusion matrix will have entries for actual classes and predicted classes. They will show how many of those signals actually belonging to class ‘a’ were predicted/recognized as belonging to class ‘a’ and if some of those signals belonging to class ‘a’ were not recognized as belonging to class ‘a’, then how many of them were mis-classified in class ‘b’ or ‘c’ or ‘d’ or ‘e’. The same logic can be placed for signals belonging to other classes, (i.e., ‘b’ or ‘c’ or ‘d’ or ‘e’). Higher the recognition capability of the classifier, lesser will be the confusion or misclassification. Another very popular method of presenting the results is in form of statistical parameters, called sensitivity (SE), specificity (SP), and overall accuracy (OA), usually given in percentage, defined below [1, 5, 10, 11]:

$$\mathrm{SE}(\%) = \frac{\text{number of correctly detected positive decisions}}{\text{number of actual positive cases}} \tag{8.13}$$

$$\mathrm{SP}(\%) = \frac{\text{number of correctly detected negative decisions}}{\text{number of actual negative cases}} \tag{8.14}$$

$$\mathrm{OA}(\%) = \frac{\text{total number of correct decisions made}}{\text{total number of actual cases considered}} \tag{8.15}$$

Generally the systems are developed on the basis of the signals taken as is, in terms of the length of the signals. However, sometimes the signals are subdivided into several segments and each segment is considered as a signal itself and the classification accuracy is reported on the basis of all these signal segments. For example, for the EEG signals in [29, 30], there are 100 EEG time-series signals in each of the five sets, with each signal comprising of 4096 samples. Some results are reported considering these 100 signals in each data set itself. However, in some cases, each signal comprising 4096 samples has been subdivided into 16 segments, each comprising 256 samples. Then, in each data set there are 1600 such signal segments and there are five such data sets each containing 1600 signal segments. These research reports, working on these signal segments, have reported their accuracy results as a percentage of the total number of such signal segments.

Table 8.1 presents several EEG-signal based methods employed for automatic detection/recognition/classification of epilepsy, in a compact form, which shows how researchers have utilized several feature extraction and classification algorithms available at their disposal, to develop schemes of their interest. Here also we would like to mention that there is huge amount of research interest in this domain and the interest is growing fast almost every day. So there is possibility that there may be a lot of schemes, which go unmentioned, compared to those schemes, which are mentioned here. We would like to remind our readers that the schemes mentioned here will act as an eye opener for interested readers to search for more such schemes already developed and reported. For many of the schemes considered here, the classification accuracy is more than 90% and sometimes even more than 99%, which should be considered as a fabulous achievement.

Table 8.1 A compact form of presentation of EEG classification schemes

Classification scheme [Reference]	Feature extraction method	Classification method	Overall classification accuracy reported (%)
[11]	DWT + Lyapunov Exponents	SVM	99.28
		PNN	98.05
		MLPNN	93.43
[5]	ApEn	ERNN	99.35–100
		PNN	98–100
[3]	DWT	MLPNN	97
		RBFN	98
[31]	DWT	Mixture of Expert Model	94.5
		MLPNN	93.2
[2]	DWT	MLPNN	93.7
[10]	DWT	Dynamic Fuzzy Neural Network	93
		MLPNN	92
[33]	DWT	ANFIS	94

8.6 Conclusion

In this chapter we have described usual feature extraction algorithms and artificial neural networks algorithms employed in the literature for classifying EEG signals on the basis of features extracted. Then we were interested in the experimental evaluation of competing EEG classification or recognition schemes. A few benchmark EEG databases are available in the internet, particularly focused on works regarding epilepsy recognition. We reported some performance measures used to compare the classification results from various methods. The accent was put on the comparison of EEG-signal based methods employed for automatic detection and classification of epilepsy.

References

1. Available: http://users.rowan.edu/~polikar/WAVELETS/WTpart1.html. Last accessed on: April 03, 2008
2. Kalayci T and Özdamar Ö (1995) Wavelet preprocessing for automated neural network detection of EEG spikes. IEEE Eng Med Biol, 160–166
3. Jahankhani P, Kodogiannis V and Revett K (2006) EEG signal classification using wavelet feature extraction and neural networks. Proc IEEE John Vincent Atanasoff 2006 International Symposium on Modern Computing
4. Chatrian E, Bergamini L, Dondey M et al. (1974) A glossary of terms most commonly used by clinical electroencephalographs. Electroenceph Clin Neurophysiol, 37:538–548
5. Srinivasan V, Eswaran C and Sriraam N (2007) Approximate entropy-based epileptic EEG detection using artificial neural networks. IEEE Trans Inf Technol Biomed, 11:288–295
6. Weng W and Khorasani K (1996) An adaptive structure neural network with application to EEG automatic seizure detection. Neural Netw, 9:1223–1240
7. Gotman J and Wang L (1991) State-dependent spike detection: Concepts and preliminary results. Electroenceph Clin Neurophysiol, 79:11–19
8. Pradhan N, Sadasivan P and Arunodaya G (1996) Detection of seizure activity in EEG by an artificial neural network: A preliminary study. Comput Biomed Res, 29:303–313
9. Acır N, Öztura İ, Kuntalp M et al. (2005) Automatic detection of epileptiform events in EEG by a three-stage procedure based on artificial neural networks. IEEE Trans Biomed Eng, 52:30–40
10. Subasi A (2006) Automatic detection of epileptic seizure using dynamic fuzzy neural networks. Expert Syst Appl, 31:320–328
11. Güler İ and Übeyli E (2007) Multiclass support vector machines for EEG-signals classification. IEEE Trans Inf Technol Biomed, 11(2), 117–126
12. Ahn C, Lee S and Lee T (1996) EEG and artifact classification using a neural network." Proc. 18th Annual International Conference of the IEEE Engineering in Medicine and Biology Society. Amsterdam, 915–916
13. Wang B, Jun L, Bai J et al. (2005) EEG recognition based on multiple types of information by using wavelet packet transform and neural networks. Proc. 2005 IEEE Engineering in Medicine and Biology 27th Annual Conference, Shanghai, China, 5377–5380, September 1–4
14. Haselsteiner E and Pfurtscheller G (2000) Using time-dependent neural networks for EEG classification. IEEE Trans Rehabil Eng, 8:457–463
15. Fukuda O, Tsuji T and Kaneko M (1996) Pattern classification of time-series EEG signals using neural networks. Proc. IEEE International Workshop on Robot and Human Communication, 217–222
16. Lu B-L, Shin J and Ichikawa M (2004) Massively parallel classification of single-trial EEG signals using a min-max modular neural network. IEEE Trans Biomed Eng, 51:551–558

17. Maiorescu V, Serban, M and Lazăr M (2003) Classification of EEG signals represented by AR models for cognitive tasks – a neural network based model. Proc. International Symposium on Signals. Circuits, and Systems, SCS 2003, 441–444
18. Haykin S (2001) Neural Networks: A comprehensive Foundation. Addison Wesley Longman Pte Ltd, Singapore
19. Bose N, and Liang P (1998) Neural Network Fundamentals with Graphs, Algorithms and Applications. Tata McGraw-Hill Publishing company limited, India
20. Hassoun M (1995) Fundamentals of Artificial Neural Networks. The MIT Press, Cambridge, Massachusetts
21. Haykin S (2001) Adaptive Filter Theory. 4th Edition, Prentice-Hall Inc., New Jersey
22. Bruhn J, Ropcke H and Hoeft A (2000) Approximate entropy as an electroencephalographic measure of anesthetic drug effect during desflurane anesthesia. Anesthesiology, 92:715–726
23. Pincas S, Gladstone I and Ehrenkranz R (1991) A regularity static for medical data analysis. J Clin Monit 7:335–345
24. Pincas S (1991) Approximate entropy as a measure of system complexity. Proc Natl Acad Sci USA, 88:2297–2301
25. Diambra L, Bastos de Figueiredo J and Malta C (1999) Epileptic activity recognition in EEG monitoring. Phys A: Stat Mech Appl 273:495–505
26. Demuth H and Beale M (2000) Neural Network Toolbox. (for use with Matlab). MathWorks, Natick, MA
27. Kumar S (2004) Neural Networks, A Classroom Approach. Tata McGraw-Hill Publishing Company Limited, New Delhi
28. Fukuda O, Tsuji T and Kaneko M (1995) Pattern classification of EEG signals using a log-linearized Gaussian mixture neural network. Proc IEEE Int Conf Neural Netw, 2479–2484
29. Andrzejak R, Lehnertz K, Rieke C et al. (2001) Indications of nonlinear deterministic and finite dimensional structures in time series of brain electrical activity: Dependence on recording region and brain state. Phys Rev E, 64, 061907
30. Online]. Available: http://www.meb.unibonn.de/epileptologie/cms/front/content.php?idcat=193&lang=3&changelang=3. Last accessed on: April 03, 2008
31. Subasi A (2007) EEG signal classification using wavelet feature extraction and a mixture of expert model. Expert Syst Appl, 32:1084–1093
32. Available: https://epilepsy.uni-freiburg.de/freiburg-seizure-prediction-project/eeg-database/. Last accessed on: April 03, 2008
33. Subasi A (2007) Application of adaptive neuro-fuzzy inference system for epileptic seizure detection using wavelet feature extraction. Comput Biol Med, 37:227–244
34. Jang J-S and Sun C (1995) Neuro-fuzzy modeling and control. Proc IEEE, 3:378–406
35. Berceikli Y (2004) On three intelligent systems: Dynamic neural, fuzzy and wavelet networks for training trajectory. Neural Comput Appl, 13:339–351
36. Berceikli Y, Oysal Y and Konar A (2004) Trajectory priming with dynamic fuzzy networks in nonlinear optimal control. IEEE Trans Neural Netw, 15:383–394
37. Available: http://www.physionet.org/physiobank/database/ Last accessed on: April 03, 2008

Chapter 9
Analysis of Event-Related Potentials Using Wavelet Networks

Hartmut Heinrich and Hartmut Dickhaus

Abstract Event-related potentials (ERPs) can be interpreted as a superposition of event-related oscillations in various EEG frequency bands. Time-frequency representations (e.g., wavelet networks) allow to describe these oscillations with a small set of parameters automatically. Using appropriate training algorithms, wavelet networks can also be applied for ERP single-trial analysis (low SNR) to investigate trial-to-trial variabilities.

9.1 Introduction

Non-invasive recordings of brain electrical fields (EEG; event-related potentials, ERPs) allow to measure neural activations associated with a variety of brain states and processes with a time resolution in the millisecond range. ERPs can be elicited in cognitive tasks, e.g., in oddball tasks or continuous performance tests. They are an important tool for studying neurocognitive functioning besides functional magnetic resonance imaging (fMRI), which provides high spatial resolution. ERP studies have significantly contributed to a better understanding of the pathophysiological processes in neuropsychiatric disorders (e.g., schizophrenia or attention-deficit/hyperactivity disorder, ADHD, [1, 17, 25]).

Usually, stimulus synchronous averaging is performed to eliminate the spontaneous part of the EEG and to extract the event-related potential. The average ERP can be described as a sequence of negative and positive deflections. These components are labelled according to their polarity and chronological order (e.g., N1, N2 or P3). They are ascribed a certain functional significance, e.g., the N1 occurring at a latency of 100–150 ms is thought to be related to early attentional orienting processing. The P3 (latency 300–600 ms) could reflect context updating and/or attentional allocation [18].

H. Heinrich (✉)
Child & Adolescent Psychiatry, University of Erlangen, Germany; Heckscher Klinikum, Munich, Germany
e-mail: hheinri@arcor.de

A. Naït-Ali (ed.), *Advanced Biosignal Processing*,
DOI 10.1007/978-3-540-89506-0_9, © Springer-Verlag Berlin Heidelberg 2009

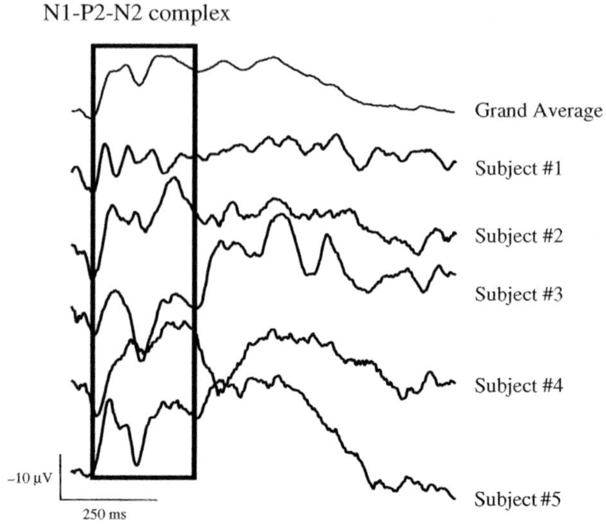

Fig. 9.1 Averaged event-related potentials (target-attended, T-A responses) of typically developing children recorded during an auditory selective attention task. In the task, 240 acoustic stimuli were presented: 96 tones with higher frequency (1500 Hz) and 144 tones with lower frequency (1000 Hz). Half of the stimuli from each type were presented to the left ear and half of them to the right ear in pseudorandom order. Children were instructed to respond to the higher tones presented to the right ear by pressing a button, i.e., the right-hand side was the attended side (= target-attended stimuli)

Traditionally, ERPs are parametrized by measuring latency and amplitude values of these peaks. In Fig. 9.1, examples of event-related potentials recorded in children during an auditory selective attention task are presented. The top trace shows the grand average, the other traces are from individual subjects.

The N1-P2-N2 complex is marked. In the grand average signal, the N1, the P2 and the N2 components can be identified quite easily. However, it is hardly not possible to determine the N1, P2 and N2 in the ERPs of individual subjects. In subject #2, there are even four negative half waves in the marked time frame. The example indicates that a reliable parametrization of the N1-P2-N2 complex in terms of latencies and maximum amplitudes measured in the averaged ERP is not possible.

9.1.1 Time-Frequency Analysis of Event-Related Potentials

Measuring latencies and amplitudes of prominent peak may be considered as a phenomenological view on the data. Alternatively, analysis could be based on an ERP model related to physiology using adequate mathematical methods.

Basar et al. [3] suggest the following ERP model:

- The EEG consists of the activity of an ensemble of generators producing rhythmic activity in several frequency ranges. Usually, these oscillators are active in a random manner; however, by application of sensory stimulation, these generators are coupled and act together in a coherent way. This synchronisation and enhancement of EEG activity gives rise to an 'evoked' or 'induced' rhythmicity.

- The superposition of event-related oscillations in various EEG frequency channels (alpha response, theta response, gamma response etc.) gives rise to the compound event-related potential.

These event-related oscillations, i.e., the relevant frequency channels may be determined using e.g., time-frequency methods like the wavelet transform [8, 16, 24]:

$$\begin{aligned} w(a,b) = \langle s; h_a \rangle &= \frac{1}{\sqrt{a}} \int s(t) h_a^* \left(\frac{t-b}{a} \right) dt \\ &= \sqrt{a} \int S(f) H^*(af) \exp(j2\pi f b)\, df \end{aligned} \tag{9.1}$$

with $s(t)$ representing the signal to be analyzed and $h_a(t)$ modified versions of the basis wavelet $h(t)$. Independent variables are the scale parameter a, which is inversely related to frequency, and the shift parameter b.

The Morlet wavelet

$$h(t) = \exp\left(-\frac{t^2}{2} + j\omega_0 t \right) \quad \text{with} \quad \omega_0 = 5.33, \tag{9.2}$$

which is a modulated Gaussian function, is frequently used as basis wavelet. From a theoretical point-of-view, it has the best time-frequency resolution and it closely resembles biological signals.

In Fig. 9.2, the continuous wavelet transform of the grand average ERP shown in Fig. 9.1 is presented. Distinct components in the time-frequency plane can clearly

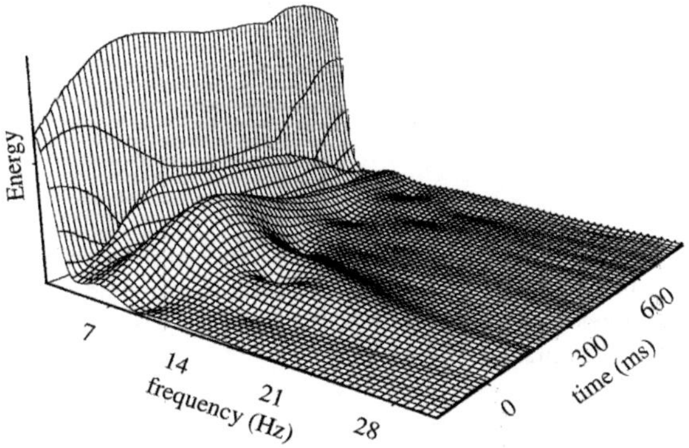

Fig. 9.2 Wavelet transform (scalogram) of the grand average ERP shown in Fig. 9.1. A Morlet wavelet was used. Different areas of the time-frequency plane contribute to the signal

be seen in the plot, e.g., a smaller one in the beta band (13–30 Hz), lasting from 0 to about 250 ms or two lower-frequency (delta, sub-delta) components lasting over the whole poststimulus period presented.

But nevertheless, parametrization has to be done by hand when using the wavelet transform. A more elegant way to determine the event-related oscillations are so-called adaptive time-frequency representations. Matching pursuits [9] and wavelet networks [23, 28] are examples for adaptive time-frequency representations which are closely related.

9.2 Wavelet Networks

9.2.1 Topology

Figure 9.3 shows the structure of a wavelet network for signal representation.[1] At the bottom, a time value t is fed into the network. The network produces an output value $\hat{s}(t)$ which is computed as the weighted sum of time-frequency components:

$$\hat{s}(t) = \sum_{k=1}^{K} w_k h\left(\frac{t - b_k}{a_k}\right) \tag{9.3}$$

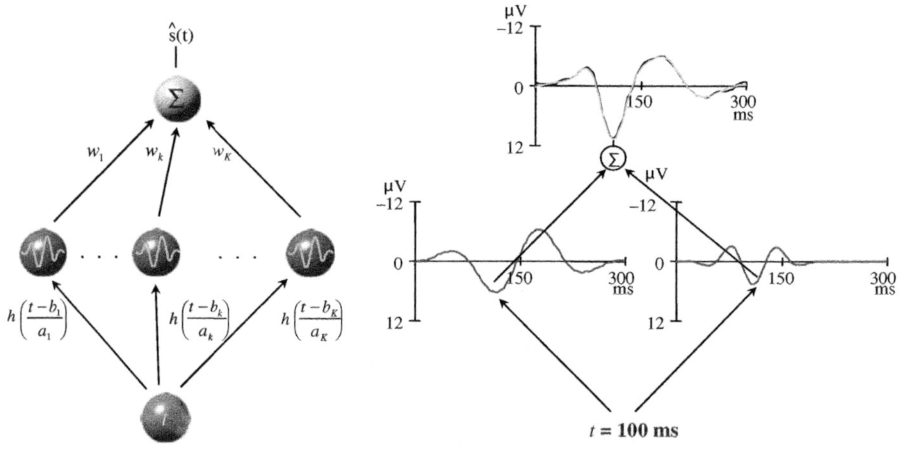

Fig. 9.3 *Left*: Topology of a wavelet network (WN). A WN can be interpreted as a one hidden-layer perceptron whose nodes in the hidden layer are represented by modified wavelet functions. *Right*: A visual evoked potential represented by a WN consisting of two nodes. Exemplarily, an input value $t = 100$ ms is illustrated

[1] In this article, only wavelet networks for signal representation are considered. For a methodological introduction and application of wavelet networks for classification we refer to Szu et al. [23] and Dickhaus and Heinrich [7].

Each wavelet node is characterized by a shift parameter b_k describing the centre of the node, a scale parameter a_k, which is responsible for the node's frequency as well as its temporal and spectral spread, and a weight w_k representing a node's contribution to the signal. $h(t)$ denotes the basis wavelet.

On the right side of the figure a wavelet network solution of an averaged visual evoked potential (VEP) is shown. The network consists of two nodes (time-frequency components). The first node lasts over the whole 300 ms post-stimulus time presented. The second, higher-frequency component is mainly located within the first 150 ms after stimulus onset. The sum of these two wavelet nodes approximates the evoked potential quite well. The wavelet nodes are modelled as time- and frequency modified versions of the Morlet wavelet:

$$\begin{aligned} \hat{s}(t) &= \sum_{k=-K,k\neq 0}^{K} w_k \exp\left[-j\omega_k\left(\tfrac{t-b_k}{a_k}\right)\right]\exp\left[-0.5\left(\tfrac{t-b_k}{a_k}\right)^2\right] \\ &= \sum_{k=1}^{K}\left[w_{\cos,k}\cos\left(\omega_k\tfrac{t-b_k}{a_k}\right)+w_{\sin,k}\sin\left(\omega_k\tfrac{t-b_k}{a_k}\right)\right]\exp\left[-0.5\left(\tfrac{t-b_k}{a_k}\right)^2\right] \end{aligned} \tag{9.4}$$

Using a variable frequency parameter ω_k, which is contrast to the wavelet transform, results in a better approximation with a smaller number of parameters.

9.2.2 Training Algorithms

Adapting a wavelet network to a given signal is a non-linear optimization problem and can be done in an iterative optimization process using a least-square (LSQ) error approach. The error E between the networks desired output $s(t)$ and its actual output $\hat{s}(t)$ has to be minimized:

$$E = \frac{1}{2}\sum_{i=1}^{N}[s(t_i)-\hat{s}(t_i)]^2 \stackrel{!}{=} \min \tag{9.5}$$

A WN can be regarded as a one hidden-layer perceptron. So, the backprop algorithm [21] or related procedures (e.g., quickprop algorithm, [10]) may be used. In each iteration, partial derivatives have to be calculated:

$$\frac{\partial E}{\partial w_{\cos,k}} = -\sum_{i=1}^{N}[s(t_i)-\hat{s}(t_i)]\cos\left(\omega_k\frac{t_i-b_k}{a_k}\right)\exp\left[-0.5\left(\frac{t_i-b_k}{a_k}\right)^2\right] \tag{9.6a}$$

$$\frac{\partial E}{\partial w_{\sin,k}} = -\sum_{i=1}^{N}[s(t_i)-\hat{s}(t_i)]\sin\left(\omega_k\frac{t_i-b_k}{a_k}\right)\exp\left[-0.5\left(\frac{t_i-b_k}{a_k}\right)^2\right] \tag{9.6b}$$

$$\frac{\partial E}{\partial \omega_k} = -\sum_{i=1}^{N} [s(t_i) - \hat{s}(t_i)] \exp\left[-0.5\left(\frac{t_i - b_k}{a_k}\right)^2\right]\left(\frac{t - b_k}{a_k}\right) \times \left(-w_{\cos,k} \sin\left(\omega_k \frac{t_i - b_k}{a_k}\right) + w_{\sin,k} \cos\left(\omega_k \frac{t_i - b_k}{a_k}\right)\right) \tag{9.6c}$$

$$\frac{\partial E}{\partial b_k} = -\sum_{i=1}^{N} [s(t_i) - \hat{s}(t_i)] \exp\left[-0.5\left(\frac{t_i - b_k}{a_k}\right)^2\right]\frac{1}{a_k} \times \left(w_{\cos,k}\left[\frac{t_i - b_k}{a_k} \cos\left(\omega \frac{t_i - b_k}{a_k}\right) + \omega_k \sin\left(\omega_k \frac{t_i - b_k}{a_k}\right)\right] + w_{\sin,k}\left[\frac{t_i - b_k}{a_k} \sin\left(\omega_k \frac{t_i - b_k}{a_k}\right) - \omega_k \cos\left(\omega_k \frac{t_i - b_k}{a_k}\right)\right]\right) \tag{9.6d}$$

$$\frac{\partial E}{\partial a_k^{-1}} = -\sum_{i=1}^{N} [s(t_i) - \hat{s}(t_i)] \exp\left[-0.5\left(\frac{t_i - b_k}{a_k}\right)^2\right](t_i - b_k) \times \left(-w_{\cos,k}\left[\frac{t_i - b_k}{a_k} \cos\left(\omega_k \frac{t_i - b_k}{a_k}\right) - \omega_k \sin\left(\omega_k \frac{t_i - b_k}{a_k}\right)\right] - w_{\sin,k}\left[\frac{t_i - b_k}{a_k} \sin\left(\omega_k \frac{t_i - b_k}{a_k}\right) + \omega_k \cos\left(\omega_k \frac{t_i - b_k}{a_k}\right)\right]\right) \tag{9.6e}$$

However, training all nodes simultaneously causes nodes to compete for high-energy signal components since the nodes cannot communicate. To avoid this 'herd effect', a recursive strategy may be applied [10]; i.e., at the kth stage of the recursive algorithm the node n_k is trained on the residual signal $e_k(t)$ which is the difference of the signal $s(t)$ and the nodes n_j ($j = 1, \ldots, k-1$) that have already been introduced. Thus, each node is focused on different signal characteristics, still possibly overlapping in the time-frequency plane. This procedure is repeated until a user-defined criterion has been fulfilled, e.g., the residual error is below a certain threshold.

Since a LSQ approach tends to neglect higher-frequency components with low energy, a criterion called 'error-of-the-gradient' may also be considered for the nodes [22]:

$$E_{grad,k} = \frac{1}{2}\sum_{i=1}^{N}\left[\left.\frac{\partial e_k(t)}{\partial t}\right|_{t_i} - \left.\frac{\partial n_k(t)}{\partial t}\right|_{t_i}\right]^2 \stackrel{!}{=} \min \tag{9.7}$$

In order to find the optimal parameter values for a signal (without noise), it is recommended to train a pool of candidate nodes, i.e., nodes with different random initial parameters spread over the time-frequency plane. The candidate node with the minimum error is chosen. After the recursive training, a final, simultaneous update of the wavelet nodes can be done.

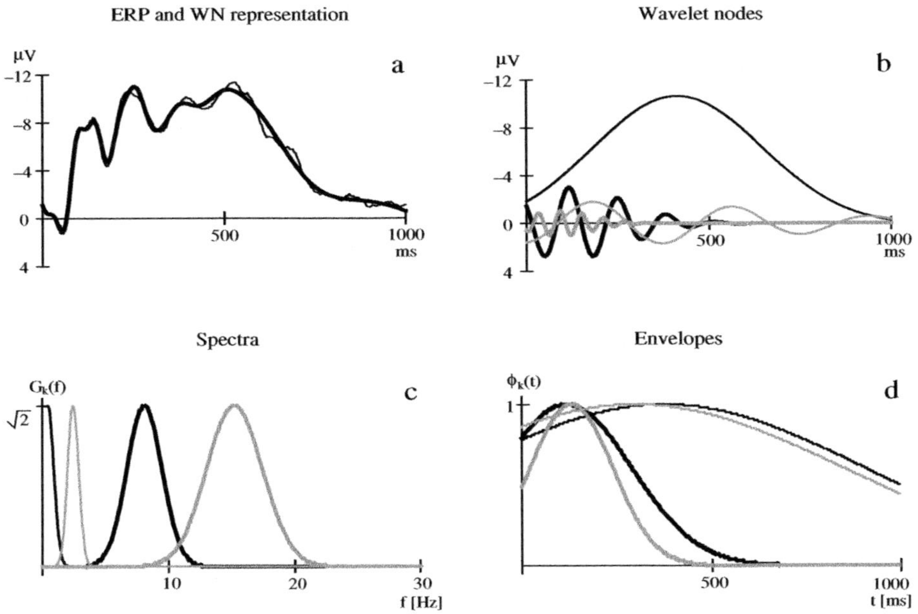

Fig. 9.4 WN representation of a grand average ERP: **(a)** grand average (*thick line*) and output a WN (4 nodes) adapted to the signal (*thick line*). **(b)** time courses of the four wavelet nodes, **(c)** the nodes' spectral functions (normalized) and **(d)** the nodes' envelopes. Filter functions and time-windows for ERP single-trial analysis may be derived from the spectral functions and envelopes in **(c)** and **(d)**, respectively

Figure 9.4a shows the WN approximation for the grand average ERP in Fig. 9.1 using the recursive training algorithm. The thin line curve represents the original channels and the thick line curve shows the WN output. The networks consists of four nodes which are plotted in Fig. 9.4b.

A dominating low-frequency component lasts over the complete 1000 ms presented. Higher-frequency components (thick-line curves) occur in the first half of the signal. Looking at the frequency characteristics of the nodes (see Fig. 9.4c), the different components can be assigned to different EEG frequency bands: the grey, thick-line component to the beta band, the black, thick-line component to the alpha band, the grey, thin-line component to the delta band and the black, thin-line component to the slow-delta band.

If noise is present in the signals, the algorithm has to be modified to achieve a reliable parametrization.

In [13], we introduced an algorithm adapted to low signal-to-noise ratios (SNR) so that it can be applied in ERP single-trial analysis:

- It is assumed that the time-frequency energy distribution of an ERP in a single trial extends over the same parts of the time-frequency plane as the distribution of an averaged ERP. Hence, WNs with the same number of nodes are trained for single trials.

- Further, each node has to represent similar time-frequency characteristics as the corresponding node of an averaged ERP. This can be achieved by projecting the residual signal $e_k(t)$ by means of filtering and tapering onto the relevant part of the time-frequency plane. We suggest to define Gaussian functions as tapering windows $\phi_k(t)$ and filter functions $G_k(f)$ and impulse response $g_k(f)$, respectively:

$$\phi_k\left(t_j\right) = \exp\left[-0.25\left(\frac{t_j - b_{REF_k}}{a_{REF_k}}\right)^2\right] \tag{9.8a}$$

$$g_k\left(t_j\right) = \frac{2\Delta t}{a_{REF_k}\sqrt{\pi}}\exp\left[-0.5\left(\frac{t_j - b_{REF_k}}{a_{REF_k}}\right)^2\right]\cos\left(\omega_{REF_k}\frac{t_j - b_{REF_k}}{a_{REF_k}}\right)$$

$$\circ\!\!-\!\!\bullet \quad G_k\left(f_j\right) = \sqrt{2}\exp\left[-2\pi^2\left(a_{REF}f_j - \frac{\omega_{REF_k}}{2\pi}\right)^2\right] + \sqrt{2}\exp\left[-2\pi^2\left(a_{REF}f_j + \frac{\omega_{REF_k}}{2\pi}\right)^2\right] \tag{9.8b}$$

The parameter ϕ_{REFk}, a_{REFk}, b_{REFk} can be chosen according to the WN parameters obtained for a representative signal (e.g., the grand average). Applying this projection operation, we obtain:

$$x_k\left(t_j\right) = \phi_k\left(t_j\right)\left(g_k\left(t_j\right)^* e_k\left(t_j\right)\right) = \phi_k\left(t_j\right)\left(\sum_m g_k\left(\tau_m\right)e_k\left(t_j - \tau_m\right)\right), \tag{9.9}$$

where '$*$' denotes convolution.

After preprocessing, a node is trained as described above. The Gaussian functions in Eqs. 9.8a and 9.8b were chosen for the following reason: If a node's output in a Morlet WN $n_k t$ is multiplied by its slightly modified envelope $\phi_k(t)$ and filtered by a normalized version of the real part of its spectral function $G_k(f)$, the output of these processing steps is identical to the node's original time course:

$$n_k\left(t_j\right) = \phi_k\left(t_j\right)\left(g_k\left(t_j\right) * n_k\left(t_j\right)\right) \tag{9.10}$$

Thus, these filtering and tapering preprocessing steps enhance the synchronized ERP components and attenuate the spontaneous activity. The WN nodes can still vary in a considerable way. The shift of a node b_k that represents a slow wave component can differ within a several hundred milliseconds interval for example. The weights $w_{cos,k}$ and $w_{sin,k}$ are not restricted at all.

For initialisation, the WN parameters of the representative signal can be chosen. On a PC with a Pentium-IV processor (1.9 GHz), adapting a WN consisting of four wavelet nodes to an ERP single trial (1000 ms; sampling frequency: 250 Hz) takes less than two seconds (500 iterations per node; 50 iterations of simultaneous fine-tuning).

9.3 Examples of an ERP Study on Children with Attention-Deficit/Hyperactivity Disorder

In this section, examples from a clinical ERP study on children with attention-deficit/hyperactivity disorder (ADHD) are presented. Children with ADHD are characterized by developmentally inappropriate levels of inattention, hyperactivity and impulsivity. ERP studies have contributed to a better understanding of the pathophysiological background of this common child and adolescent psychiatric disorder (prevalence: 3–5%), [1, 20].

25 typically developing children and 25 children with ADHD (age: 8-15 years) were included in the study. ERP signals were elicited using the auditory selective attention task which is described in the legend of Fig. 9.1. In all examples, target-attended (T-A) ERPs were considered. More details about the study can be found in [12–14].

9.3.1 Analysis of Averaged ERPs

In Fig. 9.5, the WN solutions for the averaged ERPs shown in Fig. 9.1 are presented. In each subject, a WN consisting of four nodes accounts for more than 95% of the signal energy. The time courses of the wavelet nodes are plotted on the right part of Fig. 9.5. For each subject, components with comparable time-frequency characteristics are obtained.

Visual inspection of the control group's and the ADHD group's grand average indicated that the N1 amplitude over left fronto-central electrodes could be larger for typically developing children in comparison to children with ADHD. But statistics did not reveal any effect for the N1 amplitude. The wavelet node related to the alpha band ('alpha node'; see black, thick-line curves in Fig. 9.4) surely contributes most to the N1. Calculating statistics for the amplitude of the alpha node, a trend for the group x electrode interaction was obtained.

This example illustrates that the WN approach allowing a more uniform parametrization is a more sensitive tool for ERP analysis than a conventional analysis.

9.3.2 ERP Single-Trial Analysis

What can the analysis of event-related potentials at the level of single trials be good for? Trial-to-trial variations are neglected, get smeared or even lost by the averaging process. But trial-to-trial variations may reflect important aspects of cognitive processing [11].

In this section, two examples using WNs for ERP single-trial analysis are presented. The first example addresses the question whether the effect obtained for the alpha wavelet node in averaged ERPs is based on an amplitude or phase effect in

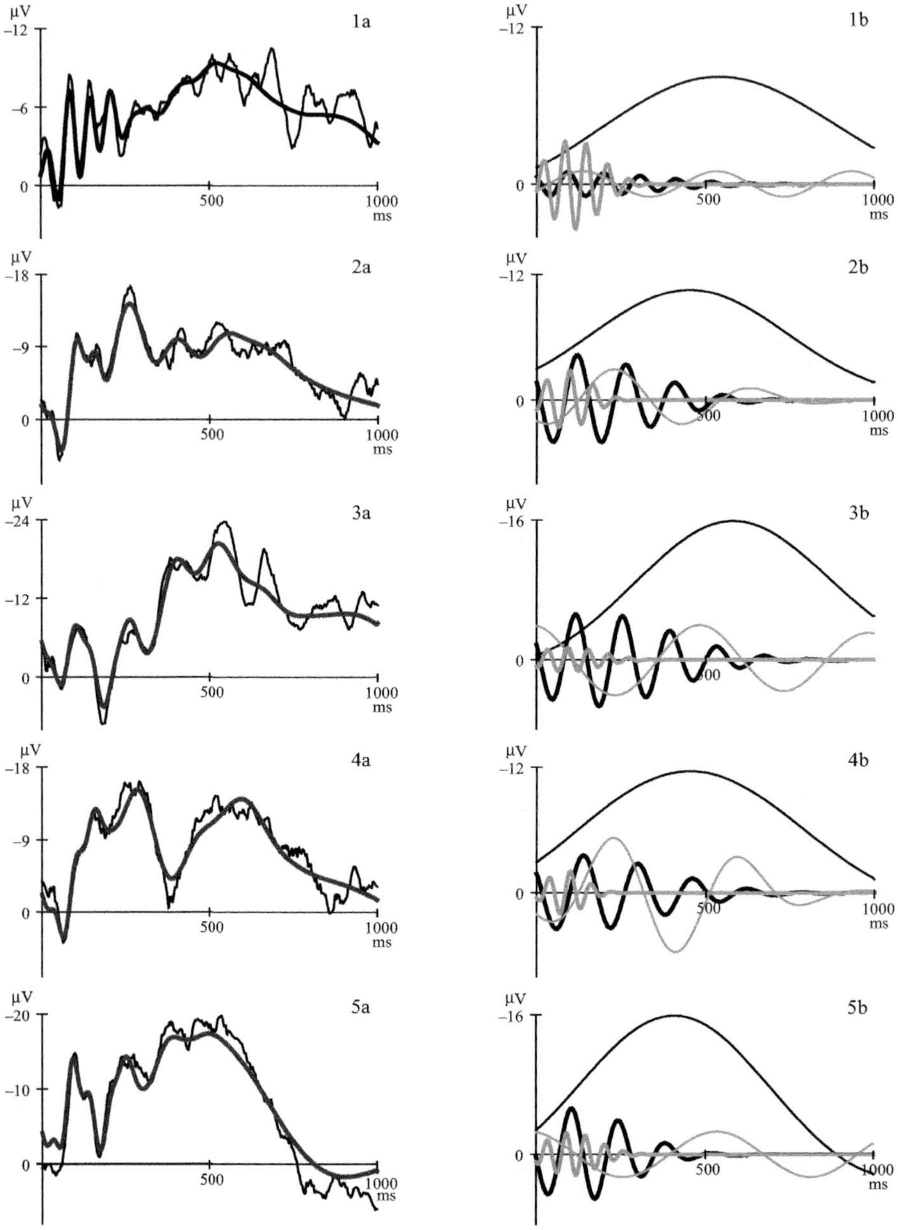

Fig. 9.5 WN solutions of the averaged ERPs in Fig. 9.1 (target-attended responses in a selective attention task; typically developing children). *Left*: ERP-signal (*thin line*) and WN output (*thick line*). *Right*: Time courses of the four wavelet nodes. For each subject, nodes with comparable time-frequency characteristics are obtained

single trials. The second example deals with time-on-task analysis in order to study the attentional problems of children with ADHD over time during completion of a cognitive task.

9.3.2.1 Amplitude vs. Phase Effect

In [13], we applied the training algorithm as described in Sect. 9.2.2 using the time-windows and filter functions shown in Fig. 9.4c,d. In the left part of Fig. 9.6, 15 consecutive single trials of a representative boy (thin line curves) and the corresponding WN estimates of the ERP component (thick line curves) are shown. Averaging the WN single-trial solutions, a signal is obtained which closely resembles the averaged ERP. This fact indicates that a WN single-sweep estimate actually represents the ERP component in a single trial.

Looking at the alpha wavelet node, which is obtained for the single trials (see Fig. 9.6 right), it is evident that the phase, e.g., represented by zero crossings, is quite stable over all trials. The training algorithm has no restrictions concerning phase. So, this finding provides further evidence that the event-related components are actually pronounced by the WN single-trial approach.

We calculated a correlation feature to parametrize the similarity of the alpha nodes over all single trials. We subjected this correlation measure reflecting phase stability as well as the amplitude of the node to statistical analysis to compare the ADHD group with the control group. Only for the correlation feature a significant group x electrode interaction was found. So, the effect observed in the averaged ERPs is not due to an increase in power in the alpha range after stimulus presentation but due to a more stable appearance of synchronized activity (phase effect).

The phase synchronisation over left fronto-central electrodes in the alpha range in typically developing children may reflect a sensori-motor integrative process [26]. This brain dynamic effect seems to be disturbed in children with ADHD.

9.3.2.2 Time-On-Task Analysis

In [14], we investigated whether time-on-task effects, i.e., systematic variations as a function of time, could be different in typically developing children and children with ADHD.

We focused on the low-delta wavelet node (see black, thin-line curve in Fig. 9.4b). At frontal sites this component represents a negative wave, at parietal sites it reflects a positive wave. Therefore, we use the terms ‘frontal negativity’ and ‘parietal positivity’.

In order to study brain-behaviour interactions, we also considered the time-on-task behaviour of measures at the behavioural level (omission errors, reaction times; see Fig. 9.7). Until the 6th T-A trial which correspond to about 45 s time on task, typically developing children and children with ADHD did not differ significantly with respect to omission errors. Then, a significant increase of omission errors was observed in the ADHD group as indicated in the intra-group comparison which also resulted in a significant inter-group effect.

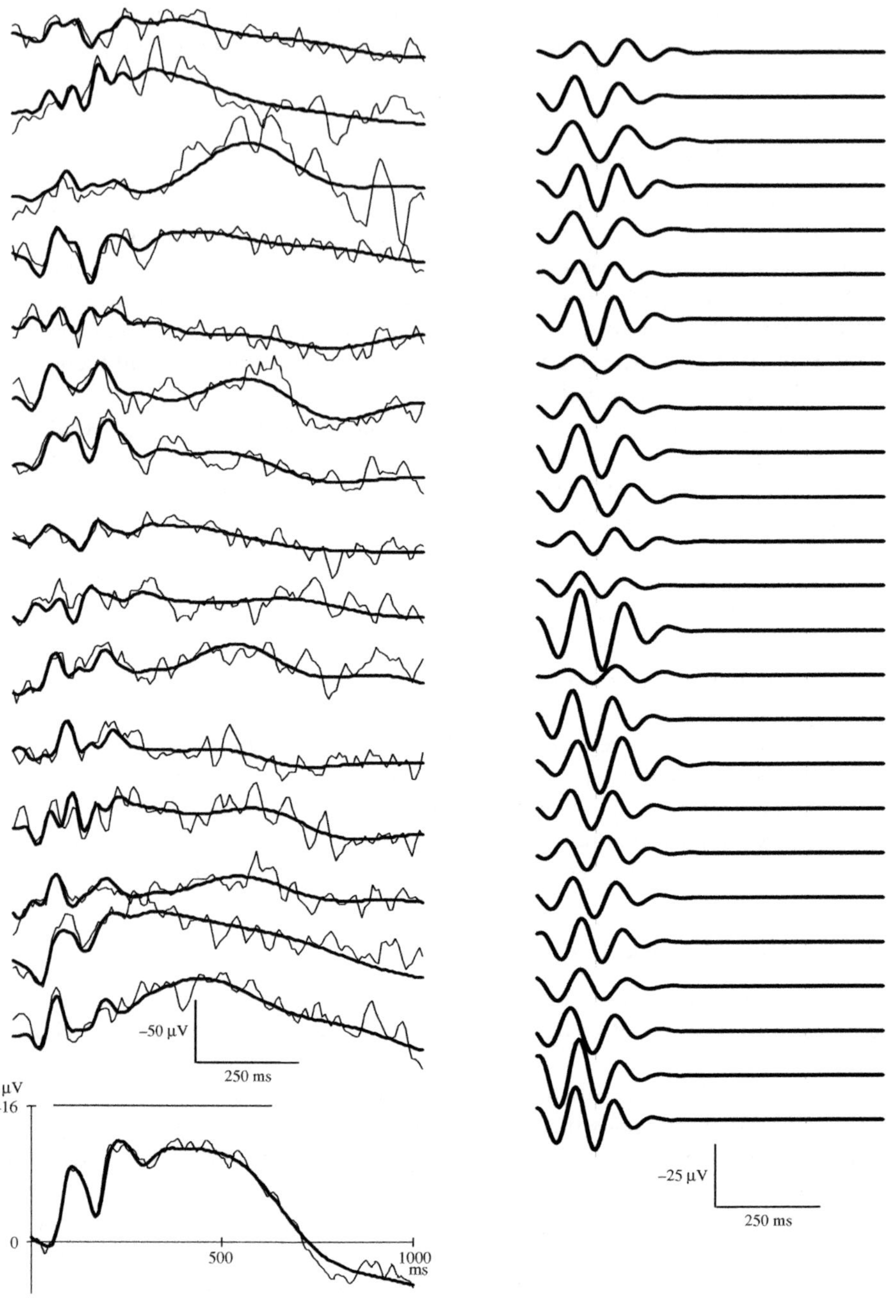

Fig. 9.6 *Left*: 15 consecutive single trials (target-attended responses, electrode Fz) of a typically developing boy recorded in a selective attention task (*thin line*) and the corresponding WN estimations (*thick line*). The average of the single-trial estimates describes the averaged ERP quite well (*bottom*). *Right*: The wavelet nodes related to the alpha frequency range, which were obtained for 25 consecutive single trials for this child, show quite a stable phase over all trials

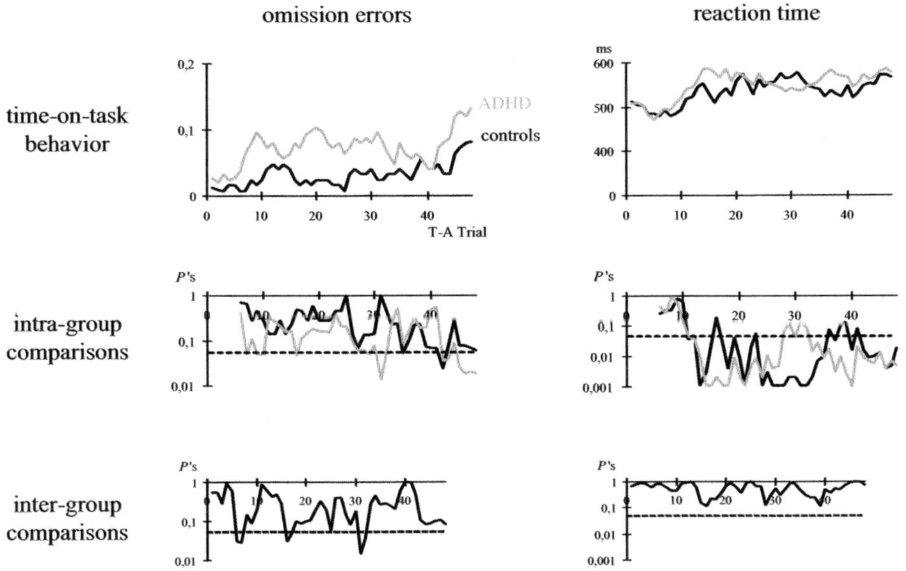

Fig. 9.7 Time-on-task analysis of omission errors (*left*) and reaction time (*right*) in typically developing children (*black curves*) and children with attention-deficit/hyperactivity disorder (ADHD, *grey curves*). *Top*: Time-on-task behaviour with respect to the 48 target-attended trials a selective attention task. Application of 10 target-attended (T-A) trials took about one minute. *Middle*: intra-group comparisons (relating a trial to a baseline value, i.e., the average of the first five trials). *Bottom*: inter-group comparisons. $P = 0.05$ level is marked by dashed lines

Between the 10th and the 40th T-A trial, smaller fluctuations in both groups did not lead to consistent effects. After the 40th T-A trial, the number of omissions was higher in both groups compared to the baseline value at the beginning of the task (intra-group comparisons: ADHD: P's < 0.05; Controls: P's < 0.1).

Results of the reaction time analysis are shown on the right. In both groups, reaction times were shortest in the beginning of the task. A significant increase of about 70 ms was observed after the 10th T-A trial. Then, reaction times did not show substantial variations. The inter-group comparisons did not reveal any significant reaction time differences between controls and children with ADHD.

In Fig. 9.8, the corresponding diagrams for the ERP measures - frontal negativity on the left side and parietal positivity on the right side - are presented.

In children with ADHD, *frontal negativity* increased significantly around the 10th T-A trial (intra-group comparison: P's < 0.01). Afterwards, smaller quasi-periodic fluctuations could be observed. Time-on-task behaviour of frontal negativity in typically developing children differed considerably from that in children with ADHD. The increase started later and lasted until the 30th T-A trial where frontal negativity was maximal, resulting in a highly significant intra-group effect (P's < 0.001).

After the 30th T-A trial, frontal negativity decreased to initial values, i.e., frontal negativity in the control group showed a quadratic course over time-on-task.

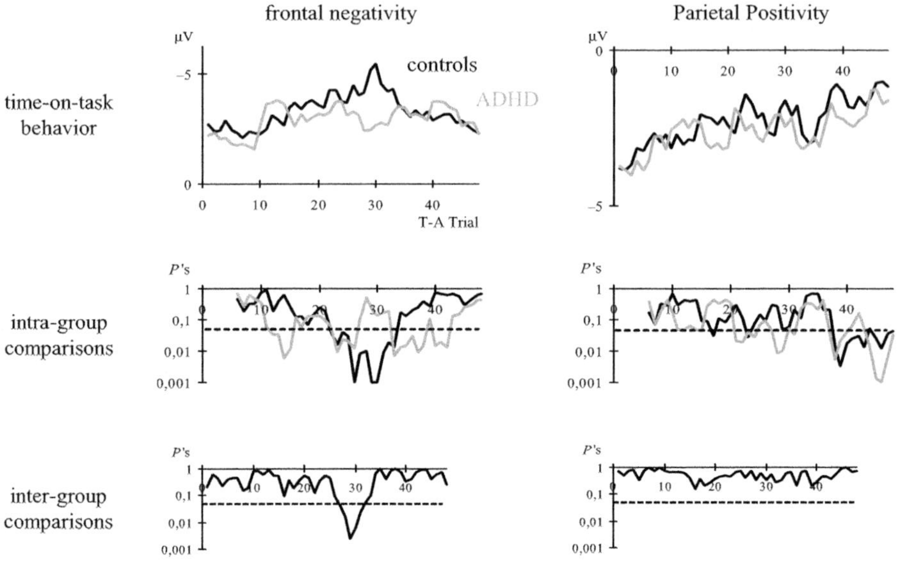

Fig. 9.8 Time-on-task analysis of frontal negativity (*left*) and parietal positivity (*right*) in typically developing children (*black curves*) and children with ADHD (*grey curves*). The composition of this figure is the same as in Fig. 9.7

Inter-group comparisons showed that around the 30th T-A trial frontal negativity was larger in controls than in children with ADHD.

A frontal negative wave Nc in children has been enhanced by processing of attended relative to nonattended stimuli [5] as well as by novel stimuli getting attention for further categorization [6]. Thus, the frontal negativity might be related to the Nc wave. So, the increase in frontal negativity, observed in both groups to T-A stimuli might indicate that more attentional resources had to be allocated for an adequate performance with increasing time on task.

In both groups, highest parietal positivity was found in the beginning of the task. An almost linear decrease of parietal positivity with smaller fluctuations riding on the line could be observed with increasing time-on-task. No group differences between controls and children with ADHD concerning time-on-task behaviour were obtained.

Summarizing the results, both behavioural and ERP measures showed distinct temporal dynamics. Time-on-task effects were not only linear, but also of higher order and started after less than one minute. Typically developing children could allocate more attentional resources during the course of the experiment. For children with ADHD, earlier time-on-task effects, i.e., an earlier increase of omission errors and frontal negativity, resulted.

In contrast to a distinct quadratic course of frontal negativity in the control group, smaller higher-order fluctuations were present in the ADHD group which could be related to shorter attention spans and a generally fluctuating cognitive behaviour, respectively [19]. So, these findings could reflect the attentional problems of children with ADHD. WN-based time-on-task analysis at the level of single trials is capable to provide a more differentiated view on the core deficits in neuropsychiatric disorders like ADHD.

9.4 Discussion and Further Perspectives

Analysing event-related oscillations with wavelet networks is a potential tool for studying neurocognitive functioning. The approach is compatible with the ERP model of [3] and, thus, closer to underlying neurophysiological processes than a phenomenologically oriented analysis of prominent peaks. But nevertheless, it has to be kept in mind that WNs are just a mathematical model.

WNs can be applied for ERP single-trial analysis to investigate trial-to-trial variabilities. In order to deal with the low SNR in ERP single trials, we propose to include preprocessing steps (i.e., projection onto a specific part of the time-frequency plane by tapering and filtering) in the recursive training process. The assumptions and constraints associated with these preprocessing steps concerning ERP generation are weaker than in other single-trial approaches [2, 15, 27]. Particularly, it is not assumed that the characteristics of the spontaneous activity does not change after stimulus presentation. However, taking advantage of an averaged signal, means that only the components which are tightly or loosely phase-locked can be estimated. The stimulus-locked, but not phase locked are dismissed.

Bénar et al. [4], who use a comparable time-frequency representation for single-trial analysis, avoid using the grand average. They introduce a cost function to minimize the dispersion of the wavelet parameters around their mean and to avoids over-fitting of noise. Another option could be to use a dictionary as it is done for matching pursuits [9].

In this article, WNs were trained for one-dimensional signals. However, the approach could easily be adapted to multichannel analysis by introducing a weight vector $\mathbf{W_k}$ into Eq. 9.3:

$$\hat{\mathbf{S}}(t) = \sum_{k=1}^{K} \mathbf{W_k} h\left(\frac{t-b_k}{a_k}\right). \tag{9.11}$$

with the vector $\hat{\mathbf{S}}(t) = (\hat{s}_1(t), \hat{s}_2(t), \ldots, \hat{s}_N(t))$ representing the network's output values for the N electrodes.

These options could help to extend and improve the wavelet network approach for ERP represention and parametrization, particularly for the analysis of single trials.

Acknowledgments Figures 9.3, 9.7 and 9.8 are reprinted from [14] with permission from Elsevier. Figures 9.4 and 9.6 are reprinted from [13] with permission from IEEE.

References

1. Banaschewski T, Brandeis D (2007) Annotation: what electrical brain activity tells us about brain function that other techniques cannot tell us – a child psychiatric perspective. J Child Psychol Psychiatry 48(5):415–435
2. Bartnik EA, Blinowska KJ, Durka PJ (1992) Single evoked potential reconstruction by means of wavelet transform. Biol Cybern 67:175–181
3. Basar E, Basar-Eroglu C, Demiralp T, Schürmann M (1995) Time and frequency analysis of the brain's distributed gamma-band system. IEEE EMB Mag 14:400–410
4. Bénar C, Clerc M, Papadopoulo T (2007) Adaptive time-frequency models for single-trial M/EEG analysis. Inf Process Med Imaging 20:458–469
5. Ciesielski KT, Courchesne E, Elmasian R (1990) Effects of focused selective attention tasks on event-related potentials in autistic and normal individuals. Electroenceph clin Neurophysiol 75:207–220
6. Courchesne E (1978) Neurophysiological correlates of cognitive development: changes in long-latency event-related potentials from childhood to adulthood. Electroenceph Clin Neurophysiol 45:468–482
7. Dickhaus H, Heinrich H (1996) Classifying biosignals with wavelet networks. IEEE EMB Mag 15(5):103–111
8. Dickhaus H, Khadra L, Brachmann J (1994) Time-frequency analysis of ventricular late potentials. Meth Inf Med 33:187–195
9. Durka PJ, Blinowska (1998) In pursuit of time-frequency representation of brain signals. In: Akay M (ed.): Time-frequency and wavelets in biomedical engineering. IEEE Press, New York
10. Fahlman SE (1988) An empirical study of learning speed in back-propagation networks. Tech Report CMU-CS-88-162. School of Computer Science, Carnegie Mellon University, Pittsburgh
11. Ford JM, White P, Lim KO, Pfefferbaum A (1994) Schizophrenics have fewer and smaller P300s: a single trial analysis. Biol Psychiatry 35:96–103
12. Heinrich H, Dickhaus H (1998) Analysis of evoked potentials using wavelet networks. In: Akay M (Hrsg.): Time frequency and wavelets in biomedical signal processing. IEEE Press, New York
13. Heinrich H, Dickhaus H, Rothenberger A, Heinrich V, Moll GH (1999) Single-sweep analysis of event-related potentials by wavelet networks-methodological basis and clinical application. IEEE Trans Biomed Eng 46(7):867–879
14. Heinrich H, Moll GH, Dickhaus H, Kolev V, Yordanova J, Rothenberger A (2001) Time-on-task analysis using wavelet networks in an event-related potential study on attention deficit hyperactivity disorder. Clin Neurophysiol 112(7):1280–1287
15. Liberati D, Bertolini L, Colombo DC (1991) Parametric method for the detection of inter- and intrasweep variability in VEP processing. Med Biol Eng Comput 29:159–166
16. Mallat S (1989) A theory for multiresolution signal decomposition: the wavelet representation. IEEE Trans Patt Anal Mach Intell 14:710–732
17. Picton TW, Hillyard SA (1988). Endogenous event-related potentials. In: Picton TW (ed.), Handbook of electroencephalography and clinical neurophysiology (Vol. 3). Elsevier, Amsterdam
18. Polich J (1998) P300 clinical utility and control of variability. J Clin Neurophysiol 15:14–33
19. Rothenberger A (1995) Electrical brain activity in children with hyperkinetic syndrome: evidence of a frontocortical dysfunction. In: Seargant J (ed.): European approaches to hyperkinetic disorder. Trümpi, Zürich
20. Rothenberger A, Döpfner M, Sergeant J, Steinhausen HC (Eds.) (2004). ADHD beyond core symptoms – not only a European perspective. Eur Child Adolesc Psychiatry 13, Suppl. 1
21. Rumelhart D, Hinton GE, Williams RJ (1986) Learning representations by back-propagating errors. Nature 323:533–536

22. Sri-Jayantha M, Stengel RF (1988) Determination of nonlinear aerodynamic coefficients using the estimation-before-modeling method. J Aircraft 25:796–804
23. Szu HH, Telfer B, Kadambe S (1992) Neural network adaptive wavelets for signal representation and classification. Optical Engineering 31:1907–1916
24. Unser M, Aldroubi A (1996) A review of wavelets in biomedical applications. Proc IEEE 84:626–638
25. van der Stelt O, Belger A (2007) Application of electroencephalography to the study of cognitive and brain functions in schizophrenia. Schizophr Bull 33(4):955–970
26. Yordanova JY, Kolev VN (1996) Developmental changes in the alpha response system. Electroenceph Clin Neurophysiol 99:527–538
27. Yu K, McGillem CD (1983) Optimum filters for estimating evoked potentials waveforms. IEEE Trans Biomed Eng 30:730–737
28. Zhang Q, Benveniste A (1992) Wavelet Networks. IEEE Trans NN 3:889–898

Chapter 10
Detection of Evoked Potentials

Peter Husar

Abstract In biosignal processing it is exceedingly difficult to detect evoked potentials in an objective way. The main problem is that neither the biosignal nor the surrounding EEG fulfils any necessary condition concerning the properties of signal and noise known from the detection theory. Further, there is a strong nonlinear and time-variant relation between the stimulation sequence and the evoked response. Thus, for a reliable detection, optimization of both the stimulating series and the stimulus response is necessary. Although evoked responses to standard stimuli are well known in healthy subjects and can be found easily, responses of ill subjects are possibly completely different if present at all. Thus, correlation detector or matching filter will not work in pathological cases. The only way for detection is the energy detector and related methods. This chapter will show exemplarily on visual evoked potentials (VEP) how the stimulation and the detection can be combined suitably for detection. Several methods for an efficient enhancement of the SNR[1] are presented. One of the main tasks is to show how the signal properties of signal and noise can be improved in the sense of detection theory.

10.1 Evoked Potentials

10.1.1 Transient Evoked Potentials, tEP

The abbreviation tEP stands for transient evoked potentials that have sufficient time to come to an end. A typical course of a multichannel tEP is shown in Fig. 10.1.

It is assumed that the sensory system returns to its rest position between two stimuli. Therefore, all responses have theoretically the same initial conditions and consequently they should exhibit identical signal parameters. However, this assumption does not apply for real sensory systems. Thus, restrictions are to be considered for

P. Husar (✉)
Technische Universität Ilmenau, Institute of Biomedical Engineering and Informatics, Germany
e-mail: peter.husar@tu-ilmenau.de

[1] Signal-to-noise ratio

A. Naït-Ali (ed.), *Advanced Biosignal Processing*,
DOI 10.1007/978-3-540-89506-0_10, © Springer-Verlag Berlin Heidelberg 2009

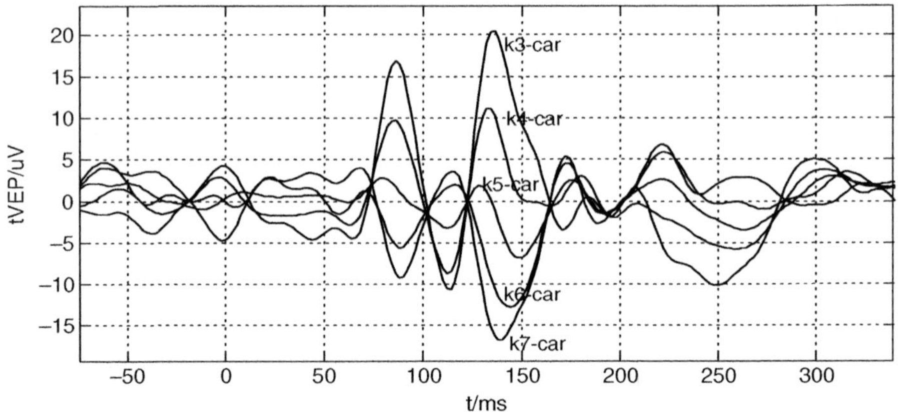

Fig. 10.1 Transient VEP: Local optical stimulation by an LED in the central visual field. The light pulse with a length of 20 ms appeared at $t = 0$. The EEG channels are centrally arranged (k_3 above Pz, k_7 above Oz) above the parietal and occipital cortex. The CAR (common average reference) was subtracted from each channel

an objective detection and analysis of the tEPs. The total period of the stimulation must not be so long that fatigue is caused. This demand means in practice that the time of investigation does not exceed a maximum period of 15 min. The temporal distance between repeated transient stimuli should not have a constant value and from a minimum distance (one second for a visual stimulation) it should vary accidentally in a broad range, if possible. This measure is important to avoid an adaptation of the sensory system and consequently the anticipative reaction of the patient. The signal shape of the tEPs is only classified and reproducible for few standardized investigation conditions. The visual stimulation techniques mostly used in neurophysiology are based on the cyclic color change of a checkerboard pattern and on a whole field flash. Numerous investigations were conducted by using these stimulation techniques. Therefore, it is possible to compare the results of physiologically and pathologically appropriate characteristics with each other relatively well. However, an exact analysis of the tEPs is not possible. Their signal shape depends on the parameters of the stimulation, the derivation and the signal processing. These dependencies are non-linear, non-monotonous and timely variable so that it is almost not possible to offer identical investigation conditions between the different laboratories. Some signal characteristics can be globally summed up for VEP: The first waves (Fig. 10.1) appear cortically after about 60 to 70 ms (N70); the maximum amplitude shows the wave (P100) between 90 and 160 ms and after this wave considerably slower ones are measured at times from 300 to 500 ms. The amplitudes of the waves are subject to strong fluctuations so that their use for diagnostic purposes is restricted. The latencies of the waves are much more stable and can be used very well as an indicator for the progression in therapies. This effect is due to the anatomy of the sensory systems: The wave latencies are mainly based on the almost constant propagation velocity of the stimuli along the nerve tracts and the locally stable condition of the latter. The amplitudes, however, clearly depend on specific derivation conditions that can never be exactly reproduced and are changing even during one measurement.

10.1.2 Steady-State Evoked Potentials, ssEP

For ssEP it is assumed that the stimulation period is so short that the sensory system cannot return to the rest position and the responses to the stimuli fuse in a periodic oscillation. The experts' opinions about the border between transient and steady-state responses differ significantly. According to my findings, sharp borders between the two conditions do not exist either in electro-neurophysiology. For VEP, the transition from transient to steady-state is smooth and is in the interval that ranges between about 150 and 500 ms of the stimulation period. Further effects known in electrophysiology are to be taken into consideration when selecting the stimulation rate or pulse width: A stimulus is principally any change of a parameter. Both the turning on and off of the light are visual stimulations. The responses to these on/off-stimulations can be proven separately, if their temporal distance is sufficiently long, about 500 ms or longer in practice. They fuse to one single response if their distance is sufficiently short, less than approximately 25 ms. If the temporal distance between the stimuli is within the specified range, they will suppress each other, and the amplitude will be reduced. Therefore, short and steep pulses are best suitable for stimulating the sensory systems. The period can be varied in the broad range according to the individual task. The conditions that apply for the signal characteristics of the ssEP are the same that apply for the tEP: The phase is more stable than the amplitude and therefore a reliable characteristic. But for higher stimulation rates, due to the 2π-periodicity, the phase measurement can result in ambiguity and consequently in false diagnostic conclusions. The amplitude is an instable characteristic and thus it can only be used as an approximated value. Compared to tEP, it must be expected for ssEP that the sensory system adjusts more rapidly to the periodic stimulation due to the considerably higher stimulation rate. This fact causes a rapid adaptation and a decrease of the amplitudes within the range of seconds. Therefore, ssEP only offer few advantages compared to tEP. The most important one is the possible use of spectral analyses for their evaluation. Whereas for tEP mostly after pathological changes the signal shape is not known in advance, for ssEP the stimulation rate indicates the point of the spectrum where the response to the stimuli is to be searched for.

10.1.3 Advanced Evoked Potentials, aEP

The comparison of tEP and ssEP can be proven more reliably because they only appear for the harmonics of the stimulation rate. However, the rapid adaptation of the sensory system to the periodic stimulus is a problem. Therefore, other possibilities for which the searched signal shape is known must be found. On the other hand, a new stimulation quality shall be presented to the sensory system continuously to prevent the adaptation. A periodic stimulation sequence with a linearly increasing stimulation rate, also known as linear chirp, is for example useful for this. From a signal-theoretical point of view, pseudo binary random sequences (PRBS) are also suited very well. The volunteer considers the sequence accidental and therefore said sequence offers continuously new information for the sensory system. To be able to use the signal-theoretical advantages of the PRBS it must be ensured that before the next stimulus the previous stimulus response has come to an end. Therefore, such stimulus sequences are only suited for analyzing early responses to stimuli (ERG, early AEP). They are not useful for cortical evoked responses because they prolong the measurement period that is anyway long. Multichannel PRBS (MLS, maximum length sequences) are suited for early responses to stimuli, for visual modes also known as multifocal stimulation (mfERG). But they are useless for analyzing cortical stimulus responses because they timely overlap and are non-linearly coupled with each other.

10.2 Basic Approaches for EP Detection

10.2.1 Preliminary Considerations

For constant or standardized investigation conditions, the expected waves with latencies of the EP are known. First, the EP can be detected by a statistic test of the expected value for a defined latency. For this purpose, a simple additive signal model according to Eq. (10.1) is assumed. Here, the index i represents a temporal or a spatial variable. As the EP have very low potentials, an increase of the SNR is required before the detection, normally by averaging Eqs. (10.2, 10.3). On the one hand, the averaging can be made synchronously with the stimulation over single realizations, similar to ensemble averaging. On the other hand, it can also be performed simultaneously over several EEG channels. This method is more identical with a true ensemble averaging. Furthermore, both approaches can be combined to the spatio-temporal averaging method. For more details, please refer to chapter "Methods of signal processing" [1].

$$\begin{array}{lll} H_0: & x_i = n_i, & i = 1, \ldots, n \\ H_1: & x_i = n_i + s_i & i = 1, \ldots, n \end{array} \tag{10.1}$$

H_0, H_1 are the zero and alternative hypotheses, n is the noise, if not otherwise indicated it is an "i.i.d.[2] process", s is the signal to be detected, i is the temporal or spatial index.

[2]Independent identically distributed

$$\bar{x} = \frac{1}{N}\sum_{i=1}^{N} x_i \tag{10.2}$$

$$\operatorname{var}(\bar{x}) = \operatorname{var}(s) + \frac{1}{N}\operatorname{var}(n) \tag{10.3}$$

The detection can be performed by a test using one sample or two samples. If one sample is used, the empiric average value will be tested for zero. If two samples are used, it will be tested for the zero difference between the spontaneous EEG and an EEG with embedded stimulation response. For example, the wave P100 of a VEP can be tested for the validity of the null hypothesis H_0: $\mu = 0$.

If a sufficient number of investigations about the amplitude of P100 have been performed, the empiric distribution can also be generated for the alternative hypothesis H_1: $\mu > 0$ (Fig. 10.2). The ROC[3] can be developed from the distributions for both hypotheses and the detection threshold can be determined according to predefined uncertainties α (false alarm) or β (miss). In practical use, the sample size must be sufficiently large for a t-test ($n > 30$) to obtain an approximately Gaussian-distributed test value according to the central limit theorem. Otherwise, a rank sum test should be used. This simple detection method gives a binary decision for the existence of an EP wave at a certain point in time. This test will be adequate, if the correct alternative hypothesis H_1 is confirmed. However, a missing wave cannot be clearly taken as granted, if the null hypothesis H_0 is accepted. Pathological changes in the sensory system normally cause a reduction of the amplitude and a longer latency of the EP waves. As the detection is a pre-stage of classification,

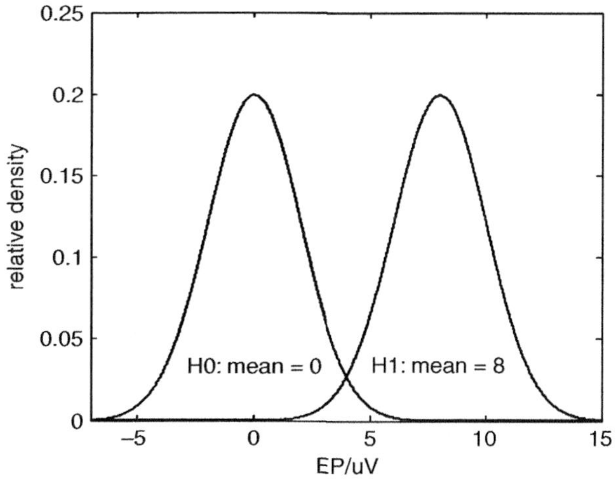

Fig. 10.2 Distributions of EEG without EP ($\mu = 0\ \mu$V) and EEG with EP ($\mu = 8\ \mu$V) at 100 ms after transient visual stimulation. Data were simulated for hypothesis testing

[3]Receiver operating characteristic

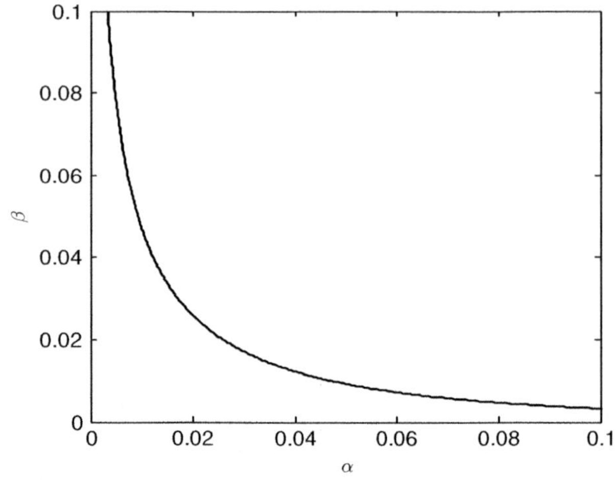

Fig. 10.3 ROC for data shown in Fig. 10.2. For standard false alarm α of 5% miss β reaches less than 1% and the test power more than 99%

pathologically changed waves should also be identified. Therefore, a statistical test for punctually selected latencies will not be sufficient. Tests with several samples or a multivariate analysis will be required. As the empiric distributions for alternative hypotheses $H_1: \mu > 0$ or $H_1: \sigma_2 > \sigma_1$ of pathologically changed waves are not known, the detection is reduced to a test of the null hypothesis. The error β (miss) is not known and cannot be determined. All the more, attention must be paid to the selection of an appropriate error probability α. If the value of α is assumed too low, the error β will increase to an unknown extent and vice versa. In the functional diagnostics based on electrophysiology false positive detections are considered more dangerous than false negative ones. Therefore, in practical signal analysis α should be considered considerably lower than 5% that are typical in biostatistics.

10.2.2 Correlation Detector

Some laboratories have databases of measured values that have been obtained under constant or standardized investigation conditions. These data can be taken to gain templates for proving EP. They could be created as spatial, temporal or spatio-temporal models. The templates are used to determine the degree of the linear connection between the template and the analyzed signal via correlation. Actually, the similarity of an individual EP and the population average is calculated. If a discrimination threshold is applied to the correlation, a correlation detector will be built. Starting from the additive signal model according to Eq. (10.1), the hypotheses for a coherent[4] detection can be expressed as follows:

[4]The signal searched for with all parameters is completely known

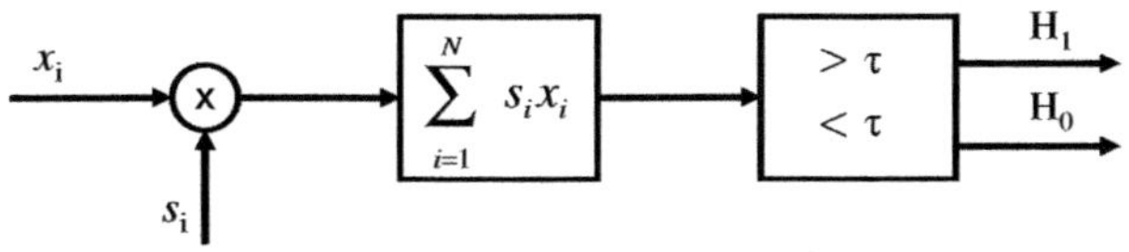

Fig. 10.4 Correlation detector (correlator, matched filter): The signal term $s_i.s_i/2$ from Eq. (10.4) has been incorporated into the threshold τ. This structure can be interpreted as a linear filter with an impulse response equal to the signal which is looked for. The output of the filter is compared with the threshold τ which works as a simple discriminator

$$\begin{aligned} H_0 &: \sum_{i=1}^{N} s_i\,(x_i - s_i/2) < \tau \\ H_1 &: \sum_{i=1}^{N} s_i\,(x_i - s_i/2) > \tau \end{aligned} \tag{10.4}$$

The structure according to Eq. (10.4) is a correlation detector for the case of a coherent signal s_i and an i.i.d. Gauss-distributed noise n_i. As the data of experimental investigations lead to a template, the coherence condition is met also in practice. But the spontaneous EEG that is considered as noise in this case does not fulfill the second assumption in any way. Consequently, it is necessary to ensure this precondition by applying suited methods in the preprocessing step. See chapter "Methods of signal processing" (prewhitening) [1]. Figure 10.4 shows a possible realization of the correlation detector. The correlator will be an optimum detector, if the requirements are met. It can prove deterministic signals up to an SNR of −20 dB and is the best choice for practical signal analysis. However, its judgment is only reliable for the rejection of the null hypothesis.

If the null hypothesis is accepted, several interpretations will be possible. Frequently an EP exists, but it not corresponds to the template, e.g., due to pathological changes. It is also very often the case that the EP exists, but it is so weak or strongly interfered that even the correlator cannot detect it. From the analyses one can conclude that the correlation decision is considered reliable after the fulfillment of the requirements for the rejection of the null hypothesis, whereas the acceptance of the null hypothesis is rather unreliable.

10.2.3 Energy Detector

If no template exists and other characteristics of the searched EP are not known either, detectors are to be used that are independent of the signal shape. If the signal shape is not known, the existence of a signal can be proven by means of a significantly increased energy compared to the existing noise. For this purpose, the following hypotheses are set up:

$$\begin{aligned} H_0 &: \quad x_i = n_i, \quad n_i N\left(0, \sigma^2\right) \\ H_1 &: x_i = n_i + s_i \;\; s_i N\left(0, \sigma_s^2\right) \end{aligned} \tag{10.5}$$

The noise n and the signal s are Gaussian distributed and i.i.d. zero-mean processes. Under these conditions the detector can be initially simplified to the following formula according to Eq. (10.5):

$$\sum_{i=1}^{N} x_i^2 \begin{matrix} < \tau : \equiv H_0 \\ > \tau : \equiv H_1 \end{matrix} \tag{10.6}$$

In practical bioanalysis, none of the mentioned conditions applies. Thus, both the noise reference and the sum signal are to be treated with a prewhitening filter. Another possibility to prove the signals by using an energy detector is the use of the periodogram as a test statistical method. By the integral transformation of the i.i.d. process into the frequency domain, a Gauss distribution of the frequency components or a χ^2-distribution of the performance is achieved for a sufficiently high N>100 [2].

$$S_x(\omega) = \frac{1}{N} \left| \sum_{k=1}^{N} x_k e^{j\omega k} \right|^2, \; -\pi \le \omega \le \pi \tag{10.7}$$

By using the periodogram, the hypotheses are as follows

$$\begin{matrix} H_0 : & S_x^0(\omega) = S_n(\omega), & n \sim N\left(0, \sigma^2\right) \\ H_1 : & S_x^1(\omega) = S_s(\omega) + S_n(\omega) & \end{matrix} \tag{10.8}$$

The energy detector is modified to

$$T_x = \frac{\left| \sum_{k=1}^{N} x_k e^{j\omega k} \right|^2}{\left| \sum_{k=1}^{N} n_k e^{j\omega k} \right|^2} \left\{ \begin{matrix} < \tau : \equiv H_0 \\ > \tau : \equiv H_1 \end{matrix} \right. \tag{10.9}$$

Consider that according to Eq. (10.8) further assumptions about the signal are not made in the detector. A noise reference that meets the condition according to Eq. (10.8) is required for calculating T_x following Eq. (10.9).

10.3 Methods for Signal Processing of EP

10.3.1 Prewhitening and Detection

From the previous theoretical consideration follows that for an unknown signal s the most reliable method of detection is the periodogram. To calculate the test statistic Tx the noise spectra $S_n(\omega)$ must be known. Real noise is not white and not i.i.d. so

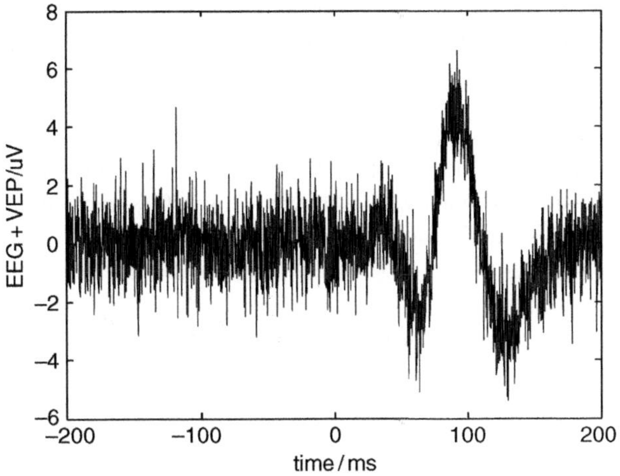

Fig. 10.5 Transient VEP embedded in white Gaussian noise. The noise reference is estimated from the pre-stimulus interval for $t < 0$. Transient visual stimulation by short flash at $t = 0$. Simulated data for detection test

that principally a prewhitening or a standardization of the sum signal $S_x(\omega)$ to the noise $S_n(\omega)$ must be performed according to Eq. (10.9) before starting the detection. The estimation of the noise reference can be based on several approaches. First it can be assumed that the spontaneous EEG just before a stimulus and after the dying out of the last response to the stimulus can be used as noise reference [2]. The test statistic Tx is calculated from the spectra of the noise reference $S_n(\omega)$ and the sum signal $S_x(\omega)$.

A typical detection situation is shown in Fig. 10.6: At lower frequencies the sum signal has a higher power so that the null hypothesis according to Eq. (10.10) can be reliably rejected. For signals with an unknown spectrum the total band width must

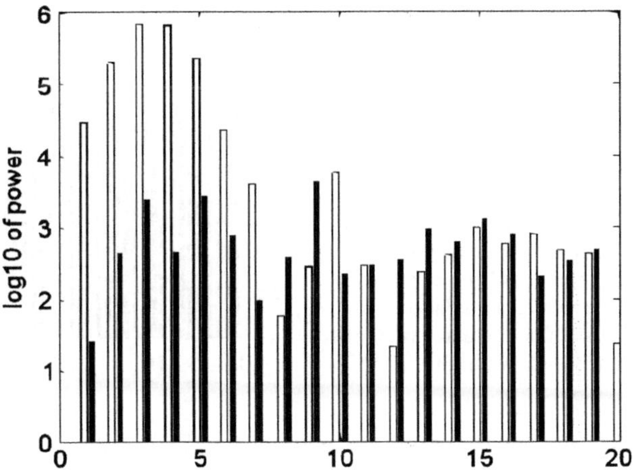

Fig. 10.6 Noise power spectrum (*black*) and power spectrum of the sum of signal and noise (*white*) estimated from the noise reference (Fig. 10.5, $t < 0$) and the sum signal ($t > 0$), respectively, vs. discrete frequency

be principally included into the determination of the test statistic T_x. This leads to the phenomenon that an evoked potential is pretended if disturbances caused by other biosignals (accidentally occurring α-wave trains of the EEG) or other technical disturbances (motion artifacts, switching peaks) occur. For this reason the *false alarm rate* α increases dramatically. A known signal spectrum is more suited for high detection reliability. A noise reference will not be necessary if the spectrum of the signal is narrow and known. This is for example the case for the Steady-State or Chirp[5]-stimulation.

$$T_x = \frac{S_x(\omega) - S_n(\omega)}{S_n(\omega)} = \left. \frac{N \sum\limits_{m=1}^{M} S_x(m)}{M \sum\limits_{n=1}^{N} S_n(n)} - 1 \right|_{H_0} \propto F_{2M,2N} \qquad (10.10)$$

But these properties can also be used for transient EPs, if it is ensured that the EEG is continuously recorded over the whole time of measurement. Then, the spectrum of the signal is to be expected at harmonics of the stimulation rate. In Fig. 10.7 this case is demonstrated for ten successive transient stimulus responses according to Fig. 10.5. Here, a substantial advantage is the fact that due to the temporal integration the non-correlated noise is reduced according to Eq. (10.3) in the Fourier transformation according to Eq. (10.7). Then, an averaging is not required for the detection.

In the real EEG, the noise is not white as in the simulated data. Therefore prewhitening must be used. If a noise reference does no exist, the noise background is estimated by smoothing the spectrum. After smoothing by a window, the spectral

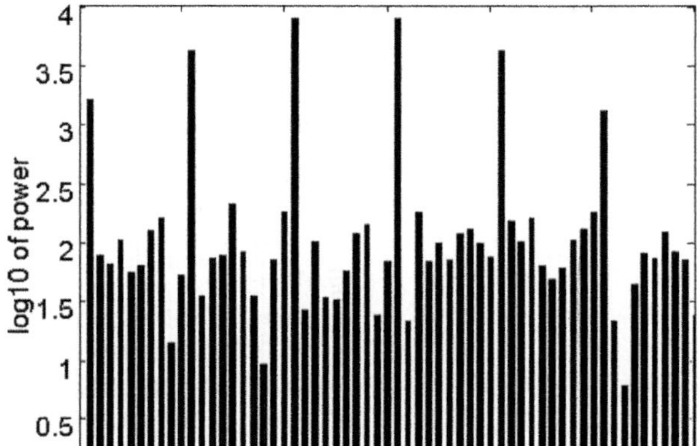

Fig. 10.7 Spectrum of signal and noise vs. discrete frequency, where the signal spectrum is known (harmonics of k = 10, 20, 30, 40, 50). No noise reference is necessary for detection. The spectrum has been computed for repeated stimulation with constant period and a train of ten EPs shown in Fig. 10.5

[5]Chirp is a pulse train with a temporal changing period

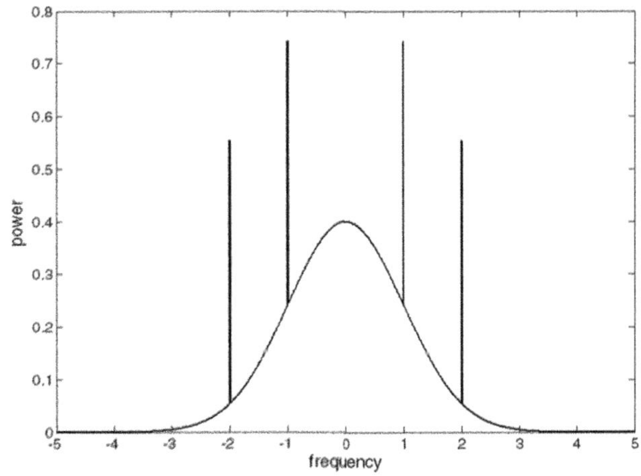

Fig. 10.8 Harmonics of stimulation rate in colored noise. This spectral shape is not appropriate for energy detection

peaks of the EP are damped. This damping appears as indentation after standardization and reduces the SNR (Fig. 10.9, on the left). It is known that the median is very robust to extreme values so that it can be used for estimating the noise. After smoothing the median of the spectrum dents do not appear in the course and the SNR of the original signal before prewhitening are maintained. (Fig. 10.9, on the right).

Decomposition methods offer another possibility for the orthogonalization of signals with the hope to separate signals from noise. A fundamental problem that could not be solved by prewhitening so far is the instationarity of the signal and the noise that exists under real conditions. The demand for stationarity results from

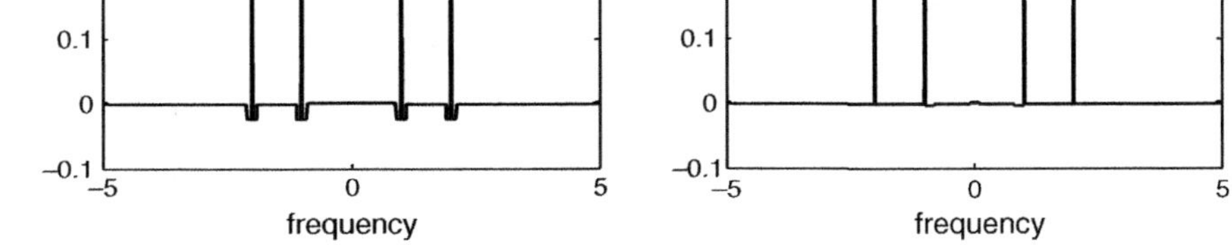

Fig. 10.9 Whitened spectrum of the signal shown in Fig. 10.8. Conventional whitening by smoothing window (*left*) and by median (*right*). Notice the differences in SNR left (7 dB and 2.7 dB) and right (10 dB and 3.2 dB)

the condition of i.i.d. processes. Up to now, the independence has certainly been established in the sense of statistics of second order (white noise), but the stationarity is still not ensured. Now, the typical representatives of this method group shall be compared with respect to their suitability for the detection of EPs [3]. On the assumption that the noise is stationary, the SVD[6] delivers excellent results. The SVD can be mathematically expressed as follows:

$$\mathbf{X} = \mathbf{U} \cdot \mathbf{S} \cdot \mathbf{V}^* \tag{10.11}$$

U and **V** are orthonormal matrices, **S** is a diagonal matrix, $\mathbf{V}^*$ is conjugate transpose of **V**. From the viewpoint of signal processing the following interpretation is useful: **X** contains the records of a multichannel signal in columns. The columns in **U** contain the orthonormal signal courses after the decomposition of **X**. The diagonal of **S** contains the ordered sequence of the signal powers of the orthogonal signals in columns of **U**. The matrix **V** contains the weights that determine the portions of individual signals in the corresponding channel. From the time courses in **U** the desired component can be selected and the singular value in **S** belonging to said component and the other components can be directly used in Eq. (10.10) for the detection test.

The procedure is explained in the following example: A 16-channels EEG contains a line disturbance (50 Hz), a transient VEP and technical noise. The sum signal

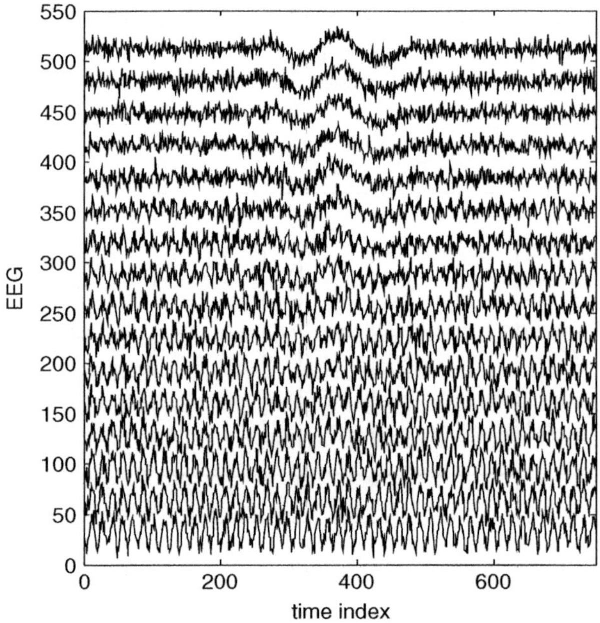

Fig. 10.10 Simulated EEG with transient VEP and line noise in 16 channels. The power of VEP decreases and the power of line noise increases from top to bottom. Stochastic noise remains constant in all channels. For illustration the channels are mathematically shifted

---

[6]Singular value decomposition

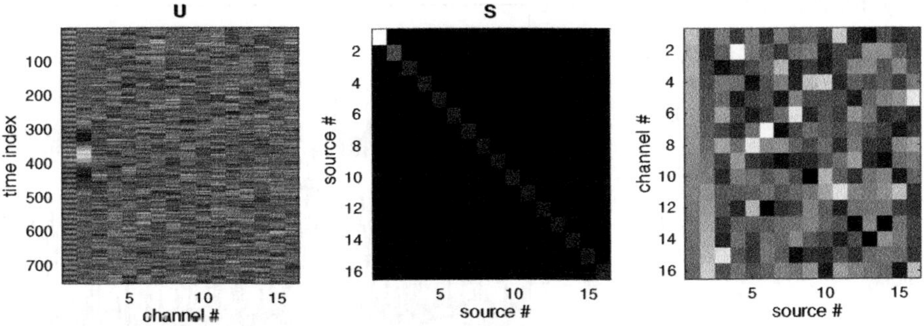

Fig. 10.11 SVD decomposition of multichannel signal shown in Fig. 10.10. The strongest component is the line noise, corresponding to the first column in **U** and the first singular value in **S**. The second strongest component is transient VEP which is looked for, corresponding to second column in **U** and second singular value in **S**. The first two columns in **V** show the weights for the first signal components in connection with channels. All other components are noise

for all channels is shown in Fig. 10.10. This signal is decomposed by the SVD; the result is illustrated in Fig. 10.11. After the SVD, orthonormal vectors that represent the signal courses are contained in the columns of the matrix U. The noise is white and normally distributed; the deterministic components do not show a normal distribution. This situation is shown in Fig. 10.12. As after the SVD the conditions to the signal characteristics for the energy detector are still not fulfilled, the periodogram according to Eq. (10.10) is used here. One channel or several channels of the matrix U, weighted with the corresponding singular values S_{ii}, is/are selected for the noise reference. The line noise must be excluded from the detection test, at best by using a notch filter already during the recording process. In analytic practice, a frequent object is also to measure the EPs, therefore the SVD will not be sufficient.

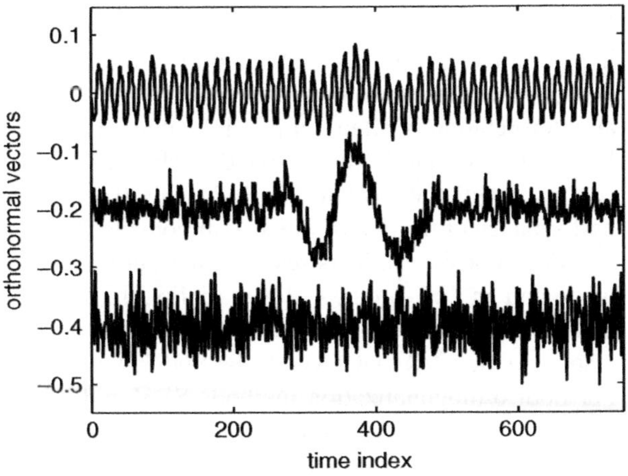

Fig. 10.12 First three components of the left matrix U after SVD. On top line noise, in the middle transient VEP, bottom Gaussian noise. The first two components are not Gaussian distributed

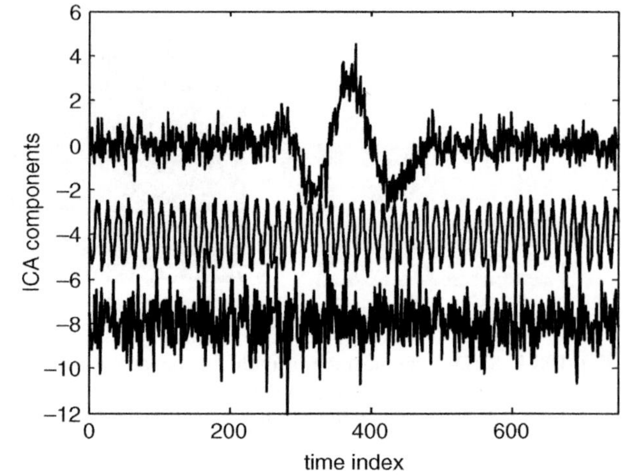

Fig. 10.13 First three independent components of simulated EEG with transient VEP, line and stochastic noise as shown in Fig. 10.10. The tVEP (*top*) is better separated from line noise (*middle*) than after SVD (Fig. 10.12). However, both deterministic components are noisy

Due to the not-normally distributed deterministic components (Fig. 10.12) they are orthogonal to each other and to the noise, but they are not independent. This can clearly be seen in the example in Fig. 10.13 in the first component that – apart from the pure line noise – partly contains the EP, too.

The ICA (Independent Component Analysis) can be used for a better separation. The principle of the ICA is based on the maximization of the not-normally distributed components of the searched signals whereas the noise is assumed to be normally distributed. The problem can be mathematically expressed as follows:

$$\mathbf{IC} = \mathbf{B} \cdot \mathbf{X} \tag{10.12}$$

IC is the matrix of the independent components, **B** is the solution of the optimization problem, and **X** is the data matrix. To solve the Eq. (10.12), several target functions and optimization algorithms that are discussed in literature can be used. Principally, the ICA achieves an independence of the components in the sense of higher moments whereas the SVD only optimizes to statistics of second order. However, it is difficult to interpret the **IC**. In case of strong and stationary disturbances an interpretation is often not possible at all. The result of the ICA of the simulated EEG with transient VEP and line disturbance is shown in Fig. 10.13. If this result is compared with the one after the SVD (Fig. 10.12), the ICA shows a better separation performance of the deterministic components. The SNR is almost identical after the two decompositions. This result is not surprising, because the stochastic noise is white and an i.i.d. process. But in reality, the spontaneous EEG is a process that is neither white nor i.i.d. In this case, the ICA delivers considerably better results with respect to the SNR that exceeds the SNR after the SVD by 10–20 db in average [3]. For both orthogonalization methods SDV and ICA the deterministic components are not-normally distributed after the decomposition so that the conventional energy detector Eq. (10.6) cannot be applied. The periodogram Eq. (10.10) is better suited because the distribution of the signal is not important here. However, the temporal integration of the periodogram causes the spectral leveling out of the signal power of transient EPs (Wiener-Kinchin theorem). Due to the time integral the increase of the SNR can be used up mostly for the ICA. The shorter the transient EP in comparison to the length of the analysis window, the higher is the SNR loss due to integration. Therefore, further methods are required for enhancing the SNR.

10.3.2 Enhancement of the SNR

The most common method for enhancing the SNR is the stimulus-synchronous averaging according to Eq. (10.12). Depending on the stimulus and the searched waves in the EP, the averaging order is between 10 and 1000. In functional diagnostics, the high averaging orders can cause unacceptable measurement times so that more effective methods are necessary. As multichannel derivations normally exist for the EEG recording, the simultaneous ensemble average can be computed at every time. But this will not be sufficient so that both approaches according to Eq. (10.13) are summarized to a spatio-temporal averager.

$$\bar{x}(k) = \frac{1}{NM} \sum_{c=1}^{N} \sum_{i=1}^{M} x_i^c(k) \tag{10.13}$$

In Eq. (10.13) c is the channel index, i is the set index, and k is the time index. The following equation applies for the indices for a repetitive simulation with a constant distance between the stimuli and a continuous recording of the data:

$$i = (k \bmod M) + 1 \tag{10.14}$$

According to Eq. (10.3) the theoretically attainable SNR is

$$SNR(\bar{x}) = \frac{\text{var}(\bar{s})}{\text{var}(\bar{n})} = NM \frac{\text{var}(s)}{\text{var}(n)} = NM \cdot SNR(x) \qquad (10.14)$$

and the enhancement of the SNR

$$\Delta SNR = NM \rightarrow \Delta SNR/dB = 10 \cdot \log_{10} NM \tag{10.15}$$

From Eq. (10.15) follows that by simultaneous averaging the otherwise required measurement time can be reduced purely by the factor N. Before using the averager according to Eq. (10.13) several conditions must be fulfilled: The signal s has a constant signal shape and simultaneously appears in all channels, the noise n is stationary and is a zero mean signal. Under real conditions, the prerequisite of simultaneous appearance is the major problem. The EP course depends on several stimulation and derivation parameters and shows significant differences between the channels even for unipolar EEG derivations. The different latencies of the EP waves

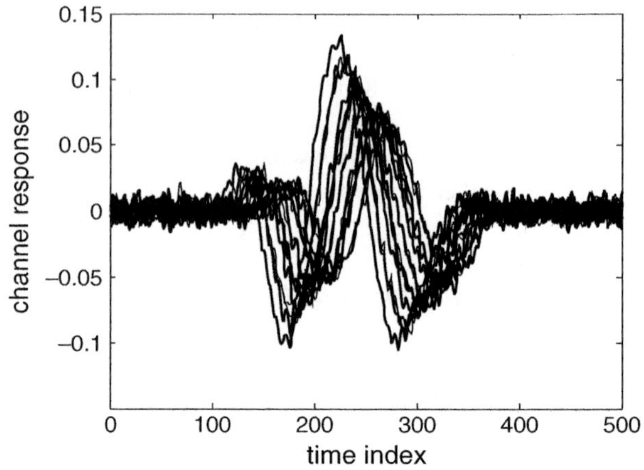

Fig. 10.14 Simulated VEP in 16 channels added to Gaussian white noise with different channel jitters. The signal shape is the same in all channels, while the amplitudes and latencies are different

generate a phase jitter that has a low pass effect and reduces the amplitude of the waves. Under unfavorable conditions, the wave can even be eliminated. The signal model according to Eq. (10.1) is extended by the jitter and processed with the simultaneous part of the averager according to Eq. (10.13) (Fig. 10.14):

$$x_i^c(k) = A_c s_c(k + \tau_c) + n_c(k) \tag{10.16}$$

$$y(k) = \frac{1}{N}\sum_{c=1}^{N} x_i^c(k) \tag{10.17}$$

The simultaneous average $y(k)$ is damped by the jitter τ_c by the factor D:

$$y_D(k) = D \cdot y(k) = \frac{1}{N}\left|\sum_{c=1}^{N} e^{-j\omega\tau_c}\right| \cdot y(k) \tag{10.18}$$

The damping factor D is between 0 and 1 and ideally it will become 1, if all time delays τ_c are zero. A typical measurement situation is demonstrated in Fig. 10.14. The channel signals show the qualitatively identical signal course, but their amplitudes and latencies are different. Therefore, the channel jitters are to be compensated to zero before applying simultaneous averaging. For this purpose, a controllable additional delay is integrated into each channel as illustrated in Fig. 10.15. The object of an optimization method is to achieve the simultaneous overlapping of the channel signals. As the signal shape is not known, a method that maximizes the SNR is used [4]. From the Eqs. (10.14) and (10.18) follows that it would be sufficient to maximize the signal power in order to maximize the SNR, because the noise power does not change. But to maximize the actual SNR a noise reference is nevertheless required, because the real EEG is non-stationary.

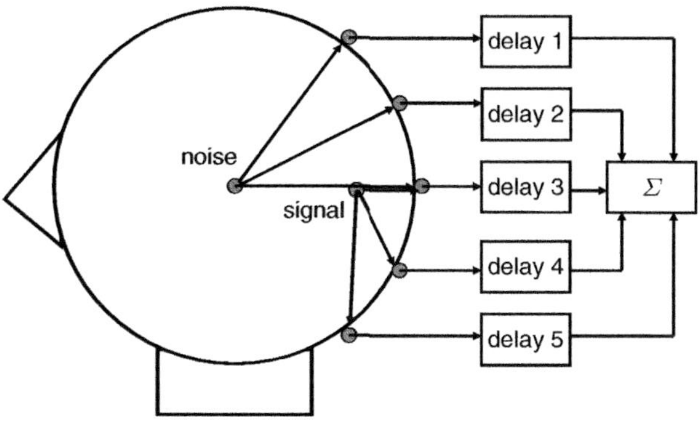

Fig. 10.15 Spatial model of signal and noise with additional channel delays. Different times of arrival are compensated by channel delays to obtain zero difference between channels

$$SNR \to \max \Rightarrow \frac{\partial SNR}{\partial \tau_c} = 0, \; \frac{\partial^2 SNR}{\partial \tau_c^2} < 0 \tag{10.19}$$

The channel delays to be determined shall compensate the jitter so that the following equation applies:

$$d_c = -\tau_c \tag{10.20}$$

The delays are recursively calculated according to Eq. (10.21) and the gradient is estimated according to Eq. (10.22).

$$d_c(k+1) = d_c(k) + \mu \hat{\nabla}(k) \tag{10.21}$$

$$\hat{\nabla}(k) = \begin{Bmatrix} \Delta SNR/\Delta\tau_1 \\ \vdots \\ \Delta SNR/\Delta\tau_N \end{Bmatrix} \tag{10.22}$$

The adaptation constant μ depends on the individual measurement conditions, the sampling rate and the derivation parameters so that it cannot be preset formally. It is empirically defined in practical analysis. The gradient according to Eq. (10.22) can be estimated in different ways. For transient EPs the following procedure is useful:

$$y(k) = \frac{1}{N}\sum_{c=1}^{N} x_i^c(k), \; y_{\Delta j}(k) = x_i^j\left(k + \Delta\tau_j\right) + \frac{1}{N}\sum_{c=1, c\neq j}^{N} x_i^c(k) \tag{10.23}$$

$$SNR(\bar{x}(k)) = \frac{\sum\limits_{k=0}^{k+M-1} y^2(k)}{\sum\limits_{k=-M}^{-1} y^2(k)} - 1, \; SNR\left(\bar{x}_{\Delta j}(k)\right) = \frac{\sum\limits_{k=0}^{k+M-1} y_{\Delta j}^2(k)}{\sum\limits_{k=-M}^{-1} y_{\Delta j}^2(k)} - 1 \tag{10.24}$$

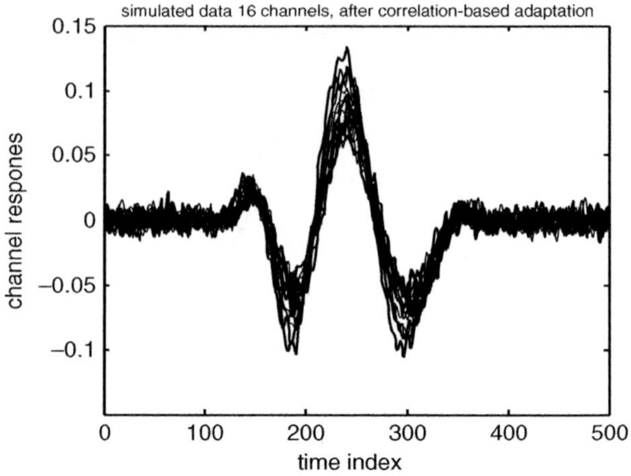

Fig. 10.16 Simulated VEP in 16 channels added to Gaussian white noise with different channel jitters after adaptation for maximal SNR. Mean enhancement of the SNR is about 3 dB compared to the VEP shown in Fig. 10.14

$$\frac{\Delta SNR}{\Delta \tau_j} = \frac{SNR\left(\bar{x}_{\Delta j}\left(k\right)\right) - SNR\left(\bar{x}\left(k\right)\right)}{\Delta \tau_j} \tag{10.25}$$

In Eq. (10.24) the individually current stimulation time is set for the time index $k = 0$ (Fig. 10.1). In this way, the actual SNR is estimated from the EEG directly before the stimulation by using the noise reference and directly after the stimulation by using the sum of the stimulus response and the noise. The gradient computed according to Eq. (10.25) can be used directly in the recursion according to Eq. (10.21). The adaptation constant μ plays a critical role in the recursion. Unlike the typical method in the theory of adaptive filtering, limits cannot be predefined for it here. The signal is not known and therefore the relation between the signal power

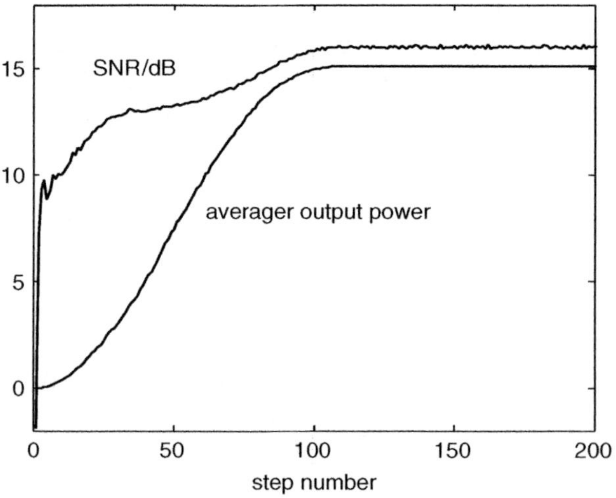

Fig. 10.17 Enhancement of the SNR and the output power of the averager with real transient VEP after completed adaptation. Notice the enhancement of more than 6 dB which is better than in simulated data

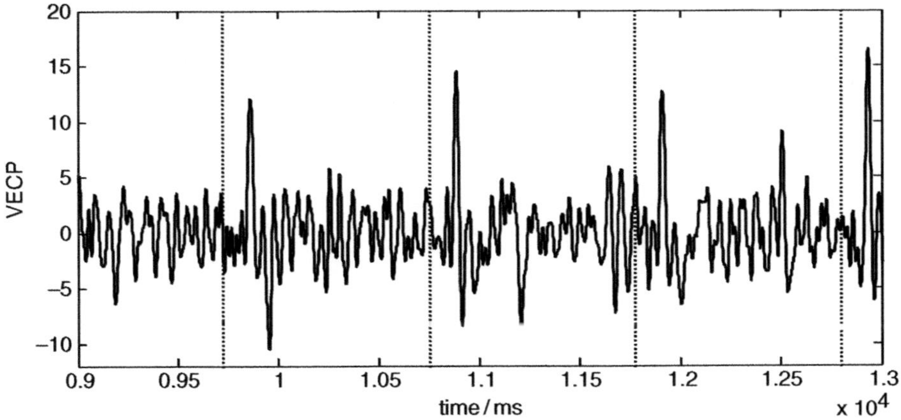

Fig. 10.18 Real VEP after completed adaptation and simultaneous averaging without stimulus related averaging Eq. (10.17), i.e., in real time across the channels. Dotted lines mark stimulus times. The stimulus has been a short (20 ms) focal (0.7° central field) light flash. The main wave P100 is clearly visible even in the strong noise from the spontaneous EEG

and the time delay are not known either. A useful adaptation constant must be determined empirically. But it is possible to show that the algorithm will converge to the maximum SNR if the constant is properly selected [4]. Compared to the dynamic change of the EP parameters, the adaptation is sufficiently quick to follow them in the analysis of EP, if the sampling rate is sufficiently high. To achieve the optimum of the simulated signal according to Fig. 10.14, 40 adaptation steps were necessary. The result is shown in Fig. 10.16. The data shown, however, are ideal simulated ones. The course of the adaptation with real VEP is illustrated in Fig. 10.17. A spot transient flash of magnitude of 2 degree was used in the central visual field.

The EEG was led via 16 channels above the occipital cortex. This result is better than the one achieved with simulated data. If the EEG is recorded continuously with a sufficient number of channels, the real-time VEP is well visible even without stimulation-related averaging; see Fig. 10.18. Although the measurement of the VEP parameters is not possible, a continuous detection in real time can be performed. Such a solution could be possible, for example, in function monitoring during neurosurgical operations. In research it would certainly be of interest to study the dynamic development of the latency. This is not possible in conventional stimulus synchronized averaging.

References

1. Poor HV (1994) An Introduction to Signal Detection and Estimation, Springer-Verlag, ISBN 0-387-94173-8
2. Liavas AP, Moustakides GV, Henning G et al (1998) A Periodogram Based Method for the Detection of Steady-State Visually Evoked Potentials, IEEE Transactions on Biomedical Engineering 242–248

3. Drozd M, Husar P, Nowakowski A et al (2005) Comparison of SVD and ICA Based Methods Using Cortical Sources Model IEEE Engineering in Medicine and Biology 51–58
4. Husar P, Berkes S, Drozd M et al (2002) An approach to adaptive beamforming in measurement of EEG. Proceedings of the European Medical and Biological Engineering Conference 1438–1439 Vienna

Chapter 11
Visual Evoked Potential Analysis Using Adaptive Chirplet Transform

Jie Cui and Willy Wong

Abstract Visual evoked potentials (VEPs) are scalp electrical signals generated in response to rapid and repetitive visual stimuli. These signals possess complex time-frequency structures and are difficult to characterize with conventional methods. In this chapter, we propose a new approach based on the adaptive chirplet transform (ACT) that can represent a complete VEP response from the transient to the steady-state portion. Our implementation involves both a non-windowed and windowed approach. The non-windowed ACT employs a coarse-refinement algorithm (MPLEM) to estimate multiple chirplets under low signal-to-noise ratio condition. We show how the chirplet parameters (i.e., time-spread, chirp rate, time-center and frequency-center) can be used to separate the transient from the steady-state portions of the response, and that as few as three chirplets are required to represent a complete VEP signal. The windowed approach is implemented by partitioning the signal into equal-length non-overlapping segments before estimating a single chirplet from each segment, resulting in significant reduction of computational time. The application of the windowed ACT to VEP analysis is also discussed.

11.1 Introduction

Visual evoked potentials (VEPs) are surface electrical potentials measured from the scalp in response to a visual signal. They are generated from the visual cortex and/or the peripheral neural pathways leading to the cortex, and are time-locked to the visual stimulus [34]. VEPs have prominent clinical significance and can help diagnose sensory dysfunctions. They are traditionally employed in tests for the integrity of the visual pathway and used as a supplement to other techniques in research into specific clinical conditions [17, 19]. It has been shown that if the repetition rate of visual stimuli is sufficiently high (usually greater than 6 Hz), responses begin to

W. Wong (✉)
Department of Electrical and Computer Engineering, University of Toronto, 10 King's College Road, Toronto, ON M5S 3G4, Canada
e-mail: willy@eecg.utoronto.ca

A. Naït-Ali (ed.), *Advanced Biosignal Processing*,
DOI 10.1007/978-3-540-89506-0_11, © Springer-Verlag Berlin Heidelberg 2009

merge and the shape of the resulting VEP becomes periodic. These responses are usually referred to as steady-state visual evoked potentials (*ss*VEPs) [33]. A variety of clinical applications require the detection of *ss*VEPs including, for instance, evaluation of optic nerve function [12, 17], objective estimation of visual acuity in infants [28, 35], detection of abnormal mental states (e.g., hysteria), and assessment of delayed neurological maturation [17, 19]. More recently, VEPs have found applications in interface design, e.g., [5, 27, 38].

Under steady-state conditions a VEP signal can be modeled in terms of a fundamental sinusoid and higher harmonics. The task of estimation and detection then is reduced to finding these periodic components embedded in background spontaneous EEG [2, 6, 21, 35, 37]. The model is, however, incomplete as it is believed that a transient component (*t*VEP[1]) first appears following the onset of the visual stimulus [34, 36]. In general, two stages of the response may be observed: (1) a transient *buildup* portion followed by (2) a steady-state portion. Even in the steady-state portion, signal energy may be distributed at frequencies other than the harmonic frequencies due to the nonlinear nature of the visual system [26]. Finally, given variability in the subject's mental state, a truly steady-state response will be hard to achieve under usual recording conditions [13]. Therefore, a linear expansion from a Fourier basis may not be the optimal way to represent the entire VEP signal. We will discuss next time-frequency representations of evoked potentials (EPs).

Techniques for characterizing sensory EPs can be broadly categorized into two classes according to the repetition rate of sensory stimuli: (1) time domain analysis to characterize the succession of positive and negative deflections in the EP response to stimuli of (usually) 6 Hz or less [18], and (2) frequency domain characterization of steady-state EPs at stimulation frequencies of 6 Hz and greater [15, 32]. Both methods have been used extensively (for a comprehensive review, see [19, 2]). Recently, the benefit of time-frequency representations has been recognized in the analysis of transient responses, since the joint time-frequency representations can reveal temporal structure of different spectral components. The effectiveness of this approach has been demonstrated in many problems of biomedical signal processing [1]. For EP analysis in particular, three categories can be identified: applications of the short-time Fourier transform (STFT), wavelet transform and the adaptive Gabor logon transform. These three methods can be seen in the context of increasing signal processing sophistication.Our work is a direct extension of this development. By introducing an additional dimension of freedom to the Gabor logon (i.e., allowing the rotation of the basis functions in the time-frequency plane) we obtain a relatively new class of functions known as the chirplet. In this chapter, we describe an application of this emerging method (the *adaptive chirplet transform*or ACT)

[1] It should be emphasized that the term "transient VEP," or *t*VEP, employed here is conceptually different from that used in traditional electrophysiological literature. It usually refers to an experimental paradigm where the potentials are evoked by visual stimuli which are sufficiently widely spaced so that the visual system can be regarded as returning to a state of rest between successive stimuli [34]. In this chapter, however, *t*VEP refers to the signal prior to the formation of steady-state VEP.

to the field of biomedical signal processing. We present two approaches, namely non-windowed ACT and windowed ACT, for time-frequency analysis of VEP signals. The goal of these approaches is to characterize the time-dependent behavior of VEP from its initial transient portion to the steady-state portion by a series of time-frequency atoms, or chirplet basis functions. The ACT technique allows us to clearly visualize, perhaps for the first time, the early moments of a VEP response.

In the next section, we introduce the adaptive chirplet decomposition and show an implementation of our algorithm. We have implemented this technique by adopting the adaptive algorithm of the matching pursuit (MP) [4]. Our approach also incorporates a maximum-likelihood estimation (MLE) algorithm to help estimate the signal under the low signal-to-noise ratio (SNR) condition that are typical of VEP measurements. We then illustrate the experimental setup and data acquisition procedure to record VEPs, and present the results of applying the non-windowed ACT methods to the data. In Section 11.3, we introduce the windowed ACT method and apply it to the VEP analysis. The optimal lower bound of the parameter estimates – Cramér-Rao lower bound (CRLB) – of a single chirplet will also be described. We show that with the windowed approach, the computational cost of chirplet analysis can be significantly reduced. Finally, we summarize our discussion and propose some directions of future research in Section 11.4.

11.2 Non-windowed ACT Method and Application

A chirp is a signal where the instantaneous frequency changes as a function of time. Chirps are found in many naturally occurring signals. It can be found in bird whistles [25], in bat echo-location signals, in human voices, in seismic signals [3], in impulsive signals dispersed by the ionosphere [31] and in EEG signals [10], among others. It is therefore desirable to have a method to analyze real signals in terms of a chirp. The basis for this approach centers around the so-called "chirplet transform".

The chirplet transform was formulated as a generalization of the wavelet transform in the early 1990s [23, 24, 25]. There was particular interest in developing the Gaussian chirplet, which is implemented as a modification to the original Gabor Logon function [16], because it has the highest joint time-frequency resolution and it is the only function whose Wigner-Ville distribution (WVD) is non-negative [4]. Another reason for the common employment of the Gaussian chirplet transform (GCT) is due to its relative simplicity in mathematical manipulation. Thus, the GCT plays a unique role in the area of time-frequency analysis.

In previous work in applying the methods of chirplet analysis to signal decomposition, e.g., [4, 22, 25, 30], assumptions were made about the signal in that the signals to be analyzed were acquired noise-free. Hence the efficacy of the approach has not been rigorously demonstrated in the case of low signal-to-noise ratio (SNR). This is of relevance to certain practical problems in biomedical signal processing. For instance, visual evoked potentials (VEPs) are usually measured under very low SNR conditions (typically SNR <-10 dB for a single trial and SNR <0 dB for an average signal of 50 trials) [9, 34]. Therefore, chirplet decomposition of a low SNR signal is in essence an estimation problem. To improve the robustness of the estimates under a decomposition process, we propose a new approach to multiple-chirplet decomposition, which consists of two basic procedures: (1) initial coarse estimates of chirplets obtained through a matching pursuit (MP) algorithm [22], and (2) refinement of the estimates by using the Logon expectation and maximization (LEM) algorithm [24]. We refer to this approach as the MPLEM algorithm. It has been shown that the MPLEM algorithm can provide more precise estimates of chirp components of a signal compared with those estimated by the MP algorithm with chirplet alone [8].

11.2.1 From "Wavelet" to "Chirplet"

The wavelet transform has been proposed to partially overcome the problem of the fixed resolution with the STFT [11]. Here we denote an arbitrary piece of sinusoid resulting from a windowing operation as a "wavelet" (Fig. 11.1), with the only constraint being that the windowing function is a Gaussian function and hence each "wavelet" is in fact a Gabor logon [23]. A family of basis "wavelet" functions can then be derived from a mother "wavelet" by applying to it two operations: scaling (or time-spread) and time translation. A "wavelet" has good time resolution but poor frequency resolution in the higher frequency bands, and vice versa for the lower frequencies. This is the reason why the wavelet transform is well-suited for analyzing signals with discontinuity or abrupt changes. However, this property also means that the wavelet transform does not provide precise estimates of the time-frequency structures for signal components that do not match the tradeoff characteristics of the wavelet signal. The wavelet is also not efficient in representing chirp-like signals either.

In order to overcome these difficulties, the logon is modified to allow for an additional degree of freedom in the time-frequency plane. By applying scaling and translation operations to a logon, we can construct a "wavelet" basis function. A further step allows this "wavelet" to rotate in the time-frequency plane (the "chirping operation" as introduced later). This is equivalent to windowing the chirp signal by using a Gaussian window and hence the resultant function is coined the "chirplet". Referring to Fig. 11.1, we can see that the relationship of a chirplet to a chirp is analogous to that of a "wavelet" to a wave.

The chirplet shown in Fig. 11.1 can also be constructed from a unitary Gaussian window $g(t) = (\pi)^{-1/4} \exp\left(-t^2/2\right)$ by applying four mathematical operations to this window, i.e.,

(1) scaling	$(\pi \Delta_t^2)^{-1/4} \exp[-(t/\Delta_t)^2/2]$,
(2) chirping	$(\pi)^{-1/4} \exp(-t^2/2) \exp(jct^2)$,
(3) time-shift	$(\pi)^{-1/4} \exp[-(t-t_c)^2/2]$,
(4) frequency-shift	$(\pi)^{-1/4} \exp(-t^2/2) \exp(j\omega_c t)$.

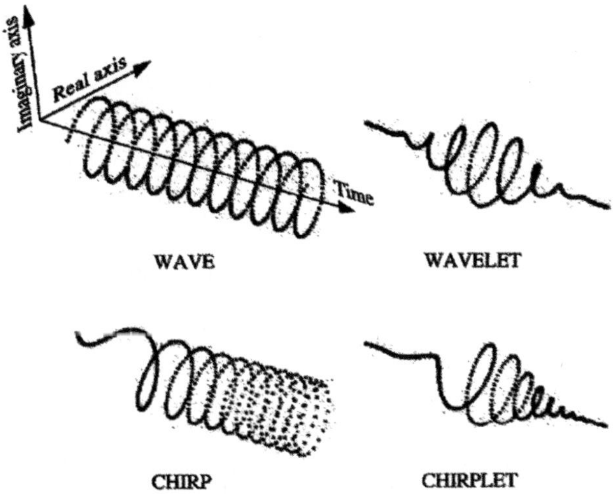

Fig. 11.1 Wave, "wavelet," chirp and chirplet revisited. The x axis corresponds to the real value of the function and the y axis to the imaginary value. Although the functions are continuous, a coarse sampling is used to enhance the 3-D appearance. Each sample is rendered as a particle in (x, y, t). WAVE – The wave appears as a 3-D helix. The angle of rotation between each sample and the next is constant, hence the frequency is constant. WAVELET – The "wavelet" is a windowed wave, where the reduction in amplitude is observed as decay toward the t axis. CHIRP – The chirp is characterized by a linearly increasing angle of rotation between one sample and the next. CHIRPLET – The chirplet is characterized by the same linearly increasing angle of rotation but first with a growing and then with a decaying amplitude. (Reproduced with permission from Mann, S.; Haykin, S., "The chirplet transform: physical considerations," IEEE Transactions on Signal Processing, vol. 43, no.11, pp. 2745–2761, Nov. 1995)

A sequential application of these operations leads to a family of wave packets with four adjustable parameters called Gaussian chirplets

$$g_{t_c,\omega_c,c,\Delta_t}(t) = \frac{1}{\sqrt{\sqrt{\pi}\Delta_t}} \exp\left\{-\frac{1}{2}\left(\frac{t-t_c}{\Delta_t}\right)^2 + j\left[c\left(t-t_c\right)+\omega_c\right]\left(t-t_c\right)\right\}, \tag{11.1}$$

where $j = \sqrt{-1}$, t_c is the time center, ω_c the frequency center, $\Delta_t > 0$ the effective time spread, and c the chirp rate that characterizes the "quickness" of frequency changes.

An alternative way to represent a signal is to use the Wigner-Ville distribution (WVD) [7]. The WVD is a time-frequency energy density that is computed by taking the correlation of a signal $f(t)$ with a time and frequency translation of itself. Later we will use the WVD to visualize the results of chirplet estimation in the time-frequency domain. The effects of the four operations on the WVD of a chirplet are shown in Fig. 11.2. It can be seen that chirplets are simply natural extension to "wavelets" by applying an additional chirping or rotational operation. Indeed, it

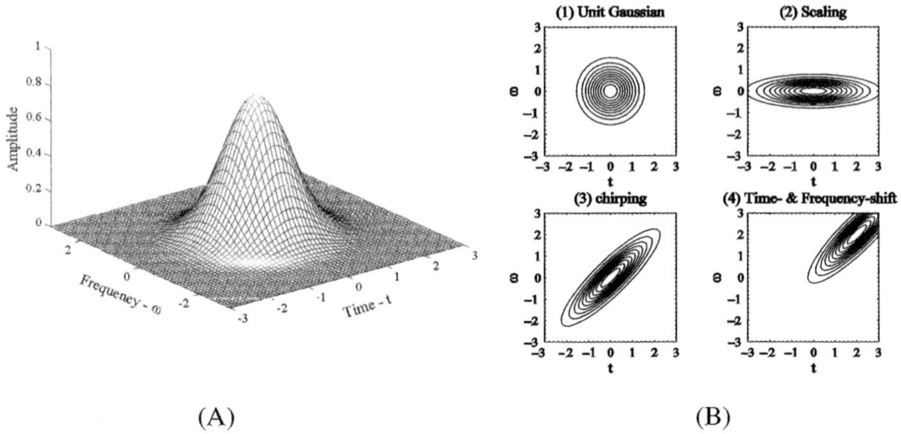

Fig. 11.2 Time-frequency distributions of Gaussian chirplets. (**A**) 3-D visualization of the WVD of the unitary Gaussian function; (**B**) (1) WVD contour of the unitary Gaussian, (2) effect of scaling, (3) effect of chirping, (4) effect of time-shift and frequency-shift

should be remembered that both the Gabor logons and "wavelets" are just special cases of chirplets, i.e., the case where the chirp rate is zero.

11.2.2 Gaussian Chirplet Transform and Adaptive Analysis

The Gaussian chirplet transform (GCT) of a signal is defined as the inner product between the signal $f(t)$ and the chirplets $g_{t_c,\omega_c,c,\Delta_t}$ defined in (11.1):

$$a_{t_c,\omega_c,c,\Delta_t} = \left\langle f, g_{t_c,\omega_c,c,\Delta_t} \right\rangle = \int_{-\infty}^{+\infty} f(t)\, g^{*}_{t_c,\omega_c,c,\Delta_t}(t)\, dt, \qquad (11.2)$$

where "$*$" denotes the complex conjugate operation. The coefficients $a_{t_c,\omega_c,c,\Delta_t}$ represent the signal's energy content in a time-frequency region specified by the chirplets. The absolute value of a coefficient is the amplitude of the projection. To simplify the notation, the set of chirplet parameters can be denoted by a continuous index set $I = (t_c, \omega_c, c, \Delta_t)$.

An arbitrary signal can then be constructed as a linear combination of Gaussian chirplets,

$$f(t) = \sum_{n=1}^{P} a_{I_n} g_{I_n}(t) + R^{P+1} f(t) = f_P(t) + R^{P+1} f(t) \qquad (11.3)$$

where I_n is the parameter set of nth chirplet, $R^{P+1} f(t)$ denotes the residue and $f_P(t)$ is defined as the Pth-order approximation of the signal. Since $g_{I_n}(t)$ has a Gaussian envelope, minimum time-frequency variance is achieved [7]. Note that the coefficient a_{I_n} is complex and hence the decomposition information at each iteration n is described by six real parameters, i.e., two from a_{I_n} and the other four from I_n.

The calculation of a_{I_n} involves selecting g_{I_n} from a predefined set of chirplets known as a *dictionary*. The approach is then to find the optimal subset of P chirplets from the dictionary so as to minimize the difference

$$\left\| f - \sum_{n=0}^{P-1} a_{I_n} g_{I_n} \right\|.$$

Unfortunately, the optimal estimation of a_{I_n} and I_n is an *NP*-hard problem [22], i.e., there exists no known polynomial-time algorithm to solve this problem. Consequently, there have been efforts to develop suboptimal techniques, e.g., [4, 30]. We will describe one such approach. The essence is to approximate the signal's energy in the time-frequency space using straight lines with arbitrary slopes (Fig. 11.3). To decompose a given signal into a sum of chirplets, two procedures are involved at each iterative step: (1) initial coarse estimates obtained using a matching pursuit (MP) algorithm and (2) progressive refinement of the estimates with the maximum likelihood estimation (MLE) procedure. The implementation of this adaptive chirplet decomposition algorithm is illustrated by the flowchart in Fig. 11.4.

The initial stage of the algorithm includes the construction of a chirplet "dictionary" and the initialization of the residue: $R^1 f = f$. The dictionary is constructed according to [4], which consists of a finite predetermined chirplets selected to cover the entire time-frequency plane. At each iteration, a single chirplet $g_{I_n}(t)$ and coefficient a_{I_n} is decided from $R^P f(t)$. This is termed "coarse estimation". The results are further optimized using a Newton–Raphson method to refine the match. The refined results are then subtracted from the signal and the steps are repeated to estimate a new chirplet from the residue $R^{P+1} f(t)$. We emphasize that the adaptive nature of the mechanism of the algorithm comes from the optimal selection of the basis functions for decomposition. The parameters of these functions are predefined

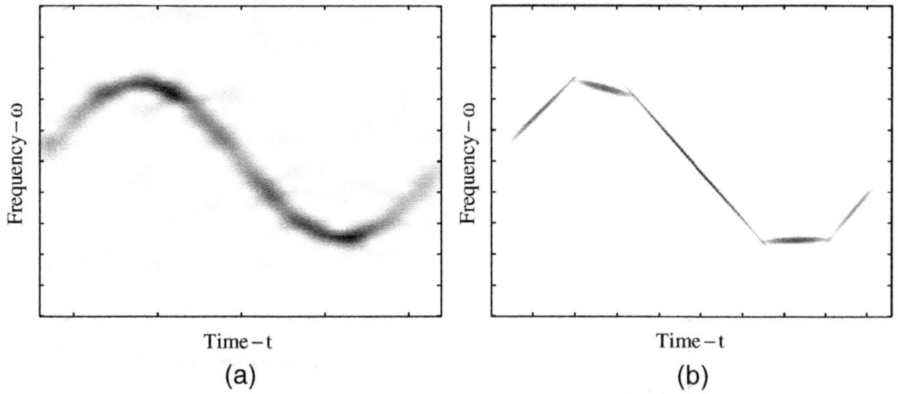

Fig. 11.3 Time-frequency plots of a signal with sinusoidal frequency modulation. (**a**) The spectrogram of the signal; (**b**) the adaptive chirplet spectrum (ACS) of the signal

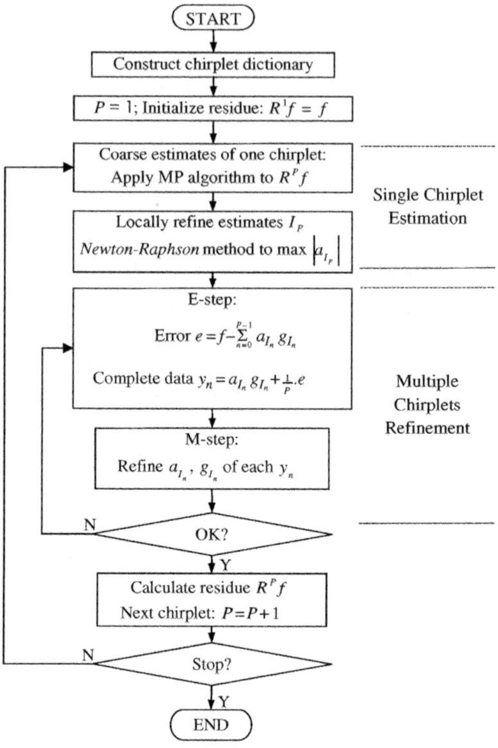

Fig. 11.4 Flowchart of the MPLEM adaptive chirplet decomposition algorithm (Reproduced with permission from Jie Cui; Willy Wong, "The adaptive chirplet transform and visual evoked potentials," IEEE Transactions on Biomedical Engineering, vol. 53, no.7, pp. 1378–1384, July 2006)

in the dictionary, which differs from the approach of adaptive filtering where the parameters are varied on a sample by sample basis.

In the case of estimating multiple chirplets, we have adopted the expectation-maximization (EM) algorithm to further refine the estimates of the chirplets. More specifically, the EM algorithm consists of two steps: an expectation step (E step) and a maximization step (M step). In the E step the complete data y_n are calculated as follows:

$$e = f - \sum_{n=1}^{P} a_{I_n} g_{I_n} \tag{11.4}$$

$$y_n = a_{I_n} g_{I_n} + \frac{1}{P} e, \quad n = 1, \ldots, P. \tag{11.5}$$

The M step applies the same algorithm employed in the estimation of a single chirplet to each of the y_n to refine the estimate of this chirplet. The EM algorithm may be repeated several times until the error in (11.4) is below a predefined level.

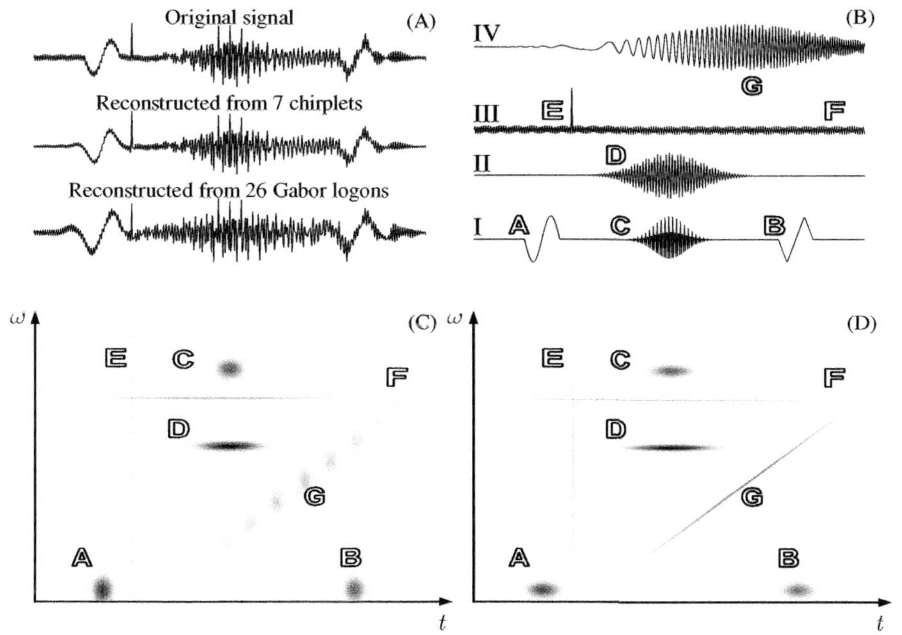

Fig. 11.5 Comparison of signal decompositions. (**A**) The waveform of signals including the original signal, its reconstruction from seven chirplets and its reconstruction from 26 Gabor logons. (**B**) Structures of the original signal. The synthetic signal is a direct sum of signals I–IV, where A = one period of a sinusoid, B = one period of a saw-tooth wave, C and D = sinusoids modulated by a Gaussian, E = delta function, F = sinusoid and G = Gaussian chirplet. (**C**) The adaptive spectrogram shows the results of decomposition using the MP algorithm with *Gabor logons*. (**D**) The ACS shows the results using the MP algorithm with *chirplets*

The results shown in Fig. 11.5 illustrate the performance of the technique. In this analysis, we show the decomposition of a complex signal into a number of Gaussian chirplets with our algorithm. The adaptive chirplet spectrogram (ACS) – which is a direct sum of the Wigner-Ville distributions of the individual chirplets – clearly shows the time-frequency structures of the signal. Note that the chirping structure "G" cannot be represented efficiently using Gabor logons.

A critical point of the analysis is the determination of the number of chirplets required to sufficiently represent the VEP signals. We did not attempt to predefine this number, but rather opted to continue the decomposition until a specific stopping criterion is satisfied. One method of approach is to calculate the coherent coefficients (cc) of the extracted chirplets, which is defined as the ratio of the energy of the projection to the energy of the residue,

$$cc_n = \frac{|a_{I_n}|^2}{||R^n f||^2} \quad n = 0, \ldots, P-1 \tag{11.6}$$

where $|a_{I_n}|^2$ is the energy of the projection and $\|R^n f\|^2$ is the energy of the residue $R^n f$. The more coherent a signal is with respect to the dictionary, the larger the cc values. Therefore, a small cc value indicates low correlation between the signal and the dictionary. A threshold based upon the cc value can be chosen as a stopping criterion.

11.2.3 VEP Analysis with Non-windowed ACT

We next apply the proposed non-windowed chirplet transform to the analysis of VEP signals. The VEPs used in this analysis were obtained through experiments involving a matrix of moving bars aligned to visual fixation crosses. It is believed that the VEP's are induced by the motion of the bars and are locked to the oscillating frequency of the stimuli. Experimental method follows procedure outlined in [29]. All stimuli were presented on an LCD monitor. VEP signals were recorded via three gold-cup electrodes, including the active electrode placed on the scalp at the Oz position, the reference placed on the left ear lobe and the third attached to right ear lobe to serve as ground. The signals were amplified (1000 $\times$), bandpass filtered (0.01–40 Hz), and passed through an A/D converter (sampled at 240 Hz) before being streamed to the harddisk. The data were then processed offline to obtain five averaged signals (denoted as $D_1 - D_5$, 1200 points each). A single trial consisted of 5 s of data. Following a 1 s prestimulation period, 16 vertical bars began oscillating in the horizontal direction with temporal frequency of 3.0 ± 0.1 Hz for 4 s. The onset of the stimuli was at the end of the first second. Five healthy adult subjects participated in this study. The subjects were instructed to pay attention to the moving bars on the screen. Eye movements were minimized by having the subjects fixate on the center cross. They were also asked to withhold eye blinks during each trial and to keep their facial muscles relaxed. Each subject sat for a total of 50 trials. They initiated the trials with a button when they were ready. The subjects were previously trained before the start of the session to ensure that they understood the task.

An averaged signal was obtained from the 50 trails of a single subject. In total, five averaged signals (denoted D_1–D_5) were obtained from the five subjects. For each of the averaged signals, ten Gaussian chirplets were estimated with the algorithm described in Fig. 11.4. We believe that this number of chirplets is sufficient for representing the VEP of interest as the residue after ten iterations was virtually indistinguishable from white noise (Fig. 11.9, discussed below). As an example, Table 11.1 summarizes the estimated parameters of the ten chirplets extracted from D_1 as well as the coherence coefficients cc_n. In general, the chirplets extracted first have higher amplitudes and higher cc. Particularly, we found that the amplitudes of the first three chirplets are significantly higher than those of the remaining chirplets. The first chirplet, c_1, represents the steady-state component of the VEP signal, as it has a long time-spread ($\hat{\Delta}_t$) and near zero chirp rate. The remaining two chirplets c_2 and c_3 have negative chirp rates, indicating that the instantaneous frequency of the prominent early components decreased with time.

Table 11.1 Ten Chirplets estimated from signal D1

Chirplets†	$\hat{a}$	$\hat{\phi}$(rad)	$\hat{t}_c$(s)	$\hat{f}_c$(Hz)	$\hat{c}$ (Hz/s)	$\hat{\Delta}_t$(s)
C_1	198.97	−2.18	3.56	5.59	−0.01	1.02
C_2	174.84	2.50	1.82	6.00	−1.35	0.26
C_3	131.53	1.40	1.47	7.75	−50.07	0.05
C_4	94.40	1.10	2.02	15.16	17.79	0.41
C_5	88.66	0.32	1.25	18.26	1.10	0.69
C_6	74.14	−2.23	4.80	8.38	−8.30	0.20
C_7	85.68	1.25	3.89	11.30	0.12	0.76
C_8	75.19	0.18	1.70	26.33	−0.42	1.01
C_9	66.76	−0.78	0.24	10.88	3.04	0.16
C_{10}	67.07	−2.77	1.02	34.37	2.05	0.78

†The ten chirplets are labeled as $c_1 - c_{10}$. The estimated parameters are in reference to Eq. (11.1). 'cc_n' is the coherent coefficient. (Reproduced with permission from Jie Cui; Willy Wong, "The adaptive chirplet transform and visual evoked potentials," IEEE Transactions on Biomedical Engineering, vol 53, no.7, pp. 1378–1384, July 2006)

To facilitate the determination of the stopping criterion, the mean and standard deviation (SD) of *cc* values of the ten chirplets from all five signals are summarized in Fig. 11.6. It can be seen that the values decrease monotonically: the entire curve is characterized by two line segments with a breakpoint at 0.10. The first line segment has a steep slope followed by a second segment which does not change appreciably. The first segment shows that three chirplets is generally sufficient to characterize the entire VEP signal. We therefore choose a stopping criterion of 0.10 for our data.

Figure 11.7 shows the visualization of the results in Table 11.1 using ACS and compares it to the conventional spectrogram calculated with STFT. The vertical dotted lines at 1 s represent the onset of visual stimuli. Figure 11.7B shows the ACS of the ten chirplets accompanied by the reconstructed signal shown directly below

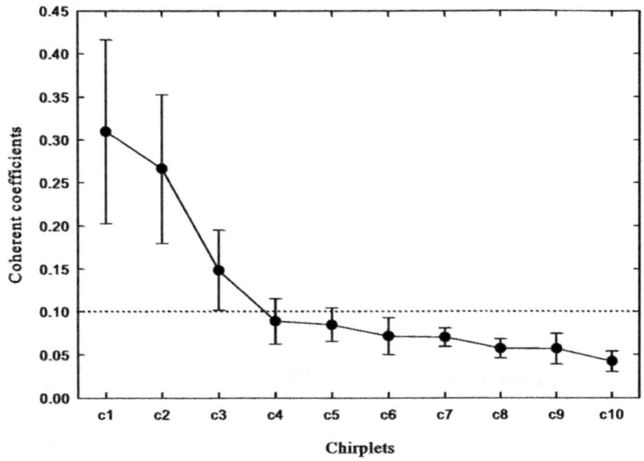

Fig. 11.6 Statistic of the coherent coefficients (Reproduced with permission from Jie Cui; Willy Wong, "The adaptive chirplet transform and visual evoked potentials," IEEE Transactions on Biomedical Engineering, vol. 53, no.7, pp. 1378–1384, July 2006)

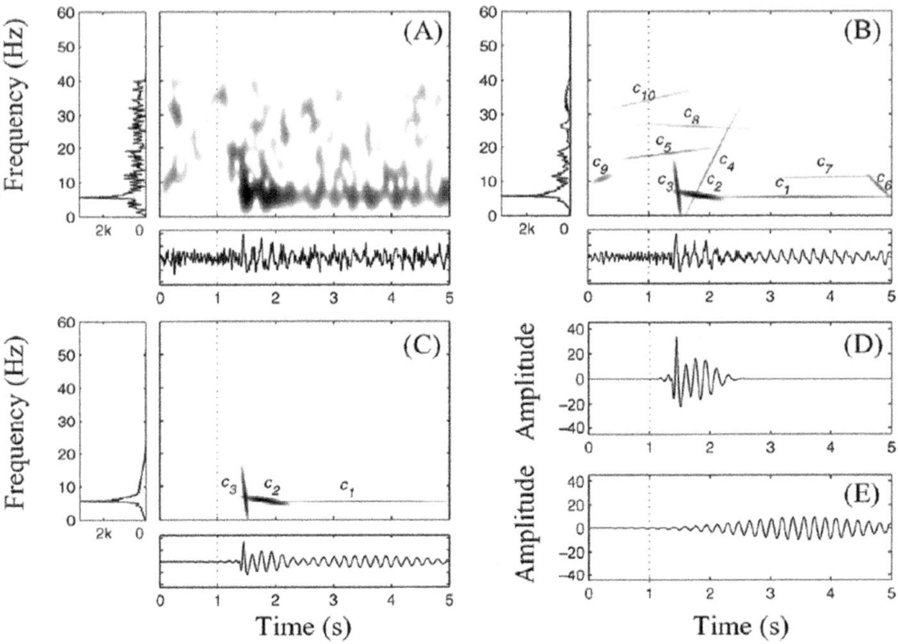

Fig. 11.7 Example: time-frequency analysis of signal D_1. (**A**) Spectrogram of D_1; immediately below is the waveform of D_1; the signal's spectrum is shown on the *left*. (**B**) ACS of the ten chirplets with the reconstructed signal (*below*) and its spectrum (*left*). (**C**) ACS of the first three chirplets (c_1–c_3) with the reconstructed signal (*below*) and its spectrum (*left*). (**D**) Reconstructed signal showing only c_2 and c_3 representing *t*VEP. (**E**) Reconstructed signal from c_1 only showing *ss*VEP (Reproduced with permission from Jie Cui; Willy Wong, "The adaptive chirplet transform and visual evoked potentials," IEEE Transactions on Biomedical Engineering, vol. 53, no. 7, pp. 1378–1384, July 2006)

and the spectrum on the left. It can be seen that the reconstructed signal provides a "less noisy" waveform. Chirplets c_1–c_3 are shown separately in Fig. 11.7C. These three chirplets show a typical representation of VEPs. In Fig. 11.7D, E, as will be discussed later, the signal was partitioned into its transient and steady-state components by assuming that the steady-state portion is constructed solely from $c_1$and the transient from c_2 and c_3.

The experimental results of the other four subjects are shown in Fig. 11.8. The ACS shows the chirplets with *cc* values above 0.10. Note that the results generally mirror the results shown for D_1. The statistics of the extracted parameters are summarized in Table 11.2. It can be seen that the values of the time-spread and the time-center are significantly different for the transient and steady-state components. ssVEPs have much wider time-spread (typically, $\hat{\Delta}_t \geq 1$) and a later time of appearance (typically, $\widehat{t_c} \geq 3.21$). Moreover both the chirp rate and the variance are close to zero at steady-state. Thus, we adopted the following criteria to classify the different components: a chirplet is characterized as steady-state if: (1) its time-spread

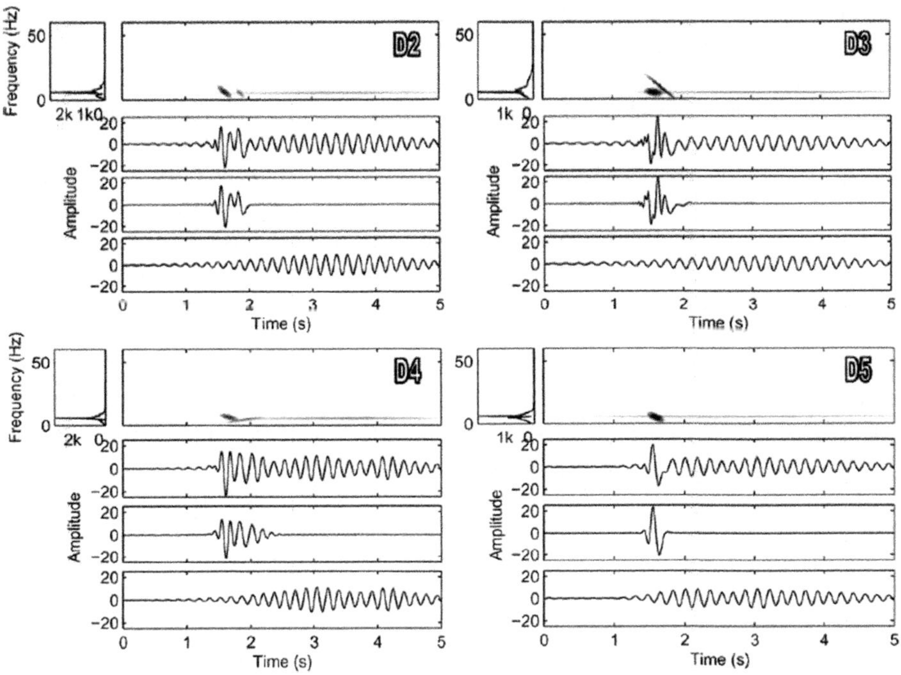

Fig. 11.8 ACS's of chirplets with *cc* values above 0.10 from signals D_2–D_5 (averaged signals from subjects 2–5). Spectrum of the reconstructed signal is shown on the left. The reconstructed signals, the *t*VEP and *ss*VEP components are shown immediately below the spectrogram (Reproduced with permission from Jie Cui; Willy Wong, "The adaptive chirplet transform and visual evoked potentials," IEEE Transactions on Biomedical Engineering, vol. 53, no. 7, pp. 1378–1384, July 2006)

is wider than 1 s; (2) its time-center is later than 2.21 s after the stimulus onset; (3) the chirp rate is less than 0.03 Hz/s. The remaining chirplets are then classified as transient responses.

In general, we concluded from our data that a 1200-point, 5 s, VEP response can be well represented by as few as three chirplets. This represents a total of 18 parameters (4 parameters per chirplet plus amplitude and phase). We can, thus, achieve a very sparse representation of the original VEP signal. Furthermore, the conventional

Table 11.2 Statistics of the parameters of transient and steady-state VEPs

	*t*VEP	*ss*VEP
$\hat{t}_c$(s)	1.69 ± 0.14	3.31 ± 0.20
$\hat{f}_c$(Hz)	6.34 ± 1.35	5.74 ± 0.16
$\hat{c}$(Hz/s)	-13.45 ± 15.98	-0.01 ± 0.02
$\hat{\Delta}_t$(s)	0.12 ± 0.08	1.14 ± 0.14

information of EP analysis such as amplitude and latency of the signal has been retained and can be retrieved easily from the reconstructed signal.

The ACS approach described in this chapter allows us to visualize the time-frequency structure of the VEP response at higher resolution than previously possible. Spectrograms that have been constructed using STFT will invariably involve smoothing of some sort yielding an overall lower resolution picture. Although the spectrogram can show some of the salient time-frequency structures of the VEP response, most of the detail is lost due to smearing. However, as we have shown with the ACS, e.g., Fig. 11.7B, C, the resulting time-frequency decomposition provides a clear picture of the underlying process. The estimated parameters obtained from the decomposition analysis provide detailed information about the local time-frequency structures of the signal not easily obtainable from the standard spectrogram alone.

By adopting the MLE algorithm, our approach assumes a signal model that the VEP signal can be represented by a weighted sum of chirplets in additive Gaussian white noise. To verify our model, we employed a statistic measure of whiteness (U), proposed by Drouiche [14], to test the whiteness of the residual signals after each iterative steps. Specifically, we test against the null hypothesis (H_0) that the residue is white. If U is larger than the critical value calculated according to a predefined threshold of significant level α, then the residue is not white . Typically, α is set at 5% and $t_\alpha = 1.65$. The mean and SD values of the whiteness measure are shown in Fig. 11.9. The measures were calculated after each of the iterations and then averaged across subjects. At a $p = 0.05$ significance level, we see that the residue is white after six iterations. That is, the VEP can be completely represented by as little as six chirplets. However, the first three estimated chirplets with the values above

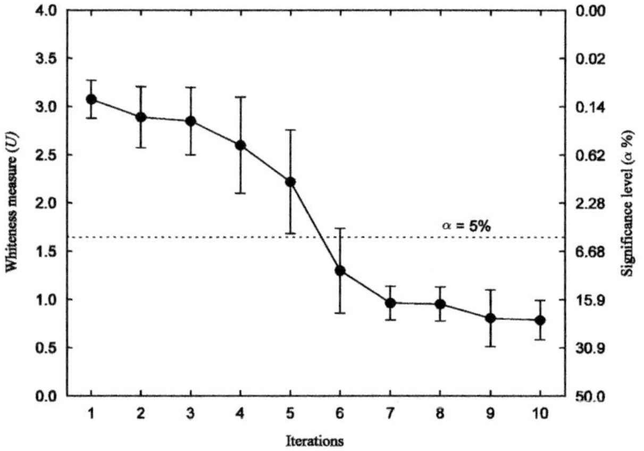

Fig. 11.9 Mean and SD values of the whiteness measure of signal residues. For each of the five signals D_1–D_5 (averaged signals from subjects 2–5), the whiteness measures (U) of signal residues after each of the iterations were calculated and the mean and SD values were obtained across the subjects (Reproduced with permission from Jie Cui; Willy Wong, "The adaptive chirplet transform and visual evoked potentials," IEEE Transactions on Biomedical Engineering, vol. 53, no. 7, pp. 1378–1384, July 2006)

Table 11.3 CRLBs of the estimates

$\hat{\theta}$	$\hat{t}_c$	$\hat{\omega}_c$	$\hat{c}$	$\hat{\Delta}_t$	$\hat{\phi}$	$\hat{A}$	$\hat{\sigma}^2$
CRLB[†]	$\frac{\Delta_t^2}{\xi}$	$\frac{1+4c^2\Delta_t^4}{\xi\Delta_t^2}$	$\frac{4}{\xi\Delta_t^4}$	$\frac{\Delta_t^2}{2\xi}$	$\frac{3+4\omega_c^2\Delta_t^2}{4\xi}$	σ^2	$\frac{\sigma^4}{N}$

[†] $\xi = A^2/2\sigma^2$

0.10 (Fig. 11.6) may be regarded as the principal components with respect to a given dictionary, since they appear to characterize most of the time-frequency variation of a VEP signal. Moreover, these three chirplets possess on average more than half of the energy in the VEP signals (cf. Table 11.3 in [9]).

11.3 Windowed ACT Method and Applications

One limitation of the non-windowed method comes from the difficulty in analyzing long-time signals. This is due to the time complexity of the analysis, which means that longer signals will incur a higher time cost. In situations where processing time is limited (e.g., in real-time applications), the non-windowed ACT approach may not be suitable. To reduce computational cost in the non-windowed ACT, one approach is the windowed ACT. In this method, the signal is partitioned into non-overlapping and equal length segments using rectangular truncation. A single chirplet is then estimated from each segment and hence the entire signal is approximated by a sequence of non-overlapping chirplets.

11.3.1 Optimal Window Length

With shorter data segments, the computational time can be significantly reduced. However, there is a tradeoff in terms of how much window length can be reduced while keeping relatively good time-frequency resolution for the estimation of local signal structures. Generally speaking, the window length of data segment must be short enough for the chirplet to adequately "sample" the time-dependent behavior of the signal spectrum, yet minimize the error in the estimate. We will describe next the theoretical minimum variance for chirplet estimation and how the minimum window length can be obtained.

First, we focus on the lower bound of window length. In windowed ACT analysis, one and only one chirplet is estimated from each of the segment with the MP algorithm. Consequently, we use a single chirplet model to represent the windowed signal, i.e., $f(t) = a \cdot g(t) + W(t)$, where $a = Ae^{-j\phi}$ is a chirplet coefficient with amplitude A and phase ϕ, $g(t)$ a chirplet defined in Eq. (11.1) (with the subscript I omitted) and $W(t)$ a process of Gaussian white noise (GWN) with zero mean and variance σ^2. In the MP algorithm, the four estimates of the chirplet parameters are obtained from

$$\begin{bmatrix} \hat{t}_c \\ \hat{\omega}_c \\ \hat{c} \\ \hat{\Delta}_t \end{bmatrix} = \underset{t_c, \omega_c, c, \Delta_t}{\arg\max} \left| \langle f, g_I \rangle \right|^2 \tag{11.7}$$

where g_I is one of the predefined chirplets contained in the "dictionary" [22]. Subsequently, the estimated chirplet can be found by substituting (11.7) into (11.1). The other estimates are found as

$$\hat{A} = |\hat{z}|, \hat{\phi} = -\measuredangle\hat{z}, \hat{\sigma}^2 = \frac{\|f\|^2 - |\hat{z}|^2}{2L}, \tag{11.8}$$

where L is the size of $f(t)$ and $\hat{z} = \langle f, g \rangle$. The minimum variance of these estimates can be obtained by Cramér-Rao lower bounds (CRLBs), for which derivations of the analytical expressions are provided next.

CRLBs can be obtained from the principal diagonal of the inverse Fisher information matrix [20]. We begin by presenting the approximations used in the derivation. When the signal is well within the recording interval and is adequately sampled, the moments of the chirplets can be approximated by the moments of the normal distribution:

$$\sum_{t=0}^{N-1} (t - t_c)^l \left| g\left(t - t_c\right) \right|^2 \approx \begin{cases} 1, & l = 0 \\ 0, & l = 1, 3 \\ \Delta_t^2/2, & l = 2 \\ 3\Delta_t^4/4, & l = 4, \end{cases} \tag{11.9}$$

$$\sum_{t=0}^{N-1} \frac{\partial \left| g\left(t\right) \right|^2}{\partial t_c} = 2\text{Re}\left\langle \vec{g}, \frac{\partial \vec{g}}{\partial t_c} \right\rangle \approx 0 \quad \text{and} \quad \sum_{t=0}^{N-1} \frac{\partial \left| g\left(t\right) \right|^2}{\partial \Delta_t} = 2\text{Re}\left\langle \vec{g}, \frac{\partial \vec{g}}{\partial \Delta_t} \right\rangle \approx 0, \tag{11.10}$$

where $\vec{g} = \{g(t)\}_{t=0}^{L-1}$. The seven deterministic but unknown parameters are $\vec{\theta} = \left\{\sigma^2, A, \Delta_t, c, t_c, \omega_c, \phi\right\}$ and the log likelihood function is

$$\begin{aligned} l\left(\vec{f};\vec{\theta}\right) &= \log p\left(\vec{f};\vec{\theta}\right) = -L\log\left(2\pi\sigma^2\right) - \tfrac{1}{2\sigma^2}\sum_{t=0}^{L-1} \left| f\left(t\right) - ag\left(t\right) \right|^2 \\ &= -L\log\left(2\pi\sigma^2\right) - \tfrac{A^2}{2\sigma^2} - \tfrac{1}{2\sigma^2}\left\|\vec{f}\right\|^2 + \tfrac{1}{\sigma^2}\text{Re}\left\langle \vec{f}, a\vec{g} \right\rangle. \end{aligned} \tag{11.11}$$

The Fisher information matrix $\left[I(\vec{\theta})\right]_{i,j} = -E[\partial^2 l(\vec{f};\vec{\theta})/\partial\theta_i\partial\theta_j], i, j = 1, 2, \cdots, 7$ is then found to be

$$I(\vec{\theta}) = \frac{A^2}{\sigma^2} \times \begin{bmatrix} \frac{N}{A^2\sigma^2} & 0 & 0 & 0 & 0 & 0 & 0 \\ 0 & \frac{1}{A^2} & 0 & 0 & 0 & 0 & 0 \\ 0 & 0 & \frac{1}{\Delta_t^2} & 0 & 0 & 0 & 0 \\ 0 & 0 & 0 & \frac{3\Delta_t^4}{16} & -\frac{\Delta_t^2\omega_c}{4} & 0 & \frac{\Delta_t^2}{4} \\ 0 & 0 & 0 & -\frac{\Delta_t^2\omega_c}{4} & \frac{1+2\omega_c^2\Delta_t^2+4c^2\Delta_t^4}{2\Delta_t^2} & -c\Delta_t^2 & -\omega_c \\ 0 & 0 & 0 & 0 & -c\Delta_t^2 & \frac{\Delta_t^2}{2} & 0 \\ 0 & 0 & 0 & \frac{\Delta_t^2}{4} & -\omega_c & 0 & 1 \end{bmatrix} \quad (11.12)$$

where $E[\bullet]$ denotes the expectation operation. From here the CRLBs of the estimates of $\vec{\theta}$ can be readily found. They are summarized in Table 11.3.

The performance of the estimators was evaluated by computer simulation (see details in [8]). We found that at the noise level found in our recordings of VEPs (SNR $\approx$ 0 dB) the window length should be not lower than 100 points. This determines the lower bound of the window length.

Next, we consider the upper bound on window length. A segment must be short enough to adequately sample the energy density curves of a signal in the time-frequency plane. Since most of the variation occurs in the transient, the choice of upper bound will be determined primarily by the transient portion of the signal. Thus, we propose the following method to estimate the duration of *t*VEP: (1) Reconstruct the *t*VEP signal from the chirplets extracted by the non-windowed ACT; (2) find the envelope of the *t*VEP; (3) locate the maximum of the envelope ($A > 0$) and set a threshold of $A/100$; (4) finally, define the duration time as the time interval between the instants at which the first point and the last point on the envelope exceeds this threshold. According to this procedure, we found that the average duration of our data was 300 ± 38 points. Considering the lower limit imposed by the SNR condition, we therefore choose 100 points (416.7 ms at sampling rate of 240 Hz) to allow the highest possible time-resolution.

11.3.2 VEP Analysis with Non-windowed ACT

We applied the non-windowed ACT to the same signals analyzed earlier. Each of the five averaged signals (D_1–D_5, 1200 points each) was divided into twelve 100-point segments with one chirplet estimated from each segment. This resulted in 12 chirplet atoms used to characterize the entire VEP response. The results of this procedure can be visualized again with Wigner-Ville distributions (WVDs). Figure 11.10 shows the results for all five signals D_1–D_5, where the individual atoms are labeled as A_1–A_{12}. The results show a similar pattern as before, that the signal before 1.25 s is well represented by three atoms, A_1–A_3 with lower amplitudes and higher frequency-centers, followed by two atoms with relatively stronger amplitudes and shorter time-spreads. Their chirp rates are generally negative, indicating downward transition of the instantaneous frequencies. The reminder of the signal can be reconstructed from a series of seven atoms denoted by A_6–A_{12}. Most of these atoms have wide

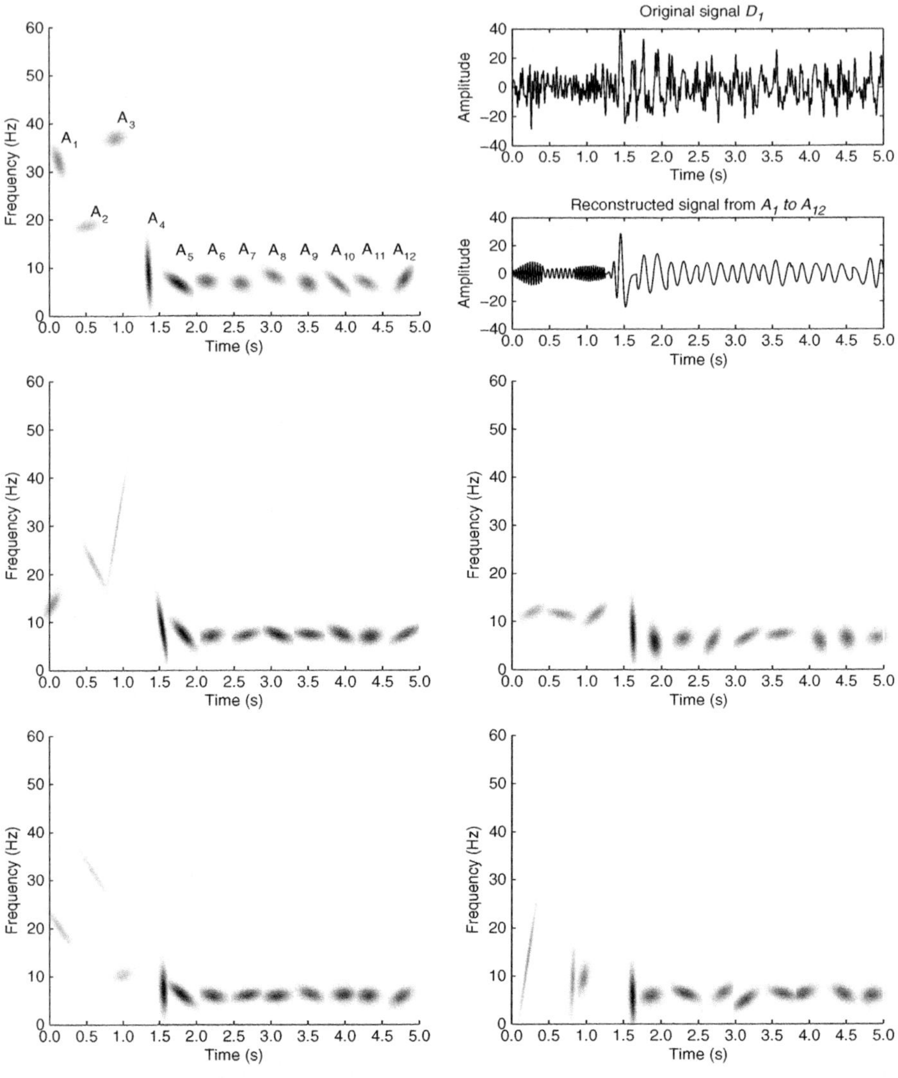

Fig. 11.10 Time-frequency structures of signals D_1–D_5. For each signal, 12 chirplet atoms have been estimated, which are labeled as A_1–A_{12} (panel D_1). As an example, the original signal D_1 and the reconstructed signal are shown on the right (Reproduced with permission from Jie Cui; Wong, W., "Investigation of short-term changes in visual evoked potentials with windowed adaptive chirplet transform," IEEE Transactions on Biomedical Engineering, vol. 55, no. 4, pp. 1449–1454, April 2008)

time-spreads and small chirp rates. Their frequency-centers are close to 6 Hz. These observations can be further confirmed by the statistics of the estimates (Fig. 11.11). For the sake of clarity, the standard deviations (SD) of the estimates are not shown in the figure, but are summarized in Table 11.4 instead. In panel (A), it can be seen

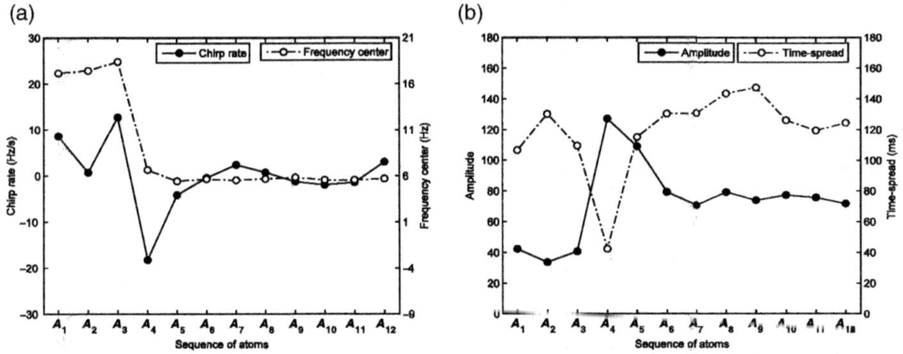

Fig. 11.11 Averaged parameters of the atoms A_1–A_{12}. (**a**) Averaged chirp rates and frequency-centers. (**b**) Averaged amplitudes and time-spreads (Reproduced with permission from Jie Cui; Wong, W., "Investigation of short-term changes in visual evoked potentials with windowed adaptive chirplet transform," IEEE Transactions on Biomedical Engineering, vol. 55, no. 4, pp. 1449–1454, April 2008)

that a clear transition of central frequencies can be observed between A_3 and A_4. In particular, the chirp rate of A_4 has a large negative value, indicating a sharp decrease in instantaneous frequency. The frequency-centers of the remaining chirplets are close to 6 Hz and their chirp rates are approximately zero. In panel (B) a prominent character is that A_4 acquires significantly shorter time-spread and higher amplitude.

Although the non-windowed approach of ACT gives a more compact representation of the VEP (panel (B) of Fig. 11.7), it should be noted that the non-windowed ACT demands much more computational time than its windowed counterpart. In practice, the non-windowed ACT also requires the entire signal before commencing analysis. The windowed method, on the other hand, reduces waiting time by requiring only one data segment, which is more suitable for situations of real-time

Table 11.4 SD of the parameter estimates of chirplets, including chirp rate (c), frequency-center (f_c), amplitude (a) and time-spread (Δ_t)

Atoms	SD(c) (Hz/s)	SD(f_c) (Hz)	SD(a)	SD(Δ_t) (s)
A_1	24.66	8.14	16.76	30.07
A_2	17.66	8.96	12.31	56.25
A_3	18.93	12.18	15.06	24.99
A_4	11.58	0.79	23.90	10.09
A_5	3.12	0.69	30.47	28.54
A_6	1.69	0.35	11.58	16.16
A_7	2.83	0.62	11.78	27.92
A_8	3.40	0.80	12.39	11.87
A_9	1.32	0.29	12.17	24.45
A_{10}	3.03	0.63	7.71	22.28
A_{11}	1.67	0.29	5.76	16.99
A_{12}	2.24	0.43	20.73	20.23

processing. The chirplets estimated by windowed ACT reveals a pattern similar to that of the non-windowed approach (cf. Figs. 11.7 and 11.8). The windowed approach will find applications where fast time computation and long-time signal monitoring are necessary.

11.4 Summary and Applications of ACT to Other Bio-signals

In this chapter, we have described the adaptive chirplet transform (ACT) and its two types of implementations, i.e., the non-windowed and windowed ACT. These two approaches were highlighted and illustrated through the analysis and estimation of visual evoked potentials. The ACT partially overcomes the disadvantages of conventional time-frequency methods and provides a parsimonious and high-resolution representation of VEP signals.

The non-windowed ACT was implemented by an iterative refinement algorithm. We showed that a complete VEP signal can be estimated by as few as seven iterations. Statistical analysis showed that a coherent coefficient (*cc*) value of 0.10 is a suitable stopping criterion for our data, as the chirplets with higher *cc* values reflect the primary time-frequency variations in the signals. A prominent advantage of the chirplet representation is its ability to separate the transient portion (*t*VEP) from the

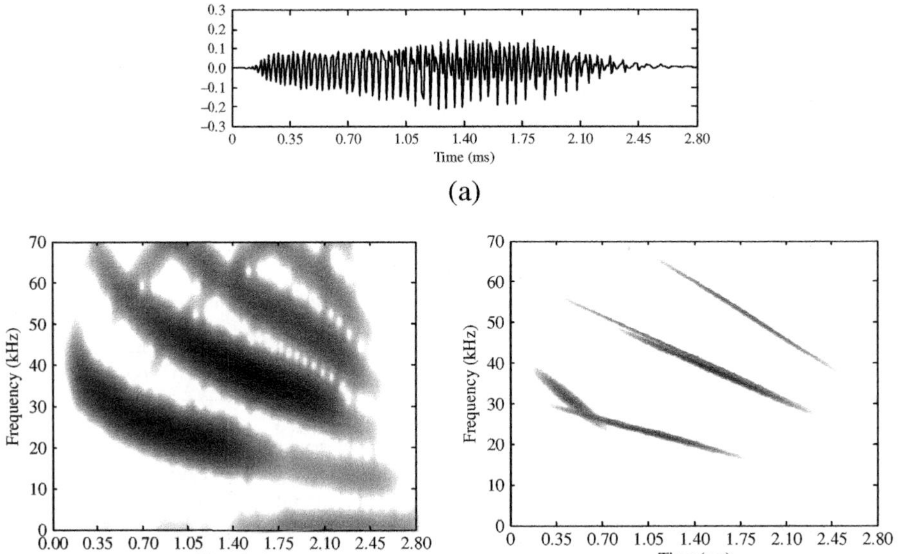

Fig. 11.12 Chirplet representations of bio-acoustical signals. (**a**) Time domain representation of the large brown bat echo-location signal (sampled at 0.14 MHz); (**b**) Spectrum of the signal (calculated with a 0.45 ms Gaussian window); and (**c**) Adaptive chirplet spectrum (ACS) representation of the signal (represented by five chirplets)

steady-state portion (*ss*VEP) as was demonstrated in Figs. 11.7 and 11.8. With the aid of the adaptive chirplet spectrogram (ACS), we also achieved a clearer visualization of the results than by conventional methods.

The windowed ACT was developed to reduce time cost and with the ultimate goal of real-time application. In this method, the data were windowed into equal-length non-overlapping segments and only one chirplet was estimated from each segment. We derived analytical expressions for the Cramér-Rao lower bounds (CRLBs) for the estimates of a single chirplet in additive Gaussian noise (Table 11.3). The selection of the optimal window length was also discussed in detail. We found that the lower bound of window length is determined by the signal-to-noise (SNR) ratio of the signal and the upper bound by the duration of the transient portion of the VEPs. We concluded that the windowed method could adequately characterize the entire VEP response from *t*VEP to *ss*VEP, and reveal a pattern similar to that found

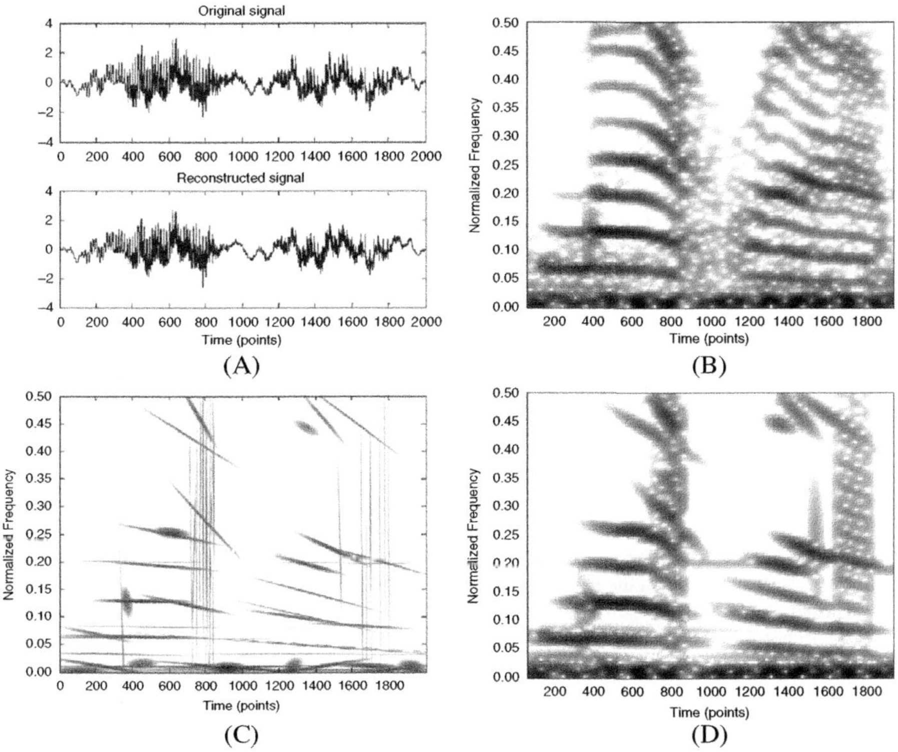

Fig. 11.13 Chirplet representation and compression of speech signal. (**A**) The waveforms of an acoustic speech signal of the spoken word "Matlab". The *upper panel* shows the original signal and *lower panel* shows the reconstructed signal from 60 chirplets; (**B**) Spectrum of the original signal; (**C**) ACS of the speech signal (represented by 60 chirplets); and (**D**) Spectrum of the reconstructed signal. Note that the original signal required 2000 real values for storage, while 60 chirplets required only 360 real values

by the non-windowed ACT. However, the time cost of the windowed method is significantly lower than that of the non-windowed method.

We emphasize that both the non-windowed and windowed approaches provide a unified representation of the *complete* VEP response (both transient and steady-state VEPs). We believe that the ACT method will be especially suitable for characterizing biomedical signals that consist of complicated time-frequency components.

As a new tool of time-frequency analysis, the ACT has potential applications in the analysis of a variety of signals that involve chirping components. Besides EEG signals, chirps exist in many other naturally occurring signals. One example is given in Fig. 11.12, where we show a chirplet decomposition of a bat echo-location ultrasound signal. The chirplets were estimated with the non-windowed ACT method and visualized with the ACS method. Note again that the compact representation achieved with the ACT approach results in high level of signal compression. Next we show an example of speech decomposition using chirplets. Figure 11.13 shows a chirplet analysis of a female utterance of the word "Matlab". The signal length was 2000 points sampled at 3.71 kHz. 60 chirplets were extracted and a reconstructed speech signal was obtained from these chirplets. There is minimal perceptual difference between the original and the reconstructed signals. The waveforms of the signals are shown in Panel (A) of Fig. 11.13. The corresponding ACS and spectrogram obtained by STFT are shown in Panels (B, C, D). Note that one chirplet can be described by six real values, i.e., two for the complex coefficients (11.2) and four for chirplet parameters, $I = (t_c, \omega_c, c, \Delta_t)$ in (11.1). Therefore, the 60 chirplets required only 360 real values for the entire data recording, showing an approximately 80% reduction in data size when compared with the 2000 real values of the original signal.

References

1. Akay, M. (1998). *Time-frequency and wavelets in biomedical signal processing*. New York: IEEE Press.
2. Aunon, J. I., McGillem, C. D., & Childers, D. G. (1981). Signal processing in evoked-potential research – averaging and modeling. *Crc Critical Reviews in Bioengineering,* 5(4), 323–367.
3. Boushash, B., & Whitehouse, H. J. (1986). *Seismic applications of the Wigner-Ville distribution.* Paper presented at the Proceedings of IEEE International Conference on Circuits System.
4. Bultan, A. (1999). A four-parameter atomic decomposition of chirplets. *IEEE Transactions on Signal Processing,* 47(3), 731–745.
5. Cheng, M., Gao, X. R., Gao, S. G., & Xu, D. F. (2002). Design and implementation of a brain-computer interface with high transfer rates. *IEEE Transactions on Biomedical Engineering,* 49(10), 1181–1186.
6. Coelho, F., Simpson, D., & Infantosi, A. (1995). *Testing recruitment in the EEG under repetitive photo stimulation using frequency-domain approaches*. Paper presented at the IEEE-EMBC and CMBEC, Montreal, Canada, 905–906.
7. Cohen, L. (1995). *Time-frequency analysis*. Englewood Cliffs, N.J: Prentice Hall PTR.
8. Cui, J. (2006). *Adaptive chirplet transform for the analysis of visual evoked potentials.* Unpublished PhD Dissertation, University of Toronto, Toronto.
9. Cui, J., & Wong, W. (2006). The adaptive chirplet transform and visual evoked potentials. *IEEE Transactions on Biomedical Engineering,* 53(7), 1378–1384.

10. Cui, J., Wong, W., & Mann, S. (2005). Time-frequency analysis of visual evoked potentials using chirplet transform. *Electronics Letters,* 41(4), 217–218.
11. Daubechies, I. (1992). *Ten lectures on wavelets.* Philadelphia, PA: Society for Industrial and Applied Mathematics.
12. Desmedt, J. E. (1990). *Visual evoked potentials.* Amsterdam, New York, Oxford: Elsevier Science Publishers B.V. (Biomedical Division).
13. Di Russo, F., & Spinelli, D. (2002). Effects of sustained, voluntary attention on amplitude and latency of steady-state visual evoked potential: A costs and benefits analysis. *Clinical Neurophysiology,* 113(11), 1771–1777.
14. Drouiche, K. (2000). A new test for whiteness. *IEEE Transactions on Signal Processing,* 48(7), 1864–1871.
15. Fricker, S. J. (1962). Narrow-band filter techniques for detection and measurement of evoked responses. *Electroencephalography and Clinical Neurophysiology,* 14(3), 411–421.
16. Gabor, D. (1946). Theory of communication. *Journal of IEE,* 93(26), 429–457.
17. Halliday, A. M. (1993). *Evoked potentials in clinical testing.* Edinburgh: Churchill Livingstone.
18. Harding, G. F. A. (1974). The visual evoked response. In M. J. Roper-Hall (Ed.), *Advances in Ophthalmology* (pp. 2–28). Basel: S Kaerger AG.
19. Heckenlively, J. R., & Arden, G. B. (2006). *Principles and practice of clinical electrophysiology of vision* (2nd ed.). Cambridge, Mass.: MIT Press.
20. Kay, S. M. (1993). *Fundamentals of statistical signal processing: Estimation and detection theory.* Englewood Cliffs, N.J.: Prentice-Hall PTR.
21. Liavas, A. P., Moustakides, G. V., Henning, G., Psarakis, E. Z., & Husar, P. (1998). A periodogram-based method for the detection of steady-state visually evoked potentials. *IEEE Transactions on Biomedical Engineering,* 45(2), 242–248.
22. Mallat, S. G., & Zhang, Z. (1993). Matching pursuit with time-frequency dictionaries. *IEEE Transactions on Signal Processing,* 41(12), 3397–3415.
23. Mann, S., & Haykin, S. (1991, June 1991). *The chirplet transform: A generalization of Gabor's logon transform.* Paper presented at the Vision Interface, Calgary, Canada.
24. Mann, S., & Haykin, S. (1992). Adaptive chirplet transform - An adaptive generalization of the wavelet transform. *Optical Engineering,* 31(6), 1243–1256.
25. Mann, S., & Haykin, S. (1995). The chirplet transform - physical considerations. *IEEE Transactions on Signal Processing,* 43(11), 2745–2761.
26. Mast, J., & Victor, J. D. (1991). Fluctuations of steady-state VEPs – Interaction of driven evoked-potentials and the EEG. *Electroencephalography and Clinical Neurophysiology,* 78(5), 389–401.
27. Middendorf, M., McMillan, G., Calhoun, G., & Jones, K. S. (2000). Brain-computer interfaces based on the steady-state visual-evoked response. *IEEE Transactions on Rehabilitation Engineering,* 8(2), 211–214.
28. Norcia, A. M., & Tyler, C. W. (1985). Spatial-frequency sweep VEP – visual-acuity during the 1st year of life. *Vision Research,* 25(10), 1399–1408.
29. Pei, F., Pettet, M. W., & Norcia, A. M. (2002). Neural correlates of object-based attention. *Journal of Vision,* 2(9), 588–596.
30. Qian, S., & Chen, D. P. (1994). Signal representation using adaptive normalized Gaussian functions. *Signal Processing,* 36(1), 1–11.
31. Qian, S., Dunham, M. E., & Freeman, M. J. (1995). Transionospheric signal recognition by joint time-frequency representation. *Radio Science,* 30(6), 1817–1829.
32. Regan, D. (1966). Some characteristics of average steady-state and transient responses evoked by modulated light. *Electroencephalography and Clinical Neurophysiology,* 20(3), 238–248.
33. Regan, D. (1972). *Evoked potentials in psychology, sensory physiology and clinical medicine.* London: Chapman and Hall.
34. Regan, D. (1989). *Human brain electrophysiology: Evoked potentials and evoked magnetic fields in science and medicine.* New York: Elsevier.

35. Tang, Y., & Norcia, A. M. (1995). Application of adaptive filtering to steady-state evoked response. *Medical & Biological Engineering & Computing,* 33(3), 391–395.
36. Van Der Tweel, L. H. (1964). Relation between psychophysics and electrophysiology of flicker. *Documenta Ophthalmologica,*, 18 287–304.
37. Victor, J. D., & Mast, J. (1991). A New Statistic for Steady-State Evoked-Potentials. *Electroencephalography and Clinical Neurophysiology,* 78(5), 378–388.
38. Wolpaw, J. R., Birbaumer, N., McFarland, D. J., Pfurtscheller, G., & Vaughan, T. M. (2002). Brain-computer interfaces for communication and control. *Clinical Neurophysiology,* 113(6), 767–791.

Chapter 12
Uterine EMG Analysis: Time-Frequency Based Techniques for Preterm Birth Detection

Mohamad Khalil, Marwa Chendeb, Mohamad Diab, Catherine Marque and Jacques Duchêne

Abstract The global aim of this work is to detect preterm deliveries using the uterine electromyography signal. For this purpose, two steps are required: the first step aims to detect all events in this signal and to identify these events by allocating them to physiological classes: contractions, foetus motions, Alvarez or Long Duration Low Frequency (LDBF) waves. The second step consists of the identification of contractions between normal contractions and preterm birth contractions. Detection and identification of events are based on the use of Wavelet Packet Transform (WPT) to select the *best basis* for both detection and classification goals. This is achieved by selecting the only WPs of the decomposition tree that are able to highlight changes in the recordings using a training set of signals. The selection criterion is based on the Kullback Leibler distance and the Generalised Gaussian Distribution (GGD). The detection algorithm "Dynamic Cumulative Sum" (DCS) is applied on selected packets. This combination of DCS and WP decomposition has been shown to be very efficient for the detection of both frequency and energy changes. After obtaining the uterine events, classification consists of identifying the detected events. The ratio between the intra-class variance and the total variance (sum of the inter-class and intra-class variances) is used as a criterion for the *best basis* of classification. In most cases, more than 85% of events are well detected and classified whatever the term of gestation. Finally, the classification of contractions is achieved using wavelet networks. This network is learned using normal contractions and preterm births contractions. As final results, uterine contractions are well classified whatever the term of gestation and this global system (detection, classification, identification and diagnostic) provides good results concerning the preterm births.

M.Khalil (✉)
Lebanese University, Faculty of Engineering, Section 1, El Koubbe, Tripoli, Lebanon
e-mail: m.khalil@ul.edu.lb

A. Naït-Ali (ed.), *Advanced Biosignal Processing*,
DOI 10.1007/978-3-540-89506-0_12, © Springer-Verlag Berlin Heidelberg 2009

12.1 Introduction

Preterm delivery, the primary cause of neonatal morbidity and mortality, remains a major problem in obstetrical practice [1]. In most developed countries, the rate of preterm births evolves between 5% and 9%. In France, a social politics of follow up and prevention of pregnancies has permitted the achievement of a low rate of 6%. Yet, despite a wealth of research work concerning the risk of premature birth, this rate has not changed for ten years. The early detection of possible preterm deliveries seems a determinant factor for the success of tocolytic treatments, hence for the prolongation of the *in utero* foetus development. The need for such an early detection opens the way towards the development of an ambulatory instrumentation based on uterine EMG for patient monitoring during pregnancy.

Uterine EMG has been a research subject since the 1950s [2–8]. In the uterus, action potentials usually occur in bursts whose frequency content ranges approximately from less than 1 Hz to 10 Hz. Uterine contraction is not the only event contained in surface EMG (SEMG) recordings. Other events can be of value for pre-term birth diagnosis: Alvarez (Alv) waves, foetus motions (MAF) waves and long duration low frequency (LDBF) waves. Only contractions are used to identify a risk of preterm delivery.

This chapter is divided into two parts: the first part presents new techniques to detect and identify all events present in the uterine EMG signal: Contractions (CT), Alvarez waves, foetus motions and LDBF waves [9, 10]. The second part deals with the contraction classification in order to diagnose a risk of preterm delivery (Fig. 12.1).

In terms of event detection, signal changes can be due either to a process modification (EMG background activity vs contraction) or to the appearance of superimposed events (foetus motions, Alvarez waves, LDBF waves). These events do not have standard features and are dependent upon the woman and the gestation term. The method proposed in this paper is therefore designed to detect changes with no *a priori* knowledge of event characteristics. This approach requires an assumption of piecewise stationarity and uses a multiscale decomposition based on Wavelet Packets (WP) [11]. The method consists in signal decomposition using a Wavelet Packet Transform (WPT), then in the selection of a subset of packets, from the WP tree, that highlights changes at best. The main idea behind WP selection (*best basis* identification) is to first define an index of change detectability, which is then computed on each WP of the tree. Selection of the best WP subset corresponds to the WP whose

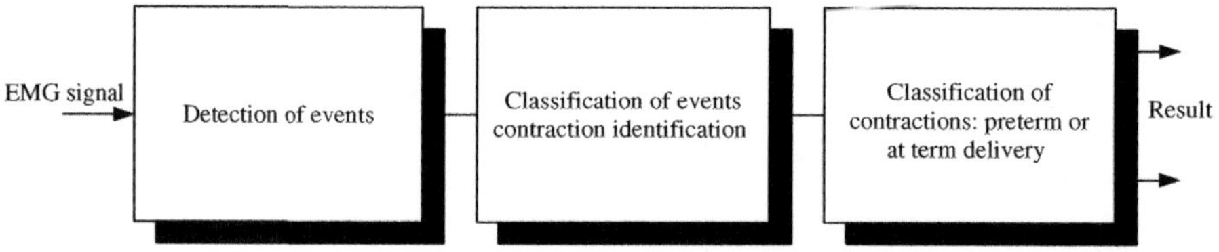

Fig. 12.1 Procedure of detection, classification and decision making

index exceeds a given threshold. The index definition is based on the distribution of the Kullback-Leiber Distance (KLD).

The detection algorithm is applied on the coefficients of the WP previously selected, in the same way as in the segmentation method proposed by Ranta et al. [12].

In WP decomposition the successive bandwidths of the scale levels decrease by a factor of two, as does the number of coefficients produced by the successive scale levels, so that a proper selection of coefficients from different scales may be used to compress or represent original signals in a compact form [13, 14].

The direct use of the wavelet coefficients without returning to the reconstructed signals is considered as an effective tool for signal processing, associating good performance and low processing time [11], and is, hence, well-adapted for detection in real time.

Detection results in a signal segmentation that isolates the various events contained in the recordings, including background activity. The next step consists of identifying those events by allocating them to physiological classes: contractions, foetus motions, Alvarez waves, or LDBF waves. The same approach by WP selection has been used for classification purposes. The idea is to keep the only WPs that optimally classify the events of a reference database (learning step). The WP selection criterion is based on the calculation of the ratio between intra-class and total variance, calculated directly from the coefficients of the WP.

In the final classification step, each event is decomposed onto the selected WP basis and the relative energy produced by each WP is computed. These energy values constitute the inputs of a classifier: several classifiers, such as K-nearest neighbors [15], the Mahalanobis distance [15], Neural Networks [16], and Support Vector Machine [17], were evaluated in the current study.

Among all classified events, only the contraction class has been used to detect the risk of preterm delivery. For that purpose, features used for classification between "efficient" and "unefficient" contractions are classically based on spectral characteristics [10, 18, 19]. Unfortunately, up to now, due to a wide inter individual variability and to the evolution of contraction characteristics along pregnancy and during parturition, there are no gold standard classes corresponding to well-identified physiological situations. Therefore, the classification process must be unsupervised, the class labeling being the ultimate step when each class is identified, comparing its features with those of the others classes. Classification of contractions is based on Wavelet Neural Network which can classify all uterine contractions, recorded in different women, according to the frequency content of their related EMG burst. Work with specialists has to be achieved later to interpret this classification on a physiological basis.

Consequently, this chapter is organized as follows. In Sect. 12.2, the detection method is presented. Wavelet Packet decomposition is shortly described and the method of WPC distribution characterization by estimating the scale and shape parameters of the Generalized Gaussian Density (GGD) is presented. The Kullback-Leibler Distance (KLD) is then described as a criterion for WP selection. Then, change detection in real time is briefly described, while the performance of the methods and examples using real datasets are also shown in this section. Supervised classification methods and corresponding results are presented in Sect. 12.3. Section 12.4 describes the contraction classification method using an artificial neural network. Finally, a conclusion and some perspectives for the near future are drawn in Sect. 12.5.

12.2 Event Detection in Uterine EMG

12.2.1 Wavelet Packet Transform (WPT)

Wavelet packet transform (WPT) is a wavelet decomposition where each level is calculated by passing details and approximations of the previous level through high and low pass filters (g and h respectively). For N decomposition levels the WPT produces 2N sets of coefficients (or nodes). Figure 12.2 represents the WPT tree. Each node $W_{j,n}$ is associated with a subspace $\Omega_{j,n}$ generated by an orthonormal basis $\{\psi_{j,n}\}_{n\in Z}$, with j being interpreted as a scale parameter, and n as a sequence parameter. The WPC at each node (j,n) are computed as [20]:

$$\left(C_{j,n}(k)\right) = \left\langle f(t), \psi_{j,n}(t - 2^j k)\right\rangle \tag{12.1}$$

where $f(t)$ is the initial signal and k is the time-localization index.

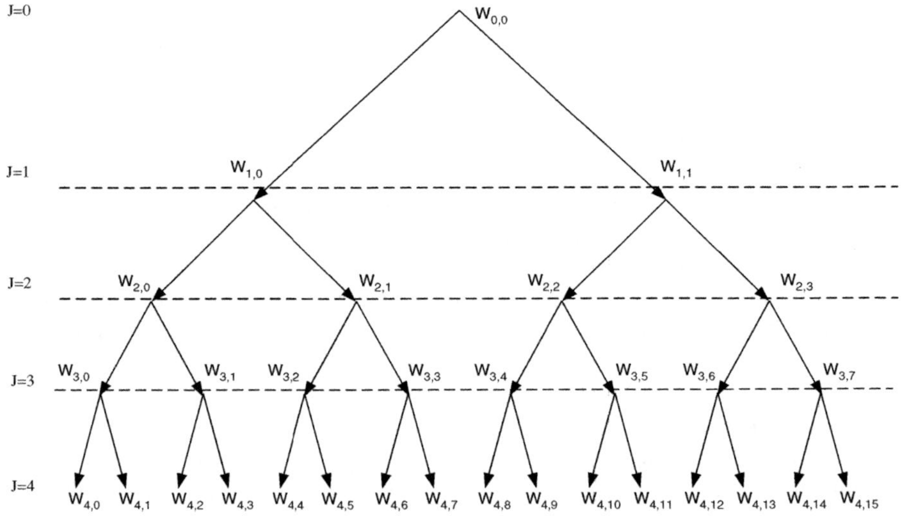

Fig. 12.2 Wavelet Packet Decomposition Tree

12.2.2 *Best Basis* Selection

12.2.2.1 Introduction

The concept of *best basis* generally relies on a selection criterion determined by analysis goals, such as compression, filtering (smoothing) [21], feature extraction and classification [22], etc. Ravier and Amblard presented a detector of transient acoustic signals combining the local wavelet analysis and higher-order statistical properties of the signals [23]. Leman and Marque used WPT and proposed a specific criterion to denoise the EHG signal [24]. Hitti and Lucas proposed a *best basis* selection method to detect abrupt changes in noisy multi-component signals [25]. They used an energy criterion to allow separation of the different frequency components of the signal from a wavelet-packet library tree. In the same way, Saito and Coifman introduced the Local Discriminant Bases (LDB) to search for a *best basis* for classification [26].

In our case the criterion has to reflect the ability of the WP subset to detect at best the various events contained in the recordings, hence to detect any change from background noise to a relevant event (and vice versa).

12.2.2.2 Criterion of the *Best Basis* Selection

The goal of this study was to define a criterion that allows the selection of the most relevant packets for a detection problem, taking into account the specificity of the application. An approach to highlight changes is to define a statistical model of the various events at each WP level (including background activity), then to select those WP for which the model differs the most from that of the background activity. This approach obviously implies the need for a measure of distance between distributions. The Kullback-Leibler Distance (KLD) has already been used in this way in previous works, (e.g., in classification problems or image comparison [26, 27]). In the present study it is proposed that we use KLD for the definition of the cost function for WP selection for the detection problem. It can be seen as a distance $D(f(X,\theta_p), f(X,\theta_q))$ between two Probability Density Functions (PDF) $f(X,\theta_p)$ and $f(X,\theta_q)$. A natural choice of D is the relative entropy (also known as cross entropy, KLD, or I divergence) between two PDFs [28]:

$$D(f(X;\theta_p),(f(X,\theta_q)) = \int f(x,\theta_p)\log\frac{f(x,\theta_p)}{f(x,\theta_q)}dx \tag{12.2}$$

In our application this distance is accessible only by estimation, as KLD computation requires the knowledge of the WPC distribution.

12.2.2.3 WPC Distribution

Previous works [27] have shown that a good approximation for the Probability Density Function (PDF) of WPC at a particular sub-band level might be achieved by adaptively varying the two parameters of a Gaussian Generalized Density (GGD) that is defined as:

$$f(x;\alpha,\beta) = \frac{\beta}{2\alpha\Gamma(1/\beta)} e^{-(|x|/\alpha)^{\beta}} \tag{12.3}$$

where $\Gamma(.)$ is the gamma function: $\Gamma(z) = \int_0^{\infty} e^{-t} t^{z-1} dt,\ z>0$. α is related to the width of the PDF peak (standard deviation *sd*), and is generally referred to as the scale parameter. β is inversely proportional to the decreasing rate of the peak, and is called the shape parameter. The GGD model is Gaussian when $\beta = 2$ and $\alpha = sd\sqrt{2}$.

The goal of this section is to verify that the events (contractions, foetus motions, Alvarez waves or LDBF waves) of the real signals, as well as the Wavelet Packet Coefficients, follow a GGD and to estimate α and β, by using the Maximum Likelihood method [27, 28, 29].

In a first step, α and β were estimated using a training test composed of 50 events of each type (contractions, foetus motions, Alvarez waves, LDBF waves) as well as 50 segments of recorded background activity (hereafter called "noise"). Two estimation methods were tested: the first one was to concatenate the whole set of events of each type and to estimate α and β only once. The second one was to estimate α and β for each segment, then to calculate their mean and standard deviation.

The adjustment quality between the distributions of the events and the GGD was performed with the Kolmogorov Smirnov test [30]. In this test, the Kolmogorov Smirnov statistic $D_{\max}$, which corresponds to the maximum distance between theoretical and sample cumulative distributions, is calculated to decide whether or not an event follows a GGD. Table 12.1 shows the estimated values of α, β and the calculated values of $D_{\max}$ (all events concatenated). In addition, Table 12.1 contains the probability $P_{D_{\max}} = P(D > D_{\max}))$ in case of GDD hypothesis.

In the case of recorded noise, the segments follow a GGD with parameters ($\alpha \approx 1.3591$ and $\beta \approx 1.7529$), which are very close to a normal distribution. Figure 12.3 shows the histogram obtained from concatenated segments of a recorded SEMG (raw signal and selected WP).

Table 12.1 Estimated values of α, β and $D_{\max}$, for the various uterine EMG events (concatenated events), $P_{D_{\max}} = P(D > D_{\max})$

	α	β	$D_{\max}$	$P_{D_{\max}}$
CT	1.0383	1.3303	0.0038	0.2000
Alvarez	1.1947	1.5473	0.0081	0.1360
MAF	0.9557	1.2337	0.0119	0.0921
LDBF	0.9868	1.2711	0.0033	0.2000
Noise	1.3591	1.7529	0.0195	0.0300

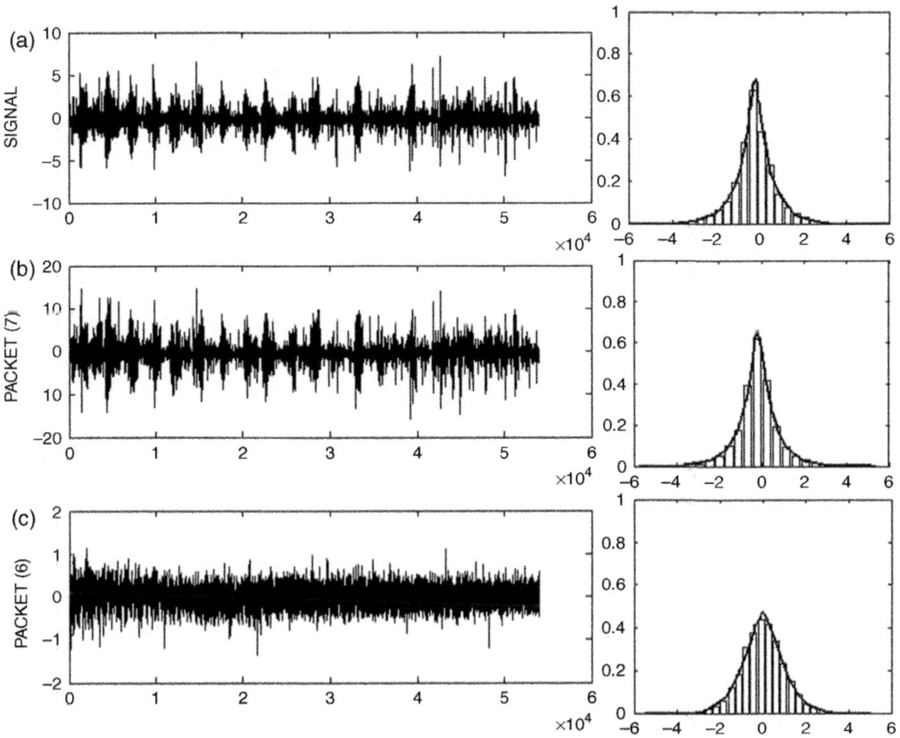

Fig. 12.3 (**a**) Uterine SEMG signal obtained from a pregnant subject with the corresponding histogram and GGD distribution. (**b**) WP 7 coefficients and the corresponding histogram and GGD distribution ($\alpha = 0.6142$, $\beta = 0.9257$). (**c**) WP 6 coefficients and the corresponding histogram and GGD distribution ($\alpha = 1.3853$, $\beta = 1.9207$)

As WPT is a linear transformation, WP coefficients exhibit the same statistical properties as the initial signal [23]. Consequently, WPC extracted from uterine events also follow a GGD. It can be extrapolated that each event contained in a recording of uterine EMG, can be statistically modelled by a Generalized Gaussian Density, whereas the associated noise can be described by a Normal (Gaussian) Distribution.

12.2.2.4 Distribution of the Estimated Kullback-Leibler Distance

As demonstrated before, the PDF of the noise WPC in each sub-band roughly follows a Gaussian distribution ($\beta = 2$). In that case the KLD becomes [31]:

$$D(f(.;\alpha_p), f(.;\alpha_q)) = K_{pq} = \log\left(\frac{\alpha_q}{\alpha_p}\right) + \frac{1}{2}\left(\frac{\alpha_p}{\alpha_q}\right)^2 - \frac{1}{2} \tag{12.4}$$

As KLD is not symmetrical, it is proposed to use: $K = K_{pq} + K_{qp}$.

If N is the length of the sequence $\mathbf{x} = \{x_1, x_2, \ldots x_N\}$, an estimation of the parameter α is given by:

$$\hat{\alpha} = \sqrt{\left(\frac{1}{N}\sum_{i=1}^{N}|x_i|^2\right)} \tag{12.5}$$

hence an estimation of K :

$$\hat{K} = \frac{1}{2}\left[\left(\frac{\hat{\alpha}_p}{\hat{\alpha}_q}\right)^2 + \left(\frac{\hat{\alpha}_q}{\hat{\alpha}_p}\right)^2 - 2\right] \tag{12.6}$$

If it is considered that only limited independent sequences $\mathbf{x}_p = \left\{x_{p1}, x_{p2}, \ldots, x_{pN}\right\}$ and $\mathbf{x}_q = \left\{x_{q1}, x_{q2}, \ldots, x_{qN}\right\}$ are available for K_{pq} estimation, then $\hat{\alpha}_p$ and $\hat{\alpha}_q$ are computed using (12.5). Hence [31]:

$$E(\hat{K}_{pq} + \hat{K}_{qp}) = E(\hat{K}) = \frac{2}{N-2} \tag{12.7}$$

12.2.2.5 Wavelet Packet Selection for Detection Purposes

If there exists a model of the distribution of K when there are no changes in a record, then a way of selecting the best WP basis would be as follows:

- Build a record including a lot of changes,
- Process it by each WP of the decomposition tree,
- For each WP output, construct the K histogram and estimate the parameters of the corresponding distribution,
- Select the WPs that exhibit the most significant difference with a distribution corresponding to "no changes in the record".

As no analytical expression of the distribution of $\hat{K}$ is available, the idea is to approach this distribution with a known distribution having at least the same general shape and expectation. Taking into account the shape of the histogram of K, an exponential distribution depending on only one parameter λ was chosen as a first approximation of the histogram. Its PDF is defined as:

$$f(x) = \lambda.e^{-\lambda.x} \text{with} E(x) = \frac{1}{\lambda} \tag{12.8}$$

An adjustment between $f(x)$ and the histogram is easily obtained by equalizing both expectations:

$$\frac{1}{\lambda} = \frac{2}{N-2} \tag{12.9}$$

In order to compare the distributions that are produced by the selection algorithm, the Kolmogorov Smirnov statistic $D_{\max}$ was used. Each WP produces a Dmax value for a given recording. After classifying those values in descending order, a threshold can be defined if there is a clear separation between WP enhancing the changes and the others (containing mostly noise) (see Fig. 12.5 and the results section). As a result, a node in the WP tree was put to "1" if the corresponding WP was selected, the others being put to "0".

The previous step identified all WP in the decomposition tree where significant activities were detected. As the tree is highly redundant, there is a need for a further step to reduce the number of nodes. The current implementation of this second step of the selection algorithm roughly follows the one proposed by Hitti and Lucas [25]. Let us define a Father Node (FN), as any connection between two branches whose ends are called Children Nodes (CN). A Father Node at level j is a Children Node for level (*j*-1). A component is the association of a FN and its two CN. Our algorithm selecting the *best basis* can be described as follows:

1. Put value "1" or "0" to each node according to the Dmax result (Fig. 12.4a)
2. Modify the node values according to the following rules:
   If all nodes in a component have the value "1", put "0*" to the FN (the whole information is contained in CN)
   If FN =1 and both CN = 0 or 0* in a component, put FN to "0" (aberrant case or information already taken into account).
   Select all CN = 1 with FN = 0 in a component (information is only detectable in the CN).
   If FN = 1 with CN1= 1 and CN2 = 0, select FN (FN and CN1 display the same information).
   If FN = 1 with CN1 = 1 and CN2= 0*, select CN1 (the only information not taken into account yet is in CN1). All these cases are displayed in Fig. 12.4b.
3. Select all nodes at "1" (Fig. 12.4c).

12.2.3 Detection from the Selected Wavelet Packets

12.2.3.1 Detection Algorithm

Any detection algorithm could work for the evaluation of the performance of the *best basis* selection algorithm. However, as a specific method, the DCS (Dynamic Cumulative sum) [32], has already been developed by our research group to be well adapted to uterine EMG recordings using the following method:

In a few words, DCS is based on local cumulative sums of likelihood ratios, computed between two locally-estimated distributions around time t. The parameters of the distributions Θ_b(before) and Θ_a(after) are estimated using two windows of identical length L before and after the current time *t*.

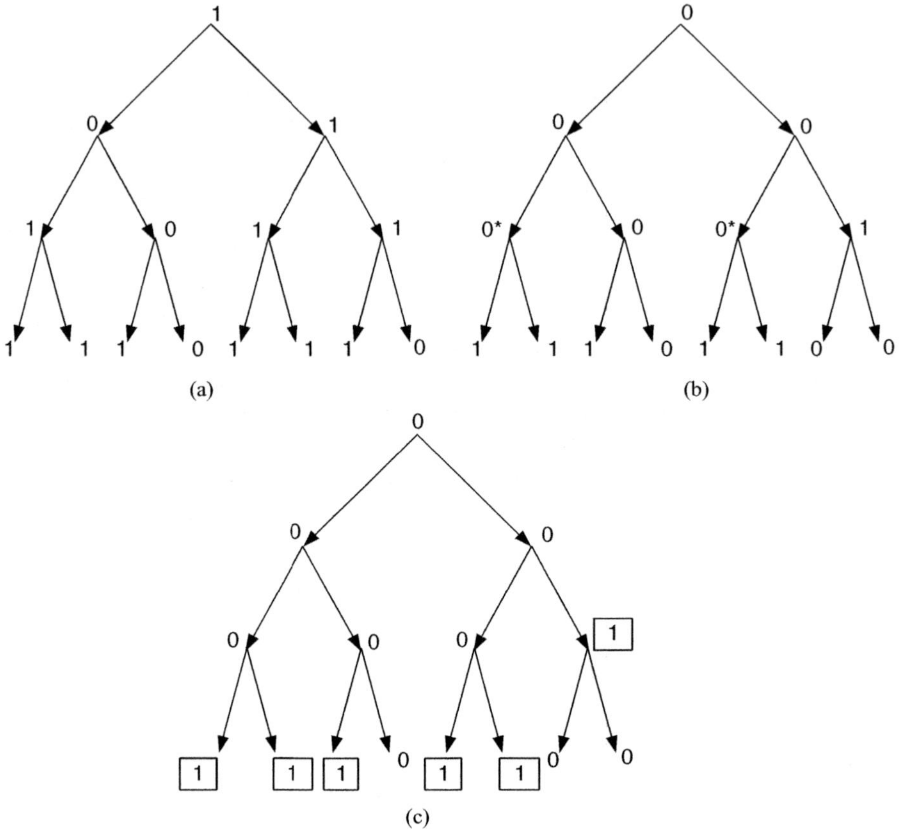

Fig. 12.4 Steps for selection of the *best basis*: (**a**) K-S result. (**b**) Modification of the node values according to the rules described in the text. (**c**) Final WP selection

After parameter estimation, DCS is defined as:

$$DCS(f_{\Theta_a^t}, f_{\Theta_b^t}) = \sum_{j=1}^{t} \log \frac{f_{\Theta_a^t}(X_j)}{f_{\Theta_b^t}(X_j)} \tag{12.10}$$

The detection function is written as:

$$d(t) = \max_{1 \le j \le t} \left[DCS\left(f_{\Theta_a^j}, f_{\Theta_b^j}\right) \right] - DCS\left(f_{\Theta_a^t}, f_{\Theta_b^t}\right) \tag{12.11}$$

Finally the stopping time is:

$$t_p = \inf\{n \ge 1 : d(t) > th\} \tag{12.12}$$

where *th* is the threshold defined from a training set (ROC curves).

12.2.3.2 Change Time Fusion

The detection algorithm is applied on every selected WP. Afterwards, it is necessary to apply a fusion algorithm in order to solve the problem of simultaneous appearance of the same change onto several WP.

The proposed fusion algorithm works as follows (all values are given in number of points):

- Each change time t_c^j detected on WP at level j has been considered as an interval $[t_c^j - 0.5, t_c^j + 0.5]$ (the time resolution at the WP level),
- Each limit of the above detection interval is transformed towards the correct position on the initial time scale, producing a detection interval on that scale [11],
- All superimposed intervals on the initial time scale are considered as indicating the same change,
- The corresponding change time is computed as the barycentre position of all superimposed intervals.

12.2.4 Results on Detection

12.2.4.1 Data Description

The first group of uterine EMG signals was defined in order to test the WP selection algorithm for detection (REELSIM). It was composed of a training set (train_REELSIM) and a test set (test_REELSIM). REELSIM consisted in a total of 200 signals, each of them containing 10 events of different types (CT, Alvarez waves, LBDF waves or fetus motions) identified by an expert. Separation between successive events was made by including white noise segments. The sampling frequency was set to Fs = 16 Hz.

The second group was constituted from real recordings (REEL), and contained 20 long duration recordings. The acquired signals were amplified and filtered between 0.2 Hz and 6 Hz to eliminate the DC component and the artefacts due to powerline interference. The sampling frequency was 16 Hz.

The third group (CLASS) was defined in order to test classification efficiency. It consisted of 100 events in each class to be considered for classification (CT, Alv, MAF, LDBF, and noise), half of them belonging to a training set (train_CLASS), the others belonging to a test set (test_CLASS).

12.2.4.2 Results

The signals of train_REELSIM were decomposed by WPT. After applying the first step of the selection algorithm, only packets 1, 3, 7 and 8 (bandwidths: [0–4], [0–2], [0–1] and [1–2] Hz) were initially selected according to the KS statistics, with a clear threshold appearing around 0.25 (Fig. 12.5). After applying all selection steps of the *best basis* (see Sect. 12.2.2.5), only packets 7 and 8 were retained.

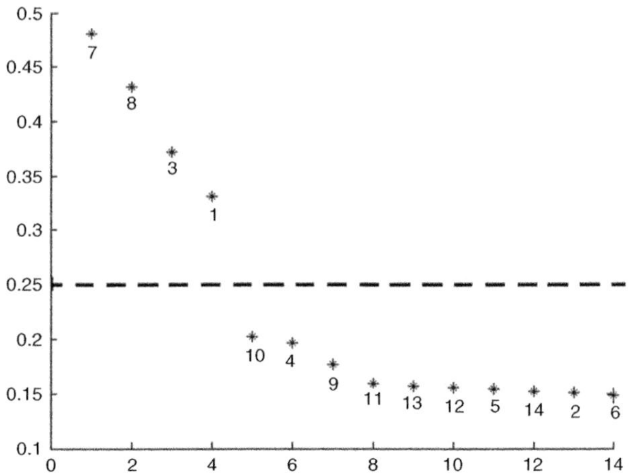

Fig. 12.5 Average of KS statistics D_{max} obtained from 100 uterine EMG signals (train_REELSIM). The numbers on the figure correspond to the WP sequential index. *X* axis: arbitrary units. *Y* axis: D_{max} values

The DCS detection algorithm was then applied by using the test_REELSIM signals on the previously selected WP before reduction (packets 1, 3, 7 and 8) and after reduction (packets 7 and 8). Table 12.2 shows the detection and false alarm probabilities after correction and fusion of the change detection times.

Table 12.2 Detection and false alarm probabilities before (WP 1, 3, 7 and 8) and after (WP 7 and 8) reduction

	Before reduction	After reduction
Detection probability	0.9989	0.9878
False alarm	0.0652	0.0545

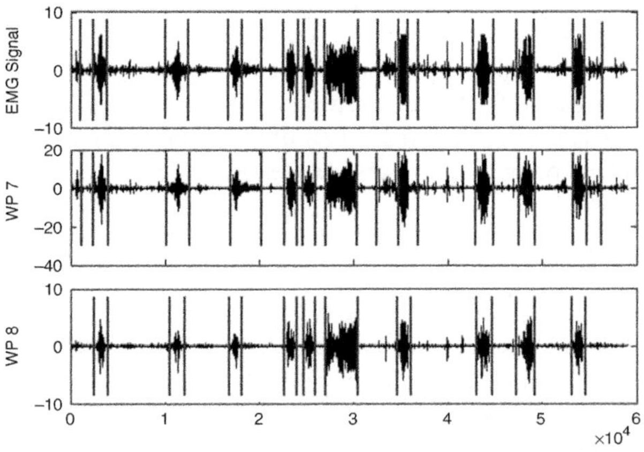

Fig. 12.6 Fusion of the change times detected and corrected on WP 7 and 8. *X* axis: number of samples. *Y* axis: amplitude in arbitrary units

For real time detection the REEL signals were used. The performance was evaluated by calculating mean (1.07 s) and standard deviation (4.76 s) of the differences between the change times estimated by the algorithm described in previously and those indicated by the expert.

The percentage of false alarms was 10%. This result was obtained by counting the change times detected by the algorithm and not indicated by the expert. The non detection rate was roughly 9.5%. It was obtained by comparison of the change times identified by the expert and not detected by the algorithm. Figure 12.6 shows the detection result for a real uterine EMG signal acquired at 32 weeks of gestation after the fusion of change detection times identified on packets 7 and 8.

12.3 Classification of Detected Events

After change detection, the problem now consists of identifying the detected events by allocating them to physiological classes: contractions, foetus motions, Alvarez waves, LDBF waves, or noise. This supervised classification has been achieved using several methods: neural network, K-nearest neighbours, Mahalanobis distance based classification and support vector machines. As wavelet packet decomposition was previously used for noise reduction and event detection, a new approach of *best basis* was also considered for classification.

12.3.1 Selection of the Best Basis for Classification

As uterine EMG is characterized by its frequency content, the relative variance (relative energy) produced on each WP by a specific event can be used as a parameter vector to characterize the event. As a consequence, a WP that produces different variance values for different events is a good candidate for event classification. A global index that could be used that way is the ratio between intra-class and total variances, computed for each WP from a reference dataset.

The intra-class variance $\hat{\Sigma}_w^n$ of a WP n is defined as:

$$\hat{\Sigma}_w^n = \frac{1}{m} \sum_{i=1}^{M} \sum_{k=1}^{m_i} \left(x_{ik}^n - \hat{g}_i^n\right)^2 \tag{12.13}$$

Where g_i is the center of gravity of the class i. The corresponding inter-class variance $\hat{\Sigma}_B^n$ is written as:

$$\hat{\Sigma}_B^n = \frac{1}{m} \sum_{i=1}^{M} m_i \left(\hat{g}_i^n - \hat{g}^n\right)^2 \tag{12.14}$$

Where g^n is the centre of gravity for all classes. The total variance $\hat{\Sigma}$ is the sum of the inter-class and intra-class variances [33]:

$$\hat{\Sigma}^n = \hat{\Sigma}_w^n + \hat{\Sigma}_B^n \tag{12.15}$$

The criterion for the classification is:

$$R^n = \hat{\Sigma}_w^n / \hat{\Sigma}^n \tag{12.16}$$

The choice of the relevant WP for classification is done by comparing R^n to a given threshold.

12.3.2 Classification Algorithms

For a given event, the input of the classification algorithm is the vector of the relative variances produced by the selected WP, possibly associated with other features, such as event duration. Several classification algorithms available through the literature have been used here for algorithm performance comparison purposes.

12.3.2.1 K-Nearest Neighbour

Let $X^n = \{\mathbf{x_1}, \ldots, \mathbf{x_j}, .., \mathbf{x_n}\}$ be the set of training data composed of n independent vectors. It is supposed that the class of each element of the training data set is known: the class of $\mathbf{x_j}$ is $w(\mathbf{x_j})$. Let x be a new event that will be put to the class of its nearest neighbours. The rule on $1NN$ (one nearest neighbour) is: $\hat{w}(\mathbf{x}) = w(\mathbf{x}_{NN})$ if:

$$d(\mathbf{x}, \mathbf{x}_{NN}) = \underset{j=1\ldots n}{Min}\ d(\mathbf{x}, \mathbf{x_j}) \tag{12.17}$$

$\mathbf{x}_{NN}$ is the nearest sample to $\mathbf{x}$ and $\hat{w}(\mathbf{x})$ is the class affected to $\mathbf{x}$.

12.3.2.2 Mahalanobis Distance Based Classification

Let $\mu_i = E(X_i)$ be the mean of class i and Σ_i the variance covariance matrix, defined as:

$$\Sigma = E[XX'] - \mu\mu' \tag{12.18}$$

The Mahalanobis distance is defined as $(X - \mu)'\Sigma^{-1}(X - \mu) = D^2$. Classification is then achieved by computing the Mahalanobis distance of X with respect to each class, and affecting X to the nearest class.

12.3.2.3 Neural Network - Feed Forward Neural Network

The Multi Layer Perceptron (MLP) is widely used to solve classification problems using supervised training, for instance, the feed forward technique, which is used to minimize error.

Such a network is based on the calculation of the output (direct calculation, weights are fixed) and adjustment of the weight by minimizing an error function. The process continues until the outputs of the network become close to those desired. The network is defined by the transfer function of the neuron, the number of layers and the number of neurons in each layer. The number of inputs is equal to the dimension of the input vectors (in this case the variances of the selected packets and the duration of the event). The number of outputs depends on the number of classes. Various transfer functions can be used as neural activation functions. The "sigmoid" functions and the "hyperbolic tangent functions" are mainly used.

12.3.2.4 Support Vector Machines

The Support Vector Machine (SVM) is a new classification method in learning theory. SVM methods were initially developed to solve classification problems, but they have been recently extended to the domain of regression problems [17]. SVM is based on the construction of an optimal hyperplane built in such a way that it maximizes the minimal distance between itself and the learning set.

To discriminate between the various uterine events, the SVM multiclass method [34] was used.

12.3.3 Classification of Uterine EMG Events

The train-CLASS and test-CLASS data sets (see Sect. 12.2.4.1) were used for that step. Each event was decomposed onto the 32 wavelet packets corresponding to four decomposition levels (level four has been experienced to be the limit where WP still contains relevant information related to events such as foetus motions and the LDBF waves).

Values of the discrimination criterion were calculated using Eq. (12.16). Figure 12.7 shows the criterion values in ascending order. Selection of the discriminant WP was made by applying a threshold and selecting those WP that exhibit the lower criterion values. WP 15, 7, 3, 1 and 16 were thus selected as the packets having the most discriminant properties. In order to demonstrate the performance associated with the choice of this threshold, the methods of K Nearest Neighbours, Mahalanobis distance, Neural Networks and Support Vector Machine were used.

Table 12.3 presents the probabilities of correct classification, when using the most discriminant WP obtained after calculation of the values of the discrimination criterion, obtained from K Nearest Neighbors, Mahalanobis distance, neural networks and SVM. The neural network method was composed of one input layer (6 neurons, 6 inputs), one hidden layer (5 neurons) and one output layer (5 neurons).

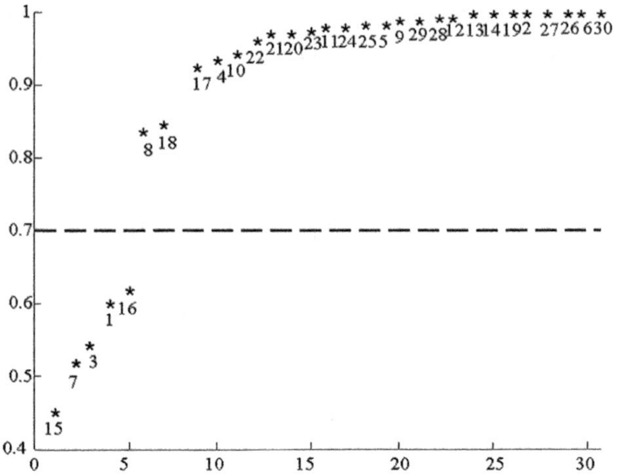

Fig. 12.7 Criterion values (R^n) for Wavelet Packets plotted in ascending order. The numbers on the figure correspond to the WP number and the dotted line is the selected threshold. *X* axis: arbitrary units. *Y* axis: values of R^n

Table 12.3 Correct classification probabilities for the four methods

Method	CT	ALV	MAF	LDBF	Noise
Mahalanobis	0.52	0.72	0.70	0.76	0.9
KNN	0.72	0.48	0.68	0.9	0.48
MLP	0.78	0.72	0.78	0.94	0.86
SVM	0.82	0.64	0.8	0.88	0.82

The activation function was the hyperbolic tangent function "tansig". At the output level, the activation function was the linear function. Thereafter, the event duration was used as an additional characteristic to improve the rate of good classification (see Table 12.4). From these results and according to the probabilities of correct classification in these tables, it can be observed that all rates of classification were improved when introducing event duration, specifically for Alvarez and MAF waves.

In SVM method, the kernel used is the RBF (Radial Basis Function) defined as follows:

$$K(u, v) = \exp(-\frac{\|u - v\|^2}{2\sigma^2}) \tag{12.19}$$

Table 12.4 Correct classification probabilities for the four methods when including event duration

Method	CT	ALV	MAF	LDBF	Noise
Mahalanobis	0.84	0.9	0.82	1.00	0.94
KNN	0.80	0.98	0.96	1.00	0.74
MLP	0.86	0.96	0.98	1.00	0.86
SVM	0.88	0.86	0.96	0.98	0.90

SVM results shown on Tables 12.3 and 12.4 are related to the values of σ Eq. (12.19) and C (regularisation parameter) [17] which gave the best classification probabilities for all events, either without (Table 12.3) or with duration (Table 12.4).

12.4 Classification of Contractions

After detection and classification of all events in the uterine EMG signals and after the identification of the uterine contractions, the aim now is to use these contractions to detect the preterm births. We use the wavelet network for this purpose.

12.4.1 Wavelet Networks (WAVNET)

The idea of combining wavelet theory with neural networks [35–37] resulted in a new type of neural network called wavelet networks. The wavelet networks use wavelet functions as hidden neuron activation functions. Using theoretical features of the wavelet transform, network construction methods can be developed. These methods help to determine the neural network parameters and the number of hidden neurons during training. The wavelet network has been applied to modeling passive and active components for microwave circuit design [38–40].

The new idea used in this work is to use directly a wavelet network applied to the parameters of the power spectral density (PSD) of the EMG, rather than to use the Wavelet decomposition, followed by a classification by neural networks. It acts as a neural network, but with activation functions similar to Wavelets (with dilation and translation parameters). This network has been mainly used for the regression. We have chosen to use it for the purpose of classification.

Wavelet networks are feedforward networks with one hidden layer, as shown in Fig. 12.8. The hidden neuron activation functions are wavelet functions.

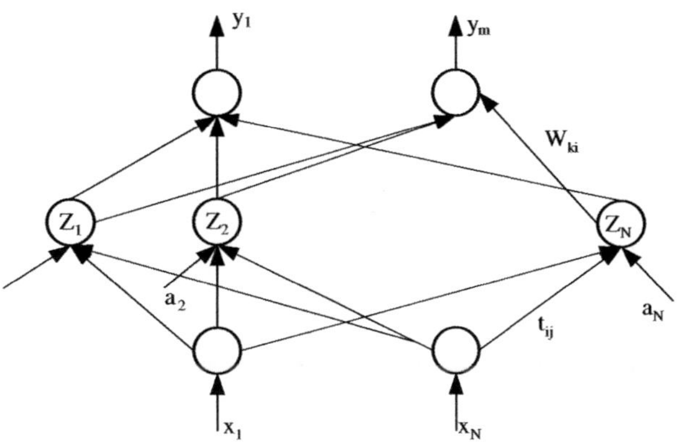

Fig. 12.8 Wavelet neural network structure

The output of the ith hidden neuron is given by

$$Z_i = \sigma(\gamma_i) = \psi\left(\frac{x - t_i}{a_i}\right), i = 1, 2, \ldots, N \qquad (12.20)$$

where N is the number of hidden neurons, $x = [x_1, x_2 \ldots x_n]^T$ is the input vector, $t_i = [t_{i1}\ t_{i2} \ldots t_{in}]^T$ is the translation parameter, ai is a dilation parameter, and $\psi(\cdot)$ is a wavelet function. The weight parameters of a wavelet network w include a_i, t_{ij}, w_{ki}, w_{k0}, $i = 1, 2, \ldots, N$, $j = 1, 2, \ldots, n$, $k = 1, 2, \ldots, m$.

The outputs of the wavelet network are computed as

$$y_k = \sum_{i=0}^{N} w_{ki} z_i,\ k = 1, 2, \ldots, m \qquad (12.21)$$

where w_{ki} is the weight parameter that controls the contribution of the ith wavelet function to the kth output.

The training process of wavelet networks is similar to that of RBF networks.

Step 1: Initialize translation and dilation parameters of all the hidden neurons, t_i, a_i, $i = 1, 2, \ldots, N$.

Step 2: Update the weights w of the wavelet network using a gradient-based training algorithm, such that the error between neural model and training data is minimized. This step is similar to MLP and RBF training.

12.4.2 Classifications of Contractions

12.4.2.1 Populations

We used 102 available subjects with known delivery terms. These contractions were extracted from 25 women. 6 women had pregnancies leading to term deliveries (TD), 19 women had pregnancies leading to preterm deliveries (PD). The contractions are divided into 5 groups according to the Registration Week of Gestation (RWG) and to the Birth Week of Gestation (BWG) (Table 12.5).

Table 12.5 Population characteristics

Group	Nb. Placenta anterior/posterior	Nb. contraction/Nb. Of women	RWG ($\bar{M} \pm SD$)	BWG ($\bar{M} \pm SD$)
G1	4/2	22/6	29±0.76	33±0.65
G2	3/1	20/4	29±0.54	31±0.59
G3	2/4	22/6	29±0.43	36±0.45
G4	2/3	22/5	25±0.62	31±0.32
G5	2/2	16/4	27±0.45	31±0.34

12.4.2.2 Results

- *Classification of groups having same RWG and different BWG*

In order to test the possibility of detecting the term of delivery at a given RWG, we have first used the groups which have the same RWG and different BWG. The first group G1 corresponds to signals recorded at 29 RWG but the corresponding women delivered at 33 BWG. The second group G2 corresponds to signals recorded at 29 RWG but the corresponding women delivered at 31 BWG. For the third group, G3, the women delivered at 36 BWG. We present the classification results as confusion matrices (Table 12.6). The obtained classification errors are 7.1% for G1/G2 and 2.3% for G2/G3 respectively.

Table 12.6 Confusion matrix explaining the result of classification (a) small difference in delivery term (2 weeks), (b) larger difference in delivery term (5 weeks)

	G1	G2		G2	G3
G1	21	1	G2	20	0
G2	2	18	G3	1	21
	(a)			(b)	

- *Classification of signals having same BWG and different RWG*

The second step was to test the possibility of differentiating contractions having the same BWG and different RWG. The aim of this study is to test the influence of pregnancy evolution on uterine EMG characteristics, for a given delivery class (preterm in this case). Three groups are used; the first G2 corresponds to signals recorded at 29 RWG with women delivering at 31 BWG. The second group G5 corresponds to women who delivered also at 31 BWG but with signals recorded at 27 RWG. For the third group, G4, the signals were recorded at 25 RWG. By applying the Wavnet, 89% of contractions are well classified when the difference between recording term is small (G2 vs G5: 2 weeks) and increases to 97%, when the difference of recording terms is increased (G2 vs G4: 5 weeks).

12.5 Discussion and Conclusion

This paper presents a method based on Wavelet Packet decomposition and wavelet network classification that was efficiently applied to uterine EMG recordings for detection and classification. The main objective was to detect and classify the relevant physiological events included in the recordings, then to discriminate between the normal (leading to term delivery) or pathological (leading to preterm delivery) contractions.

The direct use of the Wavelet Packet coefficients (WPC), as well as the procedure of WP selection aimed at reducing the WP tree to a *best basis*, produced very satisfactory results. A selection criterion based on the Kullback Leiber distance allowed a relevant WP selection related for change detection. A dynamic cumulative detection algorithm was then applied to the selected WPC and gave satisfactory segmentation results. As far as the application to uterine EMG is concerned, the proposed processing methodology is suitable for the detection of the various electrical events included in external abdominal recordings. It takes into account the non-stationary nature of the signal. Furthermore the identification of the detected events and their allocation to physiological classes (contractions, foetus motions, Alvarez waves or LDBF waves) also produced satisfactory results. The most discriminant WP obtained from the WPT were selected from a criterion well-adapted to classification problems. The ratio between intra-class and total variance was found to be a good criterion for the choice of the best discriminant Wavelet Packets.

From this point, an event was characterized by its energy computed at the level of each selected WP. An additional characteristic corresponding to the event duration was added to the inputs of the classifiers in order to improve classification performance. Four classifiers were then tested for event identification. Although good results were obtained for all methods, neural networks were more efficient when the event duration was taken into account. On average more than 85% of the events were correctly classified, regardless of the pregnancy term.

Concerning the classification of contractions, we can conclude that it is possible to distinguish between contractions recorded at the same RWG (registration week of gestation) and leading to a different BWG (birth week of gestation). Therefore, it makes it expectable to use uterine EMG as a relevant signal to detect the risk of preterm delivery.

The second approach was to classify the contractions acquired at different RWG with women having the same BWG. The outcome was successful when the difference between RWG was high enough. As a conclusion it has been shown that it is now possible to distinguish the term of gestation of women with the same pregnancy profile.

As SEMG signals can now be recorded in most anatomical, physiological and pathological conditions, the next step could be the production of a sufficiently large database to improve existing knowledge on the actual recording characteristics and their correlation to a possible diagnosis of premature birth.

References

1. Senat M V, Tsatsaris V, Ville Y et al. (1999) Menace d'accouchement prématuré. Encycl. Méd Chir, Elsevier, Paris, Urgences, p. 17
2. Dill LV and Maiden R M (1946) The electrical potentials of the human uterus in labor. Am J Obstet Gynecol 52:735–745
3. Hon E H G and Davis C D (1958) Cutaneous and uterine electrical potentials in labor. Exp Obstet Gynecol 12:47–53
4. Planes J G, Favretto R, Grangjean H et al. (1950) External recording and processing of fast electrical activity of the uterus in human parturition. Med Biol Eng Comp 22:585–591
5. Steer C M and Hertsch G J (1950) Electrical activity of the human uterus in labor: the electrohysterograph. Am J obstet Gynecol 59:25–40

6. Sureau C, Chavinié J and Cannon M (1965) L'électrpohysiologie utérine. Bull Féd Soc Gynecol Obstet 17:79–140
7. Val N, Dubuisson B and Goubel F (1979) Aide au diagnostic de l'accouchement par l'électromyogramme abdominal: sélection de caractères. Reconnaissance de formes, intelligence artificielle 3:42–50
8. Wolfs G M J A and Van Leeuwen M (1979) Electromyographic observations on the human uterus during labour. Acta Obstet Gynecol Scand Suppl 90:1–62
9. Gondry J, Marque C, Duchêne J et al. (1993) Uterine EMG Processing during pregnancy: Preliminary report. Biomed Instrum Technol 27:318–324
10. Leman H, Marque C and Gondry J (1999) Use of the EHG signal for the characterization of contraction during pregnancy. IEEE Trans on Biom Eng 46:1222–1229
11. Chendeb M, Khalil M and Duchêne J (2006) Methodology of wavelet packet selection for event detection. Signal Processing 86:3826–3841
12. Ranta R, Ranta C R, Heinrich Ch, Louis-Dorr V et al. (2001) Wavelet-based bowel sounds denoising: segmentation and characterization. Proc. of the 23rd Conference of EMBS-IEEE, Istanbul, Turkey, pp. 25–28
13. Kalayci T and Ozdamar O (1995) Wavelet preprocessing for the automatic detection of EEG spikes with neural networks. IEEE Eng Med Biol Mag 13:160–166
14. Trejo L J, and Shensa M J (1993) Linear and neural network models for predicting human signal detection performance from event-related potentials: A comparison of the wavelet transform with other feature extraction methods. Proc. of the 5th Workshop on Neural Networks, SPIE, 43–161
15. McLachlan G J (1992) Discriminant analysis and statistical Pattern Recognition, Wiley, New York
16. Turkoglu I, Arslan A and Ilkay E (2003) An intelligent system for diagnosis of the heart valve diseases with wavelet packet neural networks. Comput Biol Med 33:319–331
17. Gunn S R (1998) Support vector machines for classification and regression. Technical Report, School of Electronics and Computer Science, University of Southampton
18. Buhimschi C, Boyle M et al. (1998) Uterine activity during pregnancy and labor assessed by simultaneous recordings from the myometrium and abdominal surface in the rat. Am J Obstet Gynecol 178:811–822
19. Marque C, Terrien J, Rihana S et al. (2007) Preterm labour detection by use of a biophysical marker: the uterine electrical activity", BMC Pregnancy Childbirth 7:1–7
20. Mallat S (1999) A wavelet tour of signal processing. Academic Press, San Diego, CA
21. Coifman R R and Wickerhauser M V (1992) Entropy based algorithms for best basis selection. Proc IEEE Trans Inform Theory 38:1241–1243
22. Hsu P H (2004) Feature Extraction of Hyperspectral Images using Matching Pursuit, Geo-Imagery Bridging Continents XXth ISPRS Congress, p. 883, Istanbul, Turkey, 12–23 July
23. Ravier P and Amblard P O (2001) Wavelet packets and de-noising based on higher-order-statistics for transient detection, Signal Processing 81:1909–1926, 2001
24. Leman H and Marque C (2000) Rejection of the maternal electrocardiogram in the electrohysterogram signal. IEEE Trans Biomed Eng 47:1010–1017
25. Hitti E and Lucas M F (1998) Wavelet-packet basis selection for abrupt changes detection in multicomponent signals. Proc. EUSIPCO 1841–1844, Island of Rhodes, Greece, 8–11 September
26. Saito N and Coifman R R (1994) Local discriminant bases, A. F. Laine and M. A. Unser (Editors), Wavelet applications in signal and image processing {II}, Proc. SPIE 2303, 2–14
27. Do M N and Vetterli M (2002) Wavelet-based texture retrieval using generalized Gaussian density and Kullback-Leibler distance. Proc. IEEE Trans Image Process 11:146–158
28. Sharifi K and Leon-Garcia A (1995) Estimation of shape parameter for generalized Gaussian distributions in subband decompositions of video. IEEE Trans Circuits Syst Video Technol 5:52–56
29. Varanasi M K and Aazhang B (1989) Parametric generalized Gaussian density estimation. J Acoust Soc Amer 86:1404–1415

30. Saporta G (1990) Analyse des données et statistiques. Editions Technip, Paris
31. Chendeb M, Khalil M and Duchêne J, The use of wavelet packets for event detection. 13th Proc EUSIPCO Antalya Turkey 4–8 September
32. Khalil M and Duchene J (2000) Uterine EMG analysis: a dynamic approach for change detection and classification. IEEE Trans Biomed Eng 47:748–756
33. Dubuisson B (1990) Diagnostic et reconnaissance des formes. Editions Hermes
34. Mayoraz E and Alpaydin E (1998) Support vector machines for multi-class classification. Technical Report IDIAP
35. Zhang Q H (1997) Using wavelet network in nonparametric estimation. IEEE Trans Neural Networks 8:227–236
36. Zhang Q H and Benvensite A (1992) Wavelet networks. IEEE Trans Neural Networks 3:889–898
37. Pati Y C, and Krishnaprasad P S (1993) Analysis and synthesis of fast forward neural networks using discrete Affine wavelet transformations. IEEE Trans Neural Networks 4:73–85
38. Harkouss Y et al (1998) Modeling microwave devices and circuits for telecommunications system design. Proc IEEE Int Conf. Neural Networks, Anchorage, Alaska, 128–133, May
39. Bila S et al (1999) Accurate wavelet neural network based model for electromagnetic optimization of microwaves circuits. Int Journal of RF and Microwave CAE, Special Issue on Applications of ANN to RF and Microwave Design, 9:297–306
40. Harkouss Y et al (1999) Use of artificial neural networks in the nonlinear microwave devices and circuits modeling: An application to telecommunications system design. Int. Journal of RF and Microwave CAE, Special Issue on Applications of ANN to RF and Microwave Design, Guest Editors: Q. J. Zhang and G. L. Creech, 9:198–215

Chapter 13
Pattern Classification Techniques for EMG Signal Decomposition

Sarbast Rasheed and Daniel Stashuk

Abstract The electromyographic (EMG) signal decomposition process is addressed by developing different pattern classification approaches. Single classifier and multiclassifier approaches are described for this purpose. Single classifiers include: certainty-based classifiers, classifiers based on the nearest neighbour decision rule: the fuzzy k-NN classifiers, and classifiers that use a correlation measure as an estimation of the degree of similarity between a pattern and a class template: the matched template filter classifiers. Multiple classifier approaches aggregate the decision of the heterogeneous classifiers aiming to achieve better classification performance. Multiple classifier systems include: one-stage classifier fusion, diversity-based one-stage classifier fusion, hybrid classifier fusion, and diversity-based hybrid classifier fusion schemes.

13.1 Introduction

An electromyographic (EMG) signal is the recording of the electrical activity associated with muscle contraction. The signal recorded by the tip of the inserted needle electrode is the superposition of the individual electrical contributions of anatomical compositions called motor units (MUs), that are active during a muscle contraction, and the background interference. EMG signal analysis in the form of EMG signal decomposition is mainly used to assist in the diagnosis of muscle or nerve disorders and for the analysis of neuromuscular system.

EMG signal decomposition is the process of resolving a composite EMG signal into its constituent motor unit potential trains (MUPTs) and it is considered as a classification problem. Figure 13.1 shows the results of decomposing a 1-s interval of an EMG signal, where the classifier assigns the motor unit potentials (MUPs)

S. Rasheed (✉)
Department of Systems Design Engineering, University of Waterloo, Waterloo, Ontario N2L 3G1, Canada
e-mail: s.rasheed@ieee.org

A. Naït-Ali (ed.), *Advanced Biosignal Processing*,
DOI 10.1007/978-3-540-89506-0_13,

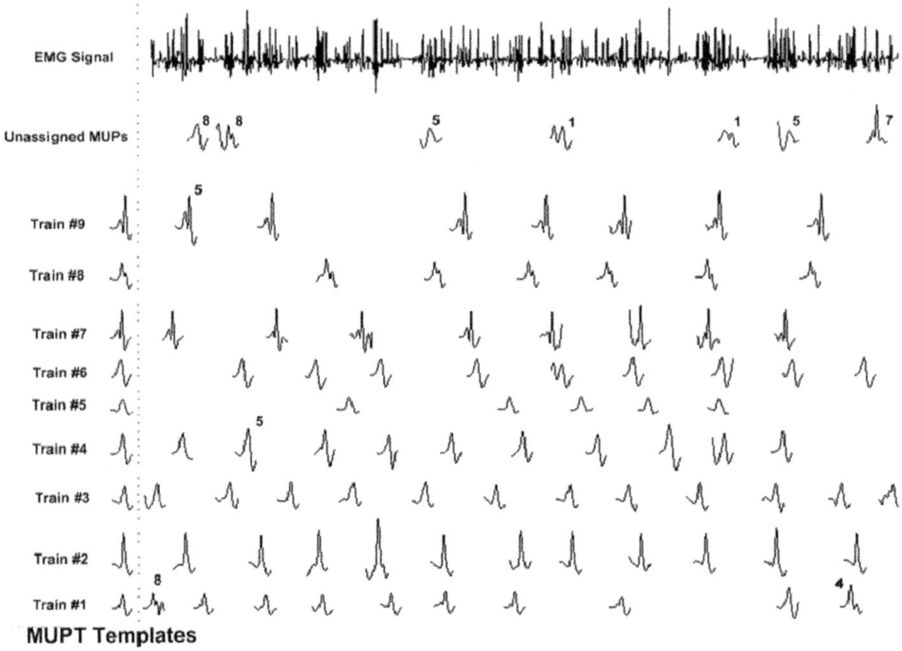

Fig. 13.1 MUPTs for a 1-s interval of an EMG signal decomposition. MUP waveforms have been expanded by a factor of 10 relative to the time scale used to depict their firing times

into their MUPTs based on a similarity criterion. Those MUPs that do not satisfy the classifier similarity criterion are left unassigned.

Automatic EMG signal decomposition techniques have been designed to follow as closely as possible the manual method [2] and a good system should do the same analysis that an electromyographer does manually [12]. This is possible only if a robust pattern recognition algorithm is developed.

Many automatic EMG signal decomposition techniques have been developed with different methodologies in the time-domain, frequency-domain, and wavelet-domain being followed for quantitative analysis [5, 13, 19, 21, 22, 27, 28, 29, 30, 32, 46, 48, 49, 53]. All of these methods use a single classifier to complete the classification task. A review of these methods can be found in [45, 47].

13.2 EMG Decomposition Process

The objective of EMG signal decomposition is often the extraction of relevant clinical information from quantitative EMG (QEMG) analysis of individual MUP waveforms and MU firing patterns. Figure 13.2 shows a flowchart depicting the major steps involved in the EMG signal decomposition process. After acquiring the EMG signal, the first task is the segmentation of the signal and detection of possible MUP

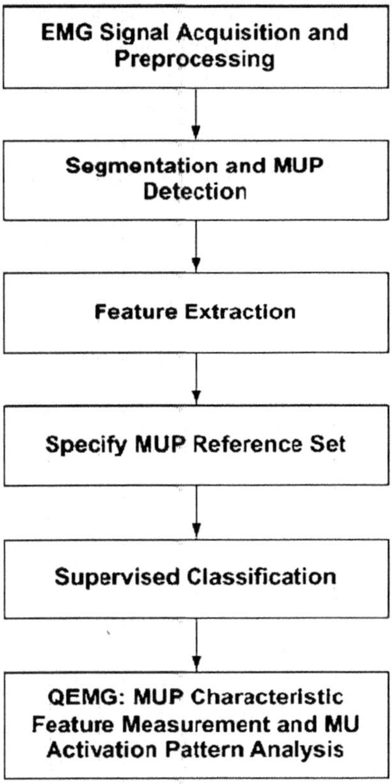

Fig. 13.2 Flowchart of major steps involved in the EMG signal decomposition process

waveforms, which is then followed by the feature extraction task and the main task of MUP classification. The classification task, which is the focus of this chapter, involves dividing detected MUPs into groups such that each set of grouped MUPs represents the activation of a single MU, and through which the activation of each active MU can be discriminated. A MUPT is the collection of MUPs generated by one motor unit positioned at their times of occurrence or separated by their interdischarge intervals (IDIs) as shown in Fig. 13.3.

The shapes and occurrence times of MUPs provide an important source of information to assist in the diagnosis of neuromuscular disorders. Some automatic EMG

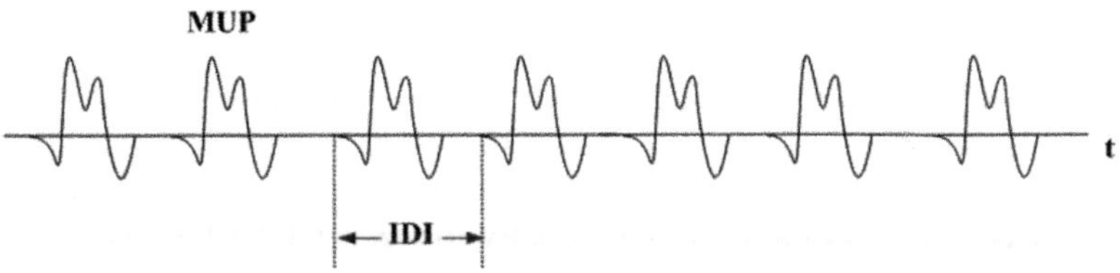

Fig. 13.3 MUPT with MUPs separated by their inter-discharge intervals (IDIs)

signal decomposition methods are designed so that the classification task considers only MUP morphological shape parameters such as: duration, amplitude, area, number of phases, number of turns, etc without evaluation of MU firing pattern or considering the variability of the MUP shape during contraction. These parameters can be used for diagnostic purposes since they reflect the structural and physiological changes of a MU. Others use MU firing patterns so that the central nervous system recruitment and control of MUs can be studied. Most of the new methods use both MUP shape parameters and either partial or full firing patterns [33].

The classification task for some of the existing decomposition methods is based on unsupervised classification, while others combine unsupervised and supervised classification methods. The unsupervised classification methods major limitation is that they only work well if there are large differences in the features of the classes involved [9] and because of the similarity between MUPs from different MUs, unsupervised classification methods often will not yield acceptable classification results [31]. Where they can result in lumping together two classes having similarly shaped MUPs into one class, or they can mistakenly separate one class into two classes [3].

On the other hand, a supervised classifier can track changing shapes over time, due to muscle fatigue and electrode or muscle movement, through updating the template of each class with each classification.

The classification task for EMG signal decomposition in this chapter is addressed using both single classifier and multi-classifier approaches. The multi-classification techniques combine the results of a set of classifiers of different kinds and based on multi-features extracted from the acquired data. The classification schemes described are based on information provided by the MUP waveform shapes and MU firing patterns.

13.3 Supervised Classification of MUPs

The task of supervised classification during the process of EMG signal decomposition is involved with the discrimination of the activation patterns of individual. MUs, active during contraction, into distinguishable MUPTs. Therefore, MUPs are most likely to belong to the same MUPT if their shapes are closely similar and if their IDIs are consistent with the discharge pattern of the considered MU. For the purpose of MUP classification, we developed single classifier and multi-classifier approaches based on the above constraints.

For each MUP classification approach, we formulated a set of firing pattern consistency statistics for detecting erroneous MUP classifications [39, 40] such that once the set of MUPTs is generated, firing pattern consistency statistics for each MUPT are calculated to detect classification errors in an adaptive fashion. This firing pattern analysis allows the algorithm to modify the threshold of assertion required for assignment of a MUP individually for each MUPT based on an expectation of erroneous assignments.

The adaptive classification process of MUPs may be modelled as a negative feedback control system depicted in Fig. 13.4 for single classifier and in Fig. 13.5 for

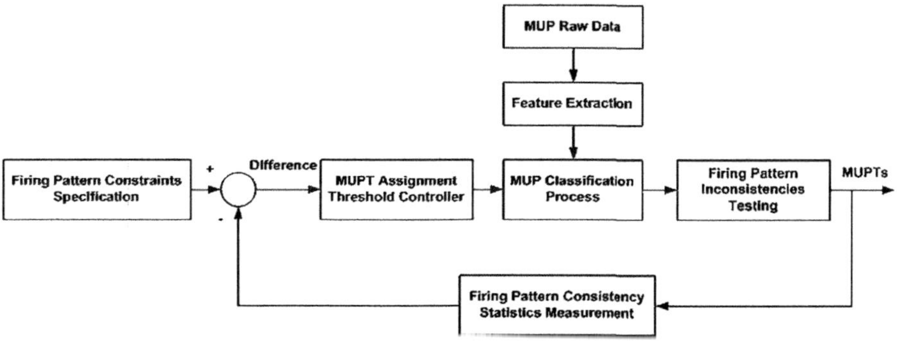

Fig. 13.4 Adaptive MUP classification using single classifiers modelled as a feedback control system

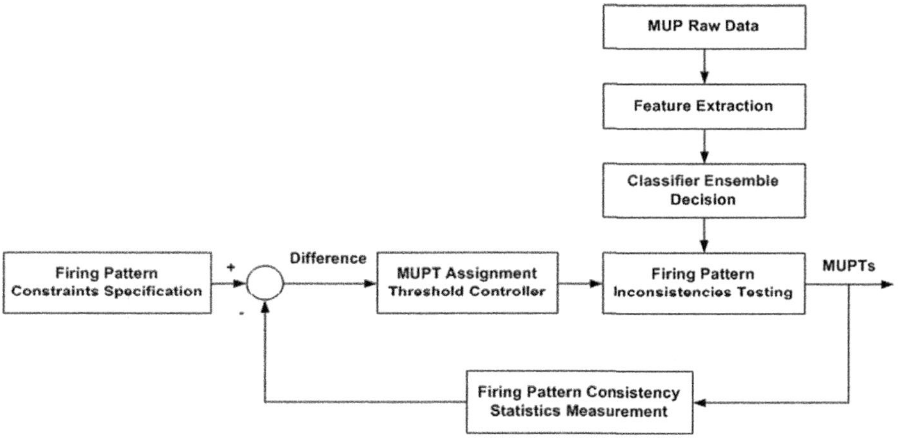

Fig. 13.5 Adaptive MUP classification using classifier fusion schemes modelled as a feedback control system

multiclassifier approaches. The MUPT assignment threshold controller is actuated by the difference between the specified firing pattern constraints and the calculated consistency statistics and if, based on the set of firing pattern statistics, it is expected that a MUPT has too many erroneous assignments, the controller increases its assignment threshold or otherwise the controller keeps it constant. This process is repeated until all the imposed firing pattern constraints for all MUPTs are satisfied.

Consider an EMG signal decomposed into M mutually exclusive sets, $\omega_i \in \Omega = \{\omega_1, \omega_2, \ldots, \omega_M\}$. Each set, ω_i, represents a MUPT into which MUPs will be classified and Ω is the set of corresponding integer labels defined such that $\Omega = \{\omega_1 = 1, \omega_2 = 2, \ldots, \omega_M = M\}$ and it provides all possible integer labels for the valid MUPTs. As some of the MUPs may not be assigned to any of the valid MUPTs, the MUP decision space set can then be extended to include $\Omega \cup \{\omega_{M+1}\}$ where ω_{M+1} = designates the unassigned category for when by some established criteria the classifier decides to not assign the input MUP.

13.4 Single Classifier Approaches

The developed single classifier approaches include certainty-based classifiers, classifiers based on the nearest neighbour decision rule, and classifiers based on MUP similarity measures. All the single classifiers take into account MUP shape and MU firing pattern information. These classifiers follow an adaptive nature for train-wise setting of the MUPT assignment threshold based on firing pattern consistency statistics.

The single classifiers are used as base classifiers in order to construct a combined classifier fusion system that usually performs better than any of the single classifiers.

13.4.1 Certainty Classifier

The Certainty classifier (CC) is a non-parametric template matching classifier that uses a certainty-based approach for assigning MUPs to MUPTs. A complete description of the CC classifier is given in [34, 47, 49] accompanied with testing and evaluation of its performance.

The CC estimates a measure of certainty expressing confidence in the decision of classifying a MUP to a particular MUPT. It determines two types of decision functions for each candidate MUP, the first is based on shape information and the second is based on firing pattern information. For a set of M MUPT class labels $\Omega = \{\omega_1, \omega_2, \ldots, \omega_M\}$, the decision functions for the assignment of MUP m_j with feature vector x, belonging to feature space X, are evaluated for only the two MUPTs with the most similar templates to MUP m_j. Each MUPT ω_i template is calculated using a labelled reference set.

The shape information decision functions include:

1. Normalized absolute shape certainty C_{ND}: represents the distance from a candidate MUP m_j to the template of a MUPT ω_i normalized by the norm of the template. For candidate MUP m_j, C^j_{NDi} is evaluated by:

$$C^j_{NDi} = max\left\{1 - \frac{r_i}{s_i}, 0\right\}, \quad i = 1, 2. \tag{13.1}$$

where, r_1 and r_2 are the Euclidean distances between MUP m_j and the closest (most similar) and second closest MUPT templates, respectively; s_1 and s_1 are the l_2 norm of the closest and second closest MUPT templates to MUP m_j, respectively.

2. Relative shape certainty C_{RD}: represents the distance from a candidate MUP m_j to the template of the closest MUPT relative to the distance from the same MUP to the second closest MUPT. For candidate MUP m_j, C^j_{RDi} is evaluated by:

$$C^j_{RDi} = 2 - i + (-1)^i \frac{r_1^2}{2.r_2^2}, \quad i = 1, 2. \tag{13.2}$$

The firing pattern information is represented by the firing certainty decision function C_{FC} with respect to the established firing pattern of the MUPT. For candidate MUP m_j, C^j_{FCi} is evaluated by:

$$C^j_{FCi} = C_f\left(I^j_{bi}, \mu_i, \sigma_i\right).C_f\left(I^j_{fi}, \mu_i, \sigma_i\right), \quad i = 1, 2. \tag{13.3}$$

where, $C_f(I, \mu, \sigma)$ is a firing time certainty function based on the deviation of an IDI, I, from the estimated mean IDI, μ, of a MUPT that has an estimated standard deviation, σ · I_{bi} and I_{fi} are the IDIs that would be created by assigning a MUP m_j to MUPT ω_i; I_{bi} is the backward IDI, the interval between MUP m_j and the previous MUP in the MUPT; I_{fi} is the forward IDI, the interval between MUP m_j and the next MUP in the MUPT.

The decision of assigning a MUP m_j to a MUPT ω_i is based on the value for which the multiplicative combination of C^j_{NDi}, C^j_{RDi} and C^j_{FCi} given by the overall certainty C^j_i in (13.4) is the greatest and if it is greater than the minimal certainty threshold (C_m) for which a classification is to be made:

$$C^j_i = C^j_{NDi}.C^j_{RDi}.C^j_{FCi}, \quad \mathrm{i} = 1, 2. \tag{13.4}$$

Otherwise, MUP m_j is left unassigned.

Certainty-based classifiers are able to track the non-stationarity of the MUP waveform shape by updating the labelled reference set once the MUP m_j to be assigned has an overall certainty C^j_i higher than an updating threshold. The reference set update is performed by the following updating rule:

$$s_{iu} = \frac{s_i + C^j_i.x}{1 + C^j_i} \tag{13.5}$$

where s_i is the moving-average template vector, s_{iu} is the updated template vector, x is the classified MUP m_j feature vector whose certainty C^j_i exceeds the updating threshold [34, 49].

The adaptive version of the Certainty classifier, the adaptive certainty classifier (ACC) uses an adaptive certainty-based approach for assigning MUPs to MUPTs.

A complete description of the ACC is given in [36, 37, 39] accompanied with testing and evaluation of its performance.

13.4.2 Fuzzy k-NN Classifier

The fuzzy k-NN classifier [24] uses a fuzzy non-parametric classification procedure based on the nearest neighbour classification rule, which bypasses probability estimation and goes directly to decision functions [9]. It assigns to MUP m_j with feature vector x, belonging to feature space X, a membership vector $(\mu_{\omega_1}(m_j), \mu_{\omega_2}(m_j), \ldots, \mu_{\omega_M}(m_j))$ for a set of M MUPT class labels $\Omega = \{\omega_1, \omega_2, \ldots, \omega_M\}$ as a function of the MUP's distance from its k nearest neighbours. The MUPT ω_i class membership of the input MUP m_j is calculated based on:

$$\mu_{\omega_i}(m_j) = \frac{\sum_{r=1}^{k} \mu_{\omega_i}(x_r) d_r^{-2}}{\sum_{r=1}^{k} d_r^{-2}} \tag{13.6}$$

where $x_1, x_2, \ldots, x_k$ denote the k nearest neighbour labelled reference MUPs of MUP m_j and $d_r = \|x - x_r\|$ is the distance between MUP m_j and its *rth* nearest neighbor x_r. The MUP m_j is assigned to the MUPT whose class label is given by:

$$\omega(m_j) = \arg\max_{i=1}^{M}(\mu_{\omega_i}(m_j) \tag{13.7}$$

The fuzzy k-NN classifier relies on the estimation of the membership functions for the set of n labelled reference MUPs $V = \{(v_1, l(v_1)), (v_2, l(v_2)), \ldots, (v_n, l(v_n))\}$ where $v_i \in X$ and $l(v_i) \in \Omega$. The fuzzy nearest neighbour labelling, known as soft labelling, assigns memberships to labelled reference patterns according to the k-nearest neighbours rule. It is required to estimate M degrees of membership $(\mu_{\omega_1}(v_j), \mu_{\omega_2}(v_j), \ldots, \mu_{\omega_M}(v_j))$ for any $v_i \in V$ by first finding the k MUPs in V closest to each labelled reference MUP v_i and then calculating the membership functions using the scheme proposed by Keller et al. [24]:

$$\mu_{\omega_i}(v_r) = \begin{cases} 0.51 + \left(\frac{k_i}{k} \cdot 0.49\right), & \text{if } l(v_r) = \omega_i \\ \left(\frac{k_i}{k} \cdot 0.49\right), & \text{if } l(v_r) \neq \omega_i \end{cases} \tag{13.8}$$

Where k_i is the number of labelled reference MUPs amongst the k closest labeled reference MUPs which are labelled in MUPT class ω_i, and r ranges from 1 to n.

The fuzzy k-NN classifier for MUP classification [40] estimates a measure of assertion expressing confidence in the decision of classifying a MUP to a particular MUPT class. It determines for each candidate MUP m_j a MUPT ω_i class membership $\mu_{\omega_i}(m_j)$ calculated from (13.6) representing the shape-based strength of membership of MUP m_j in MUPT class ω_i; and a firing assertion decision function $A^j_{FA_i}$ assessing the time of occurrence of MUP m_j with respect to the established firing pattern of MUPT class ω_i.

The firing pattern information is represented by the firing assertion decision function A_{FA}. For candidate MUP m_j and MUPT class ω_i, $A^j_{FA_i}$ is evaluated by:

$$A^j_{FA_i} = A_f\left(I^j_{bi}, \mu_i, \sigma_i\right) \cdot A_f\left(I^j_{fi}, \mu_i, \sigma_i\right) \tag{13.9}$$

where, $A_f(I, \mu, \sigma)$ is a firing time assertion function based on the deviation of an IDI, I, from the estimated mean IDI, μ, of a MUPT that has an estimated standard deviation, σ . I_{bi} and I_{fi} are the backward and forward IDIs, respectively.

The overall assertion value A_i^j for assigning MUP m_j to MUPT class ω_i is defined as:

$$A_i^j = \mu_{\omega_i}(m_j) \cdot A_{FAi}^j \tag{13.10}$$

MUP m_j is assigned to the MUPT class ω_i with the highest assertion value and if this value is above the minimum assertion value threshold (A_m) of the MUPT ω_i to which a classification is to be made, otherwise MUP m_j is left unassigned.

The adaptive fuzzy k-NN classifier (AFNNC) uses an adaptive assertion-based approach for assigning MUPs to MUPTs. A complete description of the AFNNC is given in [36, 37, 40, 42] accompanied with testing and evaluation of its performance.

13.4.3 Matched Template Filter Classifier

The basic MUP matched template filtering algorithm consists of sliding MUPT templates over the EMG signal detected MUPs, and calculating for each candidate MUP a distortion, or correlation, measure estimating the degree of dissimilarity, or similarity, between the template and the MUP. Then, the minimum distortion, or maximum correlation, position is taken to represent the instance of the template in the signal under consideration, with a threshold on the similarity/dissimilarity measure allowing for rejection of poorly matched MUPs. We used correlation measure as an estimate of the degree of similarity between a MUP and MUPT templates.

The correlation between two signals represents the degree to which signals are related, and cross correlation analysis enables determining the degree of waveform similarity between two different signals. It provides a quantitative measure of the relatedness of two signals as they are progressively shifted in time with respect to each other.

For a set of M MUPT class labels $\Omega = \{\omega_1, \omega_2, \ldots, \omega_M\}$, the correlation functions for the assignment of MUP m_j are evaluated using (13.11) and (13.12).

Two matched template filters have been investigated for supervised MUP classification. They are: the normalized cross correlation which is the most widely used correlation measure [50] and given by formula (13.11):

$$NCC_{\omega_i}^j(x) = \frac{\sum\limits_{k=1}^{n} m_j(x+k) \cdot T_i(k)}{\sqrt{\sum\limits_{k=1}^{n} m_j(x+k)^2} \cdot \sqrt{\sum\limits_{k=1}^{n} T_i(k)^2}} \tag{13.11}$$

and a pseudo-correlation [15, 16, 17] measure given by formula (13.12):

$$pC_{\omega_i}^j(x) = \frac{\sum_{k=1}^{n} (T_i(k) \cdot m_j(x+k) - |T_i(k) - m_j(x+k)| \cdot \max\{|T_i(k)|, |m_j(x+k)|\})}{\sum_{k=1}^{n} (\max\{|T_i(k)|, |m_j(x+k)|\})^2} \tag{13.12}$$

Denote p to be the matched template filter correlation coefficient such that:

$$\rho_{\omega_i}^j(x) = \begin{cases} NCC_{\omega_i}^j(x), & \text{when choosing normalized cross correlation,} \\ pC_{\omega_i}^j, & \text{when choosing pseudo correlation.} \end{cases} \tag{13.13}$$

where, m_j is the candidate MUP feature vector, T_i is the MUPT ω_i template feature vector, and x =1,2,. . .,n is the time-shifting position between the MUPT ω_i template and the candidate MUP with n being the dimension of the feature vector.

The matched template filter (MTF) classifier for MUP classification estimates a measure of similarity between a candidate MUP m_j and the MUPT templates expressing confidence in the decision of classifying a MUP to a particular MUPT. It determines, for each candidate MUP m_j, a normalized cross correlation value calculated from (13.11) or a pseudo correlation value calculated from (13.12) representing the strength of resemblance of the MUP m_j with the MUPT templates:

The MTF classifier also determines for MUP m_j a firing time similarity decision function $S_{FS_i}^j$ with respect to the established firing pattern of the MUPT. The firing pattern information is represented by the firing similarity decision function S_{FS}. For candidate MUP m_j, $S_{FS_i}^j$ is evaluated by:

$$S_{FS_i}^j = S_f\left(I_{bi}^j, \mu_i, \sigma_i\right) \cdot S_f \cdot \left(I_{fi}^j, \mu_i, \sigma_i\right) \tag{13.14}$$

where, $S_f(I, \mu, \sigma)$ is a firing time function based on the deviation of an IDI, I, from the estimated mean IDI, μ, of a MUPT that has an estimated standard deviation, $\sigma \cdot I_{bi}$ and I_{fi} are the backward and forward IDIs, respectively.

The decision of assigning a MUP to a MUPT is based on the value for which the multiplicative combination of $\rho_{\omega_i}^j(x)$ and S_{FSi}^j given in (13.15) is the greatest and if it is greater than the minimal similarity threshold (S_m) for which a classification is to be made, otherwise MUP m_j is left unassigned.

$$S_i^j = \rho_{\omega_i}^j(x) \cdot S_{FS_i}^j \tag{13.15}$$

where, S_i^j is the overall similarity associated with the classification of MUP m_j to MUPT ω_i.

The adaptive matched template filter classifier (AMTF) uses an adaptive similarity approach for assigning MUPs to MUPTs. Two types of MTF classifiers were used. One is based on the normalized cross correlation measure [50] called the adaptive normalized cross correlation classifier (ANCCC) and the other is based on the pseudo-correlation [15, 16, 17] measure called the adaptive pseudo-correlation classifier (ApCC). A complete description of the AMTF, ANCCC, and ApCC is given in [36, 37] accompanied with testing and evaluation of its performance.

13.5 Multiple Classifier Approaches

To achieve improved classification performance, multi-classifier fusion approaches for MUP classification were developed. Different classifiers typically express their decisions and provide information about identifying a MUP pattern at the abstract and measurement levels [4, 52]. At the abstract level: the classifier output is a unique MUPT class label or several MUPT class labels, in which case the MUPT classes are equally identified without any qualifying information. At the measurement level: the classifier attributes to each MUPT class label a confidence measure value representing the degree to which the MUP pattern has that label.

Different classifier fusion systems were developed that aggregate, at the abstract level and measurement level of classifier fusion, the outputs of an ensemble of heterogeneous base classifiers to reach a collective decision, and then use an adaptive feedback control system, as shown in Fig. 13.5, that detects and processes classification errors by using motor unit firing pattern consistency statistics [39, 40].

Classifier fusion system architecture belongs to the parallel category of combining classifiers. It consists of a set of base classifiers invoked concurrently and independently, an ensemble members selection module, an aggregation module that fuses the base classifier output results, and a classification errors detection module as shown in Fig. 13.6.

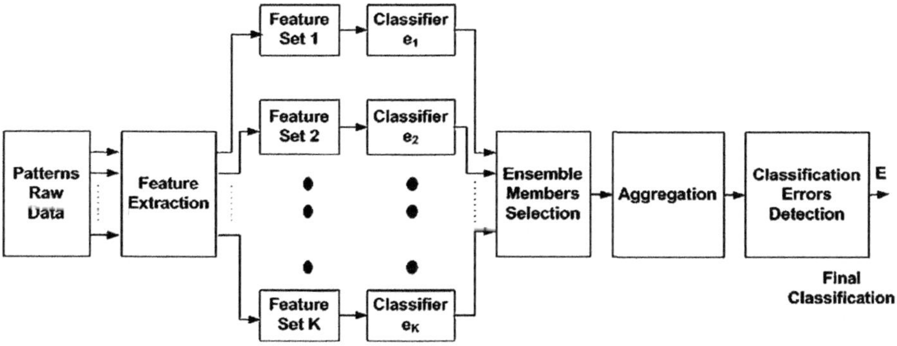

Fig. 13.6 Classifier fusion system basic architecture

13.5.1 Decision Aggregation Module

The decision aggregation module in a classifier fusion system combines the base classifier outputs to achieve a group consensus. Decision aggregation may be data independent [23], with no dependence on the data and solely rely on the output of the base classifiers to produce a final classification decision irrespective of the MUP being classified, or decision aggregation may be data dependent [23] with implicit or explicit dependency on data.

13.5.2 One-Stage Classifier Fusion

The one-stage classifier fusion system does not contain the ensemble members selection module depicted in Fig. 13.6 and it uses a fixed set of base classifiers. Choosing base classifiers can be performed directly through exhaustive search with the performance of the fusion being the objective function. As the number of base classifiers increases, this approach becomes computationally too expensive. A complete description for such a system for MUP classification is found in [36, 37, 41] accompanied with testing and evaluation of its performance.

The aggregator module in the one-stage classifier fusion system consists of one of the following classifier fusion schemes.

13.5.2.1 Majority Voting Aggregation

When classifying a MUP x at the abstract level, only the best choice MUPT class label of each classifier is used, $e_k(x)$. Therefore, to combine abstract level classifiers, a (data independent) voting method is used. The overall decision, $E(x)$, for the combined classifier system is sought given that the decision functions for the individual classifiers may not agree. A common form of voting is *majority voting* [26]. A MUP x is classified to belong to MUPT class ω_i if over half of the classifiers say $x \in \omega_i$.

13.5.2.2 Average Rule Aggregation

Average rule aggregation [1, 10, 25] is a measurement level data independent decision aggregation that does not require prior training. It is used for combining the set of decision confidences $\{Cf_{ik}(x), i = 1, 2, \ldots M; k = 1, 2, \ldots K\}$ for M MUPT classes $\{\omega_i, i = 1, 2, \ldots, M\}$ and for K base classifiers $\{e_k(x), k = 1, 2, \ldots, K\}$ into combined classifier decision confidences $\{Q_i(x), i = 1, 2, \ldots, M\}$. When using average rule aggregation, the combined decision confidence $Q_i(x)$ for MUPT class ω_i is computed by:

$$Q_i(x) = \frac{\sum_{k=1}^{K} Cf_{ik}(x)}{K} \tag{13.16}$$

The final classification is made by:

$$\omega(x) = \arg\max_{i=1}^{M}(Q_i(x)) \tag{13.17}$$

13.5.2.3 Fuzzy Integral Aggregation

Based on measurement level classifier fusion, one can train an arbitrary classifier using the $M \times K$ decision confidences $Cf_{ik}(x)$ (for all i and all k) as features in an intermediate space [10, 11]. The *Sugeno fuzzy integral* [51] approach trained by a search for a set of densities was used for combining classifiers as described in [6, 7, 36, 38].

13.5.3 Diversity-Based One-Stage Classifier Fusion

The drawback of the one-stage classifier fusion scheme described in Sect. 13.5.2 is apparent when following the *overproduce and choose* [18, 35] paradigm for ensemble members selection, where there is a need to perform an exhaustive search for the best accurate classifier ensemble. For example, if the pool of base classifiers contains 16 base classifiers and the intention is to choose an ensemble of 6 base classifiers for fusion, so there is a need to compare the performance of $\binom{16}{6} = 8008$ ensembles. Therefore, to limit the computational complexity encountered we modified the one-stage classifier fusion scheme so that the candidate classifiers chosen for fusion are based on a diversity measure.

The diversity-based one-stage classifier fusion system contains an ensemble members selection module as shown in Fig. 13.6. The ensemble choice module selects the subsets of classifiers that can be combined to achieve better accuracy. The subset giving the best accuracy can be obtained by using ensemble diversity metrics to evaluate the error diversity of the base classifiers that make up an ensemble for classifier selection purposes. The kappa statistic is used to select base classifiers having an excellent level of agreement to form ensembles having satisfactory classification performance.

The aggregator module consists of one of the classifier fusion schemes described in Sects. 13.5.2.1, 13.5.2.2, and 13.5.2.3.

A complete description for such a system for MUP classification is found in [36, 37, 45] accompanied with testing and evaluation of its performance.

13.5.3.1 Assessing Base Classifiers Agreement

The kappa statistic is used to measure the degree of decision similarity between the base classifier outputs. It was first proposed by Cohen [8] and it expresses a special type of relationship between classifiers as it quantifies the level to which classifiers agree in their decisions beyond any agreement that could occur due to chance. A value of 0.40 is considered to represent poor agreement beyond chance, values between 0.40 and 0.75 indicate fair agreement, and values beyond 0.75 indicate excellent agreement [14].

For an ensemble of K base classifiers $e_k, k = 1, 2, \ldots, K$ known to be correlated to each other as they work on the same data and are used to classify a set of N MUP patterns into M MUPT classes and the unassigned category $\omega_i \in \Omega = \{\omega_1, \omega_2, \ldots, \omega_M, \omega_{M+1}\}$ (note that ω_{M+1} represents the unassigned category). We want to estimate the strength of the association among them through measuring the degree of agreement among dependent classifiers. For $j = 1, 2, \ldots N; i = 1, 2, \ldots, M + 1$ denote by d_{ji} the number of classifiers which assign candidate MUP pattern m_j to MUPT class ω_i, i.e.,

$$d_{ji} = \sum_{k=1}^{K} T(e_k(m_j) = \omega_i) \tag{13.18}$$

where $T(e = \sigma)$ is a binary characteristic function and it equals 1 if $e = \sigma$ and 0 otherwise. Note that $\sum_{i=1}^{M+1} d_{ji} = K$ for each MUP m_j. Table 13.1 shows the per MUP pattern diversity matrix through d_{ji}.

Based on the per MUP pattern diversity matrix of K classifiers, the degree of agreement among correlated classifiers $e_k(m_j), k = 1, 2, \ldots, K$ in classifying MUP m_j is measured using the following kappa hat statistic formula for multiple outcomes (classes) and multiple classifiers [14]:

$$\widehat{\kappa} = 1 - \frac{NK^2 - \sum_{j=1}^{N} \sum_{i=1}^{M+1} d_{ji}^2}{KN(K-1) \sum_{i=1}^{M+1} \bar{p}_i \bar{q}_i} \tag{13.19}$$

where $\bar{p}_i = \frac{\sum_{j=1}^{N} d_{ji}}{NK}$ represents the overall proportion of classifier outputs in MUPT ω_i, and $\bar{q}_i = 1 - \bar{p}_i$.

The value of the kappa hat statistic $\widehat{\kappa}_i$, (for MUPT class w_i) $i = 1, 2, \ldots, M$ and the unassigned category ω_{M+1} may be measured using the following formula [14]:

Table 13.1 Per MUP pattern diversity matrix of K classifiers

MUP Pattern	MUPTω_1	MUPTω_2	...	MUPTω_M	MUPTω_{M+1}	$\sum_{i=1}^{M+1} d_{ji}^2$
m_1	d_{11}	d_{12}	...	d_{1M}	$d_{1(M+1)}$	$\sum_{i=1}^{M+1} d_{1i}^2$
m_2	d_{21}	d_{22}	...	d_{2M}	$d_{2(M+1)}$	
.	.	.	...	.	.	.
.	.	.	...	.	.	.
.	.	.	...	.	.	.
m_N	d_{N1}	d_{N2}	...	d_{NM}	$d_{N(M+1)}$	$\sum_{i=1}^{M+1} d_{Ni}^2$
Total	$\sum_{i=1}^{N} d_{j1}$	$\sum_{i=1}^{N} d_{j2}$	...	$\sum_{i=1}^{N} d_{jM}$	$\sum_{i=1}^{N} d_{j(M+1)}$	$\sum_{j=1}^{N} \sum_{i=1}^{M+1} d_{ji}^2$

$$\hat{\kappa}_i = 1 - \frac{\sum_{j=1}^{N} d_{ji}(K - d_{ji})}{KN(K-1)\bar{p}_i\bar{q}_i} \tag{13.20}$$

13.5.4 Hybrid Classifier Fusion

The hybrid classifier fusion system does not contain the ensemble members selection module depicted in Fig. 13.6 and it uses a fixed set of base classifiers. It uses a hybrid aggregation module which is a combination of two stages of aggregation: the first aggregator is based on the abstract level and the second is based on the measurement level. Both aggregators may be data independent or the first aggregator may be data independent and the second data dependent. We used as the first aggregator, the majority voting scheme behaving as a data independent aggregator, while, as the second aggregator, we used either the average rule aggregation behaving as a data independent aggregator or the Sugeno fuzzy integral as an implicit data dependent aggregator.

The hybrid aggregation scheme works as follows:

• First stage: The outputs of the ensemble of classifiers are presented to the majority voting aggregator. If all classifiers state a decision that a MUP pattern is left unassigned, then there is no chance to re-assign that MUP pattern to a valid MUPT class and it stays unassigned. If over half of the classifiers assign a MUP pattern to the same MUPT class, then that MUP pattern is allocated to that MUPT class and no further assignment is processed. For these MUP patterns, an overall confidence value is calculated for each MUPT class by averaging the confidence values given by the ensemble of base classifiers who contributed to the decision of assigning the MUP pattern. In all other situations, i.e., when half or less than half of the classifiers specify a decision for a MUP pattern to be assigned to the same MUPT class, the measurement level fusion scheme is used in the second stage to specify to which MUPT class the MUP pattern should be assigned based on which MUPT class has the largest combined confidence value. From the first stage, a set of incomplete MUPT classes are generated missing those MUP patterns that need to be assigned to a valid MUPT class in the second stage.

• Second stage: This stage is activated for those MUP patterns for which only half or less than half of the ensemble of classifiers in the first stage specify a decision for a MUP pattern to be assigned to the same MUPT class. The outputs of the ensemble of classifiers are presented to the average rule aggregator, or the trainable aggregator represented by the Sugeno [51] fuzzy integral. For each MUP pattern, the overall combined confidence values representing the degree of membership in each MUPT class are determined and accordingly, the MUP pattern is assigned to the MUPT class for which its determined overall combined confidence is the largest and if it is above the specified aggregator confidence threshold set for that MUPT class, otherwise the MUP pattern is left unassigned. The MUP patterns satisfying the assignment condition are placed in their assigned MUPT classes and thus forming a more complete set of MUPT classes.

13.5.5 Diversity-Based Hybrid Classifier Fusion

The hybrid classifier fusion scheme described in Sect. 13.5.4 uses as a classifier ensemble a fixed set of classifiers and consequently both aggregators act on the outputs of the same base classifiers.

The diversity-based hybrid classifier fusion scheme is a two-stage process and consists of two aggregators with a pre-stage classifier selection module for each aggregator. The ensemble candidate classifiers selected for aggregation are decided through assessing the degree of agreement using the kappa statistic measure given in Sect. 13.5.3.1.

The diversity-based hybrid fusion scheme works as follows:

• First stage: The ensemble candidate classifiers selected for aggregation by the first aggregator are those having the maximum degree of agreement, i.e., having the maximum value of kappa statistics $\widehat{\overline{\kappa}}$ evaluated using (13.19). The outputs of the ensemble candidate classifiers are presented to the majority voting aggregator. If all the classifiers state a decision that a MUP pattern is left unassigned, then there is no chance to re-assign that MUP pattern to a valid MUPT class and it stays unassigned. If over half of the classifiers assign a MUP pattern to the same MUPT class, then that MUP pattern is allocated to that MUPT class and no further assignment is processed. For these MUP patterns, an overall confidence value is calculated by averaging the confidence values given by the ensemble classifiers who contributed in the decision of assigning the MUP pattern. In all other situations, i.e., when half or less than half of the classifiers specify a decision for a MUP pattern to be assigned to the same MUPT class, the measurement level fusion scheme is used in the second stage to specify to which MUPT class the MUP pattern should be assigned based on which MUPT class has the largest combined confidence value. From the first stage, a set of incomplete MUPT classes are generated missing those MUP patterns that need to be assigned to a valid MUPT class in the second stage.

• Second stage: This stage is used for those MUP patterns for which only half or less than half of the ensemble classifiers in the first stage specify a decision for a MUP pattern to be assigned to the same MUPT class. The ensemble candidate classifiers selected for aggregation at the second combiner are those having a minimum degree of agreement considering only the unassigned category, i.e., having the minimum value of kappa statistics $\widehat{\kappa}$ evaluated using (13.20) for $i = M + 1$. The outputs of the ensemble classifiers are presented to the average rule aggregator or the trainable aggregator represented by Sugeno [51] fuzzy integral aggregator. For each MUP pattern, the overall combined confidence values representing the degree of membership in each MUPT class are determined and accordingly the MUP pattern is assigned to the MUPT class for which its determined overall combined confidence is the largest and if its above the specified aggregator confidence threshold set for that MUPT class. The MUP patterns whose overall combined confidence are greater than zero and the specified aggregator confidence threshold are placed in the assigned MUPT class and thus forming a more complete set of MUPT classes.

13.6 Results and Comparative Study

The single classifier and multi-classifier approaches were evaluated and compared in terms of the difference between the correct classification rate $CC_r\%$ and error rate $E_r\%$. The correct classification rate $CC_r\%$ defined is as the ratio of the number of correctly classified MUP patterns, which is equal to the number of MUP patterns assigned minus the number of MUP patterns erroneously classified, to the total number of MUP patterns detected:

$$CC_r\% = \frac{\text{number of MUPs correctly classified}}{\text{total number of MUPs detected}} \times 100 \tag{13.21}$$

The error rate $E_r\%$ is defined as the ratio of the number of MUP patterns erroneously classified to any valid MUPT class to the number of MUP patterns assigned:

$$E_r\% = \frac{\text{number of MUPs erroneously classified}}{\text{number of MUPs assigned}} \times 100 \tag{13.22}$$

The effectiveness of using single classifier and multi-classifier approaches for EMG signal decomposition was demonstrated through the analysis of simulated and real EMG signals. The EMG signal data used consisted of two sets of simulated EMG signals: an independent set and a related set, each of 10 seconds length, and a set of real EMG signals. The characteristics of the EMG signal data sets can be found in [39, 40].

The simulated data were generated from an EMG signal simulator based on a physiologically and morphologically accurate muscle model [20]. The simulator enables us to generate EMG signals of different complexities with knowledge of the signal intensity represented by the average number of MUP patterns per second (pps), the numbers of MUPT classes, and which motor unit created each MUP pattern. Furthermore, the amount of MUP shape variability represented by jitter and/or IDI variability can be adjusted.

The EMG signals within the set of independent simulated signals have different levels of intensity and each have unique MUPT classes and MUP distributions. The EMG signals within the set of related simulated signals have the same level of intensity and the same MUPT classes and MUP distributions but have different amounts of MUP shape variability and firing pattern variability.

The set of real EMG signals are of different complexities and were detected during slight to moderate levels of contraction. They have been decomposed manually by an experienced operator using a computer-based graphical display algorithm.

The classification performance results of the correctly classified, and erroneously classified MUP patterns are presented in terms of mean and mean absolute deviation (MAD) of the difference between the correct classification rate $CC_r\%$ and error rate $E_r\%$ across the used data sets.

The base classifiers used for experimentation belong to the types described in Sect. 13.4. For each kind, we used four classifiers, i.e., four ACC classifiers e_1, e_2, e_3, e_4, four AFNNC classifiers e_5, e_6, e_7, e_8, four ANCCC classifiers e_9, e_{10}, e_{11}, e_{12}, and four ApCC classifiers e_{13}, e_{14}, e_{15}, e_{16}. Classifiers e_1, e_5, e_9, e_{13} were fed with time-domain first-order discrete derivative features and use MUP patterns with sequential assignment for seeding [40]. Classifiers e_2, e_6, e_{10}, e_{14} were fed with time-domain first-order discrete derivative features and use high-certainty MUP patterns for seeding [39]. Classifiers e_3, e_7, e_{11}, e_{15} were fed with wavelet-domain first-order discrete derivative features and use MUP patterns with sequential assignment for seeding. Classifiers e_4, e_8, e_{12}, e_{16} were fed with wavelet-domain first-order discrete derivative features and use the highest shape certainty MUP patterns for seeding. The performance of the above single classifiers approaches across the three EMG signal data sets is reported in Table 13.2.

We performed two experiments using the multi-classifier approaches. In the first experiment, we used a single classifier pool containing eight base classifiers e_1, e_2, e_3, e_4, e_5, e_6, e_7, e_8 from which we selected all the eight classifiers or only six classifiers out of the eight to work as a team in the ensemble. When using only six classifiers, the number of classifier ensembles that can be created is $\binom{8}{6} = 28$

Table 13.2 Mean and mean absolute deviation (MAD) of the difference between correct classification rate CC_r and error rate E_r for the different single classifier approaches across the three EMG signal data sets

Classifier	Independent simulated signals	Related simulated signals	Real signals
e_1	81.9 (4.9)	75.0 (2.5)	78.5 (0.9)
e_2	83.9 (4.9)	76.5 (1.8)	72.0 (0.3)
e_3	82.3 (4.1)	75.6 (1.6)	71.9 (1.4)
e_4	84.7 (4.0)	76.4 (1.5)	67.1 (1.4)
e_5	85.2 (2.5)	73.7 (1.0)	80.9 (0.9)
Best single classifiers e_6	90.4 (1.7)	80.7 (2.3)	77.5 (2.6)
e_7	83.1 (2.4)	73.3 (0.5)	73.5 (0.4)
e_8	88.9 (1.5)	79.0 (1.8)	73.4 (2.4)
Average of 8 single classifiers	85.0 (3.3)	76.2 (1.6)	74.3 (1.2)
e_9	79.3 (3.2)	59.2 (0.4)	69.0 (1.3)
e_{10}	80.6 (3.1)	54.6 (2.6)	63.0 (0.7)
e_{11}	76.1 (2.7)	59.4 (1.0)	58.7 (0.4)
e_{12}	77.7 (2.7)	56.0 (0.8)	49.5 (0.3)
e_{13}	78.0 (3.4)	62.6 (0.7)	71.3 (0.4)
e_{14}	77.8 (2.8)	59.4 (0.9)	66.1 (2.7)
e_{15}	77.5 (2.9)	62.2 (0.1)	68.0 (1.5)
e_{16}	77.6 (2.2)	58.9 (0.9)	56.6 (3.6)
Average of 16 single classifiers	81.5 (0.1)	67.7 (0.4)	68.5 (0.0)

Table 13.3 Mean and mean absolute deviation (MAD) of the difference between correct classification rate CC_r and error rate E_r for the different single classifier and multi-classifier approaches across the three EMG signal data sets

Classifier	Independent simulated signals	Related simulated signals	Real signals
	Single Classifiers		
Weakest of 8 Single Classifiers e_4	84.7 (4.0)	76.4 (1.5)	67.1 (1.4)
Weakest of 16 Single Classifiers e_{12}	77.7 (2.7)	56.0 (0.8)	49.5 (0.3)
Best Single Classifier e_6	90.4 (1.7)	80.7 (2.3)	77.5 (2.6)
Average of 8 Single Classifiers	85.0 (3.3)	76.2 (1.6)	74.3 (1.2)
Average of 16 Single Classifiers	81.5 (0.1)	67.7 (0.4)	68.5 (0.0)
	One-Stage Classifier Fusion [44]		
Majority Voting (fixed of 6)	86.7 (4.3)	80.5 (2.9)	77.0 (3.9)
Average Fixed Rule (fixed of 6)	90.6 (2.1)	83.9 (0.5)	82.6 (1.6)
Sugeno Fuzzy Integral (fixed of 6)	87.6 (2.6)	81.7 (1.2)	80.4 (1.9)
	One-Stage Classifier Fusion [45]		
Majority Voting (fixed of 8)	86.0 (4.6)	79.2 (3.0)	77.3 (4.3)
Average Fixed Rule (fixed of 8)	88.0 (2.5)	82.0 (0.7)	85.1 (1.2)
Sugeno Fuzzy Integral (fixed of 8)	82.3 (2.7)	78.3 (1.9)	80.9 (2.9)
	Diversity-based One-Stage Classifier Fusion [45]		
Majority Voting 6/8	87.6 (4.2)	80.1 (2.7)	78.8 (4.8)
Average Fixed Rule 6/8	88.5 (2.2)	82.1 (1.1)	84.9 (0.8)
Sugeno Fuzzy Integral 6/8	84.6 (2.4)	80.2 (1.1)	82.0 (0.8)
	Hybrid Classifier Fusion [44]		
AMVAFR (fixed of 6)	91.8 (1.8)	84.6 (1.3)	82.7 (2.5)
AMVSFI (fixed of 6)	91.8 (1.8)	84.6 (1.3)	82.5 (1.7)
	Diversity-based Hybrid Classifier Fusion [43]		
ADMVAFR – 6/8	91.6 (1.8)	84.4 (0.7)	85.5 (0.9)
ADMVSFI – 6/8	91.2 (1.8)	84.0 (0.8)	85.2 (0.9)
ADMVAFR – 6/16	90.0 (3.2)	83.2 (0.7)	83.7 (0.6)
ADMVSFI – 6/16	89.6 (3.3)	82.5 (0.8)	82.8 (0.4)

AMVAFR, ADMVAFR – stands for Adaptive (or Diversity-based) Majority Voting with Average Fixed Rule hybrid classifier fusion scheme, respectively.

AMVSFI, ADMVSFI – stands for Adaptive (or Diversity-based) Majority Voting with Sugeno Fuzzy Integral one-stage classifier fusion scheme, respectively.

6/8, 6/16 – stands for selecting 6 base classifiers from the classifier pool containing 8 or 16 classifiers, respectively.

fixed of 6 or 8 – stands for using fixed ensemble of 6 or 8 single classifiers, respectively.

ensembles. The performance of multi-classifier approaches for this experiment is reported in Table 13.3

In the second experiment, we used a base classifier pool containing sixteen base classifiers e_1, e_2, e_3, e_4, e_5, e_6, e_7, e_8,e_9, e_{10}, e_{11}, e_{12}, e_{13}, e_{14}, e_{15}, e_{16} from which we selected six classifiers to work as a team in the ensemble for every signal at each stage aggregator of the diversity-based hybrid classifier fusion system. The number of classifier ensembles that can be created is $\binom{8}{6} = 28$ ensembles. The performance of this experiment is reported in Table 13.3.

From Table 13.3, we see that the one-stage aggregator classifier fusion and its diversity-based variant schemes have classification performance better than the average performance of the constituent base classifiers and also better than the performance of the best base classifier except across the independent simulated signals. The hybrid classifier fusion and its diversity-based variant approaches on the other hand have performances that not only exceed the performance of any of the base classifiers forming the ensemble but also reduced classification errors for all data sets studied [36, 37, 44].

The improvement in classification performance and the reduction of the classification errors using the multi-classifier approaches is a result of the complementary action of the base classifiers when working together in an ensemble whose base classifier members being selected using the kappa statistic diversity measure.

Beside improving the classification performance using multi-classifier approaches, the other reason that turned our attention toward managing the uncertainty in classifying MUP patterns during EMG signal decomposition is the inability to guarantee that a single high performance classifier will have the optimal performance and robustness across EMG signals of different complexities.

13.7 Conclusion

In this chapter, we studied the effectiveness of using different classification approaches for EMG signal decomposition with an objective to improve classification performance and robustness. To achieve this goal, we explored many classification paradigms and adapted them to the problem we are investigating, evaluated the developed classifiers using simulated and real EMG signals of different complexities, refined the misclassification in created MUPT classes through proposing a set of IDI statistics capable of detecting erroneous MUP classifications, proposed and tested a new hybrid classifier fusion approach for improving the results, and finally adopted an iterative adaptive MUP classification approach for train-wise adjustment of each MUPT class assignment threshold based on MUPT class firing pattern statistics to exclude MUP patterns causing firing pattern inconsistencies.

References

1. Alexandre L A, Campilho A C and M. Kamel (2001) On combining classifiers using sum and product rules. Pattern Recognition Letters, 22:1283–1289
2. Andreassean S (1987) Methods for computer-aided measurement of motor unit parameters. The London Symposia – Supplement 39 to Electroencephalography and Clinical Neurophysiology, 13–20
3. Atiya A F (1992) Recognition of multiunit neural signals. IEEE Transactions on Biomedical Engineering, 39(7):723–729

4. Brunelli R and Falavigna D (1995) Person identification using multiple cues. IEEE Transactions in Pattern Analysis and Machine Intelligence, 17(10):955–966
5. Chauvet E, Fokapu O, Hogrel J Y et al (2003) Automatic identification of motor unit action potential trains from electromyographic signals using fuzzy techniques. Medical & Biological Engineering & Computing, 41:646–653
6. Cho Sung-Bae and Kim J H (1995) Combining multiple neural networks by fuzzy integral for robust classification. IEEE Transactions on Systems Man and Cybernetics, 25(2):380–384
7. Cho Sung-Bae and Kim J H (1995) Multiple network fusion using fuzzy logic. IEEE Transactions on Neural Networks, 6(2):497–501
8. Cohen J (1960) A coefficient of agreement for nominal scales. Educational and Psychological Measurement, 20(2):37–46
9. Duda R O, Hart P E and Stork D (2001) Pattern Classification. John Wiley & Sons, 2nd edition
10. Duin R (2002) The combining classifier: to train or not to train? In Proceedings of the 16th International Conference on Pattern Recognition, 2:765–770
11. Duin R and Taz D (2000) Experiments with classifier combining rules. In J Kittler and F Roli, editors, Multiple Classifier Systems, Lecture Notes in Computer Science, 1857:16–29, Gagliari, Italy, Springer
12. Etawil H A Y (1994) Motor unit action potentials: discovering temporal relations of their trains and resolving their superpositions. Master's thesis, University of Waterloo.,Waterloo, Ontario, Canada.
13. Fang J, Agarwal G and Shahani B (1999) Decomposition of multiunit electromyogram signals. IEEE Transactions on Biomedical Engineering, 46(6):685–697
14. Fleiss J L, Levin B and Paik M C (2003) Statistical Methods for Rates and Proportions. John Wiley & Sons, 3rd edition
15. Florestal J R, Mathieu P A and Malanda A (2004) Automatic decomposition of simulated EMG signals. In Proceedings of the 28th Conference of the Canadian Medical and Biological Engineering Society, 29–30
16. Florestal J R, Mathieu P A and Malanda A (2006) Automated decomposition of intramuscular electromyographic signals. IEEE Transactions on Biomedical Engineering, 53(5):832–839
17. Florestal J R, Mathieu P A and Palmondon R (2007) A genetic algorithm for the resolution of superimposed motor unit action potentials. IEEE Transactions on Biomedical Engineering, 54(12):2163–2171
18. Giacinto G and Roli F (2001) An approach to the automatic design of multiple classifier systems. Pattern Recognition Letters, 22:25–33
19. Gut R and Moschytz G S (2000) High-precision EMG signal decomposition using communication techniques. IEEE Transactions on Signal Processing, 48(9):2487–2494
20. Hamilton-Wright A and Stashuk D W (2005) Physiologically based simulation of clinical EMG signals. IEEE Transactions on Biomedical Engineering, 52(2):171–183
21. Hassoun M H, Wang C and Spitzer R (1994) Nnerve: Neural network extraction of repetitive vectors for electromyography – part i: algorithm. IEEE Transactions on Biomedical Engineering, 41(11):1039–1052
22. Hassoun M H, Wang C and Spitzer R (1994). Nnerve: Neural network extraction of repetitive vectors for electromyography – part ii: performance analysis. IEEE Transactions on Biomedical Engineering, 41(11):1053–1061
23. Kamel M S and Wanas N M (2003) Data dependence in combining classifiers. In T.Windeatt and F. Roli, editors, Multiple Classifier Systems, Lecture Notes in Computer Science, 2790: 1–14 Guilford UK Springer
24. Keller J M, Gray M R, and Givens J A (1985) A fuzzy k-nearest neighbor algorithm. IEEE Trans Systems Man and Cybernetics, 15(4):580–585
25. Kittler J M, Hatef M, Duin R P V et al. (1998) On combining classifiers. IEEE Transactions in Pattern Analysis and Machine Intelligence, 20(3):955–966
26. Lam L and Suen C Y (1997) Application of majority voting to pattern recognition: an analysis of its behavior and performance. IEEE Transaction on Systems Man and Cybernetics – Part A: Systems and Humans, 27(5):553–568

27. LeFever R S and De Luca C J (1982) A procedure for decomposing the myoelectric signal into its constituent action potentials – part i: technique, theory, and implementation. IEEE Transactions on Biomedical Engineering, 29(3):149–157
28. Loudon G H, Jones N B and Sehmi A S (1992) New signal processing techniques for the decompositon of emg signals. Medical & Biological Engineering & Computing, 30(11): 591–599
29. McGill K C (1984) A method for quantitating the clinical electromyogram. PhD dissertation, Stanford University, Stanford, CA
30. McGill K C, Cummins K and Dorfman L J (1985) Automatic decomposition of the clinical electromyogram. IEEE Transactions on Biomedical Engineering, 32(7):470–477
31. Mirfakhraei K and Horch K (1997) Recognition of temporally changing action potentials in multiunit neural recordings. IEEE Transactions on Biomedical Engineering, 44(2):123–131
32. Hamid Nawab S, Wotiz R and De Luca C J (2004) Improved resolution of pulse superpositions in a knowledge-based system EMG decomposition. In Proceedings of the 26th Annual International Conference of the IEEE Engineering in Medicine and Biology Society, 1: 69–71
33. Nikolic M, Sørensen J A, Dahl K et al. (1997) Detailed analysis of motor unit activity. In Proceedings of the 19th Annual International Conference of the IEEE Engineering in Medicine and Biology Society Conference, 1257–1260
34. Paoli G M (1993) Estimating certainty in classification of motor unit action potentials. Master's thesis, University of Waterloo., Waterloo, Ontario, Canada
35. Partridge D and Yates W B (1996) Engineering multiversion neural-net systems. Neural Computation, 8:869–893
36. Rasheed S (2006) A multiclassifier approach to motor unit potential classification for EMG signal decomposition. Ph. D. dissertation, URL: http://etd.uwaterloo.ca/etd/srasheed2006.pdf, University of Waterloo, Waterloo, Ontario, Canada
37. Rasheed S (2008) Diversity-Based Hybrid Classifier Fusion: A Practical Approach to Motor Unit Potential Classification for Electromyographic Signal Decomposition. VDM Verlag Dr. Müller, Berlin, Germany
38. Rasheed S, Stashuk D and Kamel M (2004) Multi-classification techniques applied to EMG signal decomposition. In Proceedings of the 2004 IEEE International Conference on Systems, Man and Cybernetics, SMC 04, 2:1226–1231, The Hugue, The Netherland
39. Rasheed S, Stashuk D and Kamel M (2006) Adaptive certainty-based classification for decomposition of EMG signals. Medical & Biological Engineering & Computing, 44(4):298–310
40. Rasheed S, Stashuk D and Kamel M (2006) Adaptive fuzzy k-NN classifier for EMG signal decomposition. Medical Engineering & Physics 28(7):694–709
41. Rasheed S, Stashuk D and Kamel M (2008) Fusion of multiple classifiers for motor unit potential sorting. Biomedical Signal Processing and Control, 3(3):229–243
42. Rasheed S, Stashuk D and Kamel M (2008). A software package for motor unit potential classification using fuzzy k-NN classifier. Computer Methods and Programs in Biomedicine, 89:56–71
43. Rasheed S, Stashuk D and Kamel M (2009) Integrating heterogeneous classifier ensembles for EMG signal decomposition based on classifier agreement. Accepted for publication in IEEE Transactions on Information Technology in Biomedicine and now is published on-line at http://ieeexplore.ieee.org
44. Rasheed S, Stashuk D and Kamel M (2007) A hybrid classifier fusion approach for motor unit potential classification during EMG signal decomposition. IEEE Transactions on Biomedical Engineering, 54(9):1715–1721
45. Rasheed S, Stashuk D and Kamel M (2008) Diversity-based combination of non-parametric classifiers for EMG signal decomposition. Pattern Analysis & Applications, 11:385–408
46. Stashuk D W (1999) Decomposition and quantitative analysis of clinical electromyographic signals. Medical Engineering & Physics, 21:389–404
47. Stashuk D W (2001) EMG signal decomposition: how can it be accomplished and used? Journal of Electromyography and Kinesiology, 11:151–173

48. Stashuk D W and de Bruin H (1988) Automatic decomposition of selective needle-detected myoelectric signals. IEEE Transactions on Biomedical Engineering, 35(1):1–10
49. Stashuk D W and Paoli G (1998) Robust supervised classification of motor unit action potentials. Medical & Biological Engineering & Computing, 36(1):75–82
50. Stefano L D and Mattoccia S (2003) Fast template matching using bounded partial correlation. Machine Vision and Applications, 13:213–221
51. Sugeno M (1977) Fuzzy measures and fuzzy integrals - a survey. In Fuzzy Automata and Decision Processes 89–102 North-Holland, Amsterdam
52. Xu L, Krzyzak A and Suen C Y (1992) Methods of combining multiple classifiers and their applications to handwriting recognition. IEEE Transaction on Systems Man and Cybernetics, 22(3):418–435
53. Zennaro D, Wellig P, Koch V. M et al. (2003) A software package for the decomposition of long-term multi-channel EMG signals using wavelet coefficients. IEEE Transactions on Biomedical Engineering, 50(1):58–69

Chapter 14
Parametric Modeling of Some Biosignals Using Optimization Metaheuristics

Amir Nakib, Amine Naït-Ali, Virginie Van Wassenhove and Patrick Siarry

Abstract This chapter describes advanced optimization techniques that can be used for dealing with non-linear parametric models to characterize the dynamics and/or the shape of some biosignals. The class of optimization techniques that will be covered encompasses Genetic and Particle Swarm Optimization family of optimization methods. Both methods belong to a specific optimization category called "metaheuristics". Two specific examples will illustrate the use of these techniques. In the first example, Genetic Algorithms are used as an optimization technique used to model Brainstem Auditory Evoked Potentials (BAEPs). In the second example, the Particle Swarm Optimization Algorithms are used for curve fitting Event Related Potentials and the heart beat signal. As will be further discussed, metaheuristics can naturally be extended to various applications relevant to biosignal processing.

14.1 Introduction

Modeling signals or systems can efficiently analyze and parameterize specific phenomena and the closer the employed model is to reality, the more accurate the results will be. However, real systems are generally complex and, in particular, they can be highly non-linear; consequently, modeling parametrically these systems can be very difficult. The main obstacle is that, for a given observation, optimizing a non-linear model can be extremely time consuming. Hence, for dynamic systems, a real time estimation of optimal parameters is a major constraint and for this reason, choosing an appropriate optimization technique is of great importance. Among the large number of optimization techniques including, for instance, nonlinear programming, dynamic programming, or calculus of variations, one can now include the technique of metaheuristics.

A. Nakib (✉)
Université Paris 12, Laboratoire Image, Signaux et Systèmes Intelligents, LiSSi, EA 3956 61 avenue du Général de Gaulle, 94010, Créteil France
e-mail: nakib@univ-paris12.fr

A. Naït-Ali (ed.), *Advanced Biosignal Processing*,

DOI 10.1007/978-3-540-89506-0_14, © Springer-Verlag Berlin Heidelberg 2009

Metaheuristics are heuristic methods which can solve a very general class of computational problems. They can generally be applied to problems for which there is no satisfactory problem-specific algorithm or heuristic (local search). The most common use for metaheuristics addresses combinatorial and continuous optimization problems but metaheuristics can also handle problems when formulated under the same optimization constraints, for instance, in solving Boolean equations. Since the early 1980s, metaheuristics have developed significantly and have now achieved widespread success in addressing numerous practical and difficult continuous and combinatorial optimization problems. They incorporate concepts from biological evolution, intelligent problem solving, mathematical and physical sciences, neural systems, or statistical mechanics. A great deal of effort has been invested in the field of continuous optimization theory in which heuristic algorithms have become an important area of research and applications. It should be mentioned that 10 or 20 years ago, metaheuristics were considered a time consuming technique; nowadays, the phenomenal progress in high performance mono/multiprocessor platforms allows solving problems even if under real-time constraints. Metaheuristics can thus be employed as efficient tools to deal with some biomedical problems as will be shown throughout this chapter.

The current chapter is organized in three parts: in Sect. 14.2, Genetic Algorithms (GA) are introduced and applied to estimate the dynamics of Brainstem Auditory Evoked Potentials (BAEP) through a simulation example. In Sect. 14.3, the Swarm Optimization Algorithm (PSO) is described and applied for the purpose of curve fitting. Two examples will then be considered: the first example deals with a real Event Related Potentials (ERP) and the second example with a real Electrocardiographic (ECG) beat signal.

14.2 Modeling the Dynamics (Specific Case)

As mentioned previously, GA are employed to estimate the dynamics of BAEPs. First, a parametric model is defined and its optimal parameters are then defined once the convergence is reached.

14.2.1 Defining a Parametric Model

BAEPs have already been briefly mentioned in Chap. 1: they are low energy electrical signals generated by the brain stem (a subcortical structure) in response to acoustic stimulations (such as short acoustic impulses) to the ear. BAEPs are a widely used tool for the early clinical diagnosis of acoustic neuromas but they can also be used to study the effects of certain substances on the auditory system. Importantly, BAEPs are a useful tool for surgeons as they are used during surgery to help keep the auditory pathways intact. As can be seen in Fig. 14.1, a BAEP is characterized by five major consecutive waves, denoted by I, II, III, IV/V.

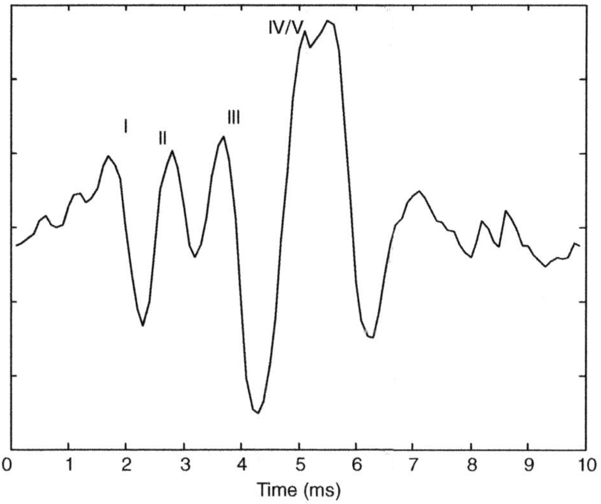

Fig. 14.1 A real processed BAEP showing the five waves I, II, III, IV/V

The classic method used to estimate the BAEPs consists in averaging the brainstem responses recorded via electroencephalography over multiple acoustic stimulations. The signal-to-noise ratio (SNR) can reach values of –20 to –30 dB. In practice, it is generally impossible to observe directly a BAEP with a single stimulation, even after the filtering process. The classical averaging technique is based on the assumption of a stationary brainstem response. In other words, it is assumed that (i) the BAEP (the signal of interest) obtained from an average of single responses is time-invariant and (ii) that the noise (the EEG) is a zero-mean stationary signal. Therefore, even if this technique seems to be simple, it does achieve excellent results in some cases especially when the subject is healthy and the recordings are carried out under appropriate conditions. In this case, 800 or more responses are generally required to extract an averaged BAEP.

In some specific cases, namely pathological, even if it is possible to reduce the effect of the noise during the averaging process, the non-stationary BAEP (both in its dynamics and its shape) leads to an unrecognizable smoothed average signal. Hence, an objective analysis becomes a critical task, in particular for the measurement of latency (taken as the time between the onset of the stimulation and the maximum amplitude value of the BAEP waveform) and of the conduction times (the latency differences or duration between peaks I and III, or between peaks I–IV.)

In the following experiment, we make the assumption that the BAEPs vary over time from one response to another according to random delays. This "jitter" or desynchronization of the signals can be of physical and/or physiological origin and is a classic problem in signal processing. Many different methods exist to deal with jitters: for instance, techniques used in radar detection or in some specific biomedical engineering applications. Unfortunately, these methods cannot be adapted to our problem given the very low SNR of the BAEPs recording method. Additionally, under the assumption of the presence of jitters in the brainstem response, a phenomenon known as "smoothing" will occur during the averaging process and is unavoidable. The distortion caused by smoothing can lead to serious consequences depending on the nature of the desynchronization (e.g. distribution and variance).

By considering this phenomenon, the corresponding model can be expressed as follows:

$$x_i(n) = s(n + d_i) + b_i(n) \tag{14.1}$$

where,

$x_i(n)$ is the signal corresponding to the ith stimulation,
$s(n)$ is the useful signal (BAEP) to be estimated,
$b_i(n)$ is the noise (mainly the EEG) during the ith acquisition,
d_i is the time delay of each signal $s(n)$ with respect to the onset of stimulation.

For L stimulations, the averaging leads to:

$$\bar{x}(n) = \underbrace{\frac{1}{L}\sum_{i=0}^{L-1} s(n + d_i)}_{A} + \underbrace{\frac{1}{L}\sum_{i=0}^{L-1} b_i(n)}_{B} \tag{14.2}$$

From (14.2) it is clear that the term B (which is related to the noise) can be neglected if the statistical average of the noise is zero or close to zero ($E\,[b(n)] \approx 0$). This situation occurs if a sufficient number of stimulations (L) is used during the acquisition process. In other words, the higher L is, the higher the SNR. Consequently, in such a situation, the averaged signal can be expressed as follows:

$$\bar{x}\,(n) = \frac{1}{L}\sum_{i=0}^{L-1} s(n + d_i) \tag{14.3}$$

The issue that now arises is the following: how is it possible to estimate the optimal delay parameters d_i? This naturally requires optimizing a criterion.

If L responses are necessary to reduce the noise during the averaging process, the optimal set of the parameters d_i can be found by maximizing the energy of the averaged signal. This occurs when BAEPs are aligned i.e. synchronized. In mathematical terms, this energy can be expressed as:

$$f_{\mathbf{d}} = \sum_{n=0}^{N-1}\left(\frac{1}{L}\sum_{i=0}^{L-1} x_i(n - d_i)\right)^2 \tag{14.4}$$

where,

$\mathbf{d} = [d_0, d_1, d_2, \ldots.., d_{L-1}]$ represents the vector of the delay parameters,
N is the number of samples for each response.

Therefore, the optimization problem leads to the minimization of the following equation:

$$J_{\mathbf{d}} = -\sum_{n=0}^{N-1}\left(\frac{1}{L}\sum_{i=0}^{L-1} x_i\,(n-d_i)\right)^2 \tag{14.5}$$

Since this criterion is neither quadratic nor convex, one of the possible optimization techniques that can be used to find the best values of d_i is a metaheuristic such as Simulated Annealing (SA) [1, 2, 3, 4] or GAs, which are under consideration in this chapter. A instance of simulation using this metaheuristic technique is now presented below.

14.2.2 Optimization Using Genetic Algorithms

14.2.2.1 What are Genetic Algorithms?

Genetic Algorithms were developed in 1979 by J. Holland and his colleagues at the University of Michigan. The underlying concept of GAs was inspired by the theory of natural selection and genetics. In order to clarify how GAs work, we provide some of their basic principles (Table 14.1). We then turn to the problem of BAEPs estimation and how it can be handled by GAs.

Let's assume that a population is made up of K individuals. In genetics, each individual is characterized by a chromosome and, at a lower level, the chromosome is made up of genes. The generation which is initially able to evolve in a given environment, in order to produce other generations – by following the rules of selection, crossover and mutation – constitutes the first generation. In the course of the process, those individuals that are, for instance, the weakest will disappear

Table 14.1 The principle of Genetic Algorithms in the context of delay estimation

Population	A set of potential solutions in a generation (m). $\mathbf{d}_1^{(m)}, \mathbf{d}_2^{(m)}, \ldots \mathbf{d}_K^{(m)}$ $\mathbf{d}_i^{(m)}$is a vector with M parameters to be determined.
Chromosome or individual	A potential solution $\mathbf{d}_i^{(m)}$: $\mathbf{d}_i^{(m)} = \left[d_{i,0}^{(m)}\ d_{i,1}^{(m)}\ \ldots\ d_{i,M-1}^{(m)} \right]^t$
Gene	An element of a potential solution. For example: $d_{i,n}^{(m)}$, $n=0,\ldots M-1$.
Reproduction	A potential solution in a generation (m–1) is maintained in the next generation (m).
Crossover	Two potential solutions of a given generation (m–1) are combined to generate two other solutions for the future generation (m). Example: $\mathbf{d}_i^{(m-1)}$ and $\mathbf{d}_j^{(m-1)}$ can produce $\mathbf{d}_i^{(m)}$ and $\mathbf{d}_j^{(m)}$
Mutation	If $\mathbf{d}_i^{(m-1)}$is a potential solution in a given generation, the mutation can occur in the following generation in order to generate $\mathbf{d}_i^{(m)}$ by modifying one of its elements.

Table 14.2 Application of GA for delay estimation

1. Choose randomly K vectors which represent potential solutions. Each solution is made up of M delays which correspond to M responses,
2. **Reproduction stage**: generate k other potential solutions using both crossover and mutation operators,
3. Evaluate the objective function for each potential solution,
4. Selection stage: take the best K solutions amongst the $K+k$ solutions so that the following generation can be produced,
5. Save the best solution,
6. If the number of maximal generations is not reached, go back to 2,
7. Solution = best point found, stop the process.

from one generation to the next. Conversely, the strongest individuals will be able to reproduce with no modifications. Importantly, the child is the clone of his parent. Additionally, some individuals from the same generation will be able to crossover so that the future generation can have similar characteristics. An individual's genes can also be changed and replaced by genes from the search space. The final algorithm is then given in Table 14.2.

14.2.2.2 An Example on Simulated BAEPs

Let's now consider a small example which reproduces the problematic evoked above, namely, the estimation of delays or jitters in each recorded response. For this purpose, one can simulate the process by using a template such as the one shown in Fig. 14.2. In this example, 200 responses were generated. The responses were then contaminated using a heavy noise, so that BAEPs became visually unrecognizable (i.e. low SNR) and delayed randomly according to a Gaussian distribution. The set

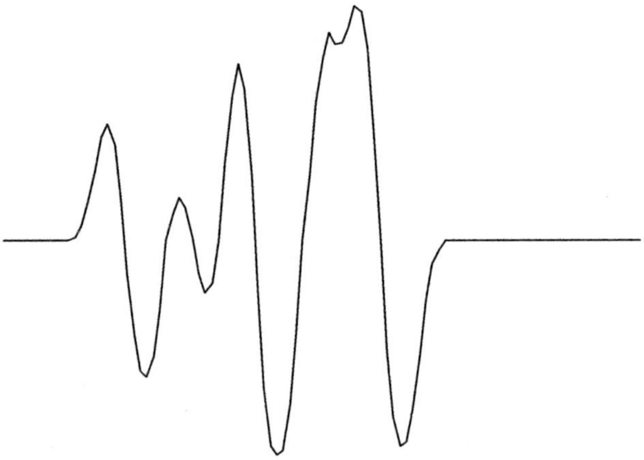

Fig. 14.2 BAEP template used for the simulation purpose

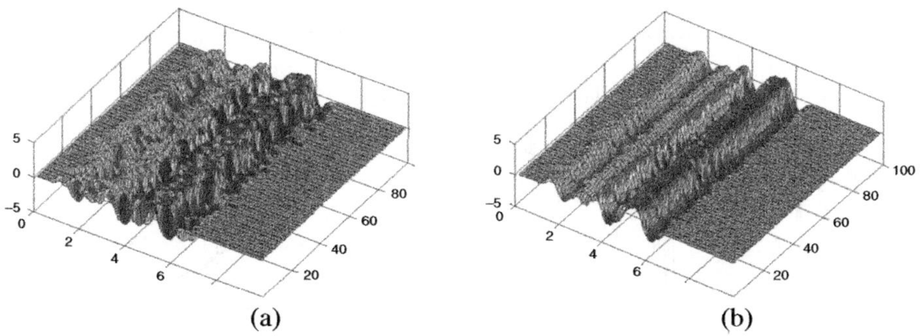

Fig. 14.3 Bidimensional representation of BAEPs (**a**) BAEPs delayed randomly according to a Gaussian distribution (**b**) aligned BAEPS after GAs convergence. For the clarity of the graphs, the noise is not represented

of responses are represented in Fig. 14.3a as a bidimensional surface. Note that in this figure, we have voluntarily removed the noise for clarity. As previously mentioned, the classic averaging method applied to these BAEPs provides a smoothed signal for which the waves are difficult to identify (Fig. 14.4a). However, after application of the GAs on this set of data, the BAEPs are properly aligned (Fig. 14.3b) and provide the expected shape of the averaged signal (see Fig. 14.4b). The estimated BAEP dynamics from successive responses is compared in Fig. 14.5a to random delays, used initially in this simulation. Both curves seem to be similar. These results are naturally obtained after GAs convergence (see Fig. 14.5b).

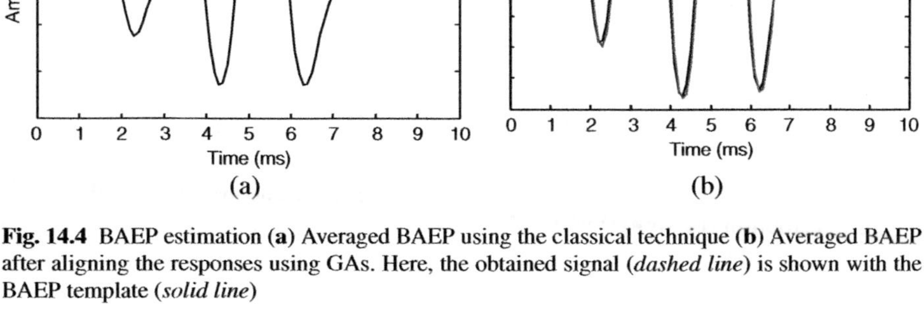

Fig. 14.4 BAEP estimation (**a**) Averaged BAEP using the classical technique (**b**) Averaged BAEP after aligning the responses using GAs. Here, the obtained signal (*dashed line*) is shown with the BAEP template (*solid line*)

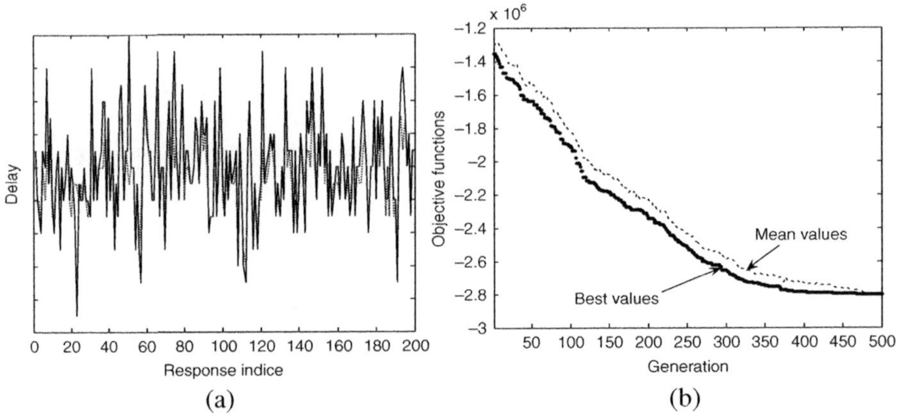

Fig. 14.5 (**a**) The estimated BAEP dynamics (delay evolution) represented by the *dashed line* compared to the random delays used for the simulation (*solid line*), (**b**) GAs convergence curve

In the next section, we present another recent metaheuristic called Particle Swarm Optimization Algorithm (PSO). The evaluation of this algorithm is achieved on a real Event Related Potential (ERP) as well as a real ECG beat, for the purpose of curve fitting.

14.3 Shape Modeling Using Parametric Fitting

Parametric fitting involves finding coefficients (parameters) for one or more models that fit a given recorded data. These data are assumed to be statistical in nature and can be divided into two components: a deterministic component and a random component (*data* = *deterministic component* + *random component*).

The deterministic component is given by a parametric model and the random component is often described as an error associated with the data (*data* = *model* + *error*). The model is a function of the independent (predictor) variable and one or more coefficients. The error represents random variations in the data that follow a specific probability distribution (usually Gaussian); these variations can come from many different sources but are always present at some level when one is dealing with measured data.

Parametric fitting are also known as regression techniques and approximate signals or value distributions using a mathematical distribution function with a limited number of free parameters. Values for these parameters are then chosen to fit the actual distribution or signal. If the model is a good fit for the distribution or the signal, this provides an accurate and compact approximation; however, since the shape of the distribution (or signal) is usually not known beforehand, this is often not the case. To overcome this inflexibility, we use a model and apply a metaheuristic to choose its coefficients.

These techniques compute their estimates by collecting and processing random samples of the data. Sampling techniques [5] offer high accuracy and probabilistically warrants the quality of the estimation. However, since the sampling itself is typically carried out at the time of the approximation, the resulting overhead prohibits the use of sampling for query optimization.

Since our main concern is a compact representation of data, and not its acquisition, we will now focus on the properties of parametric techniques which offer the important advantage of being adaptive.

In the sections, which follow we will show how one can apply a curve fitting method using a PSO. A parametric model is first presented followed by optimization by PSO. The performances are reported for real ERP and real ECG beat.

14.3.1 Defining a Parametric Model

Let's consider a signal $s(n)$ corrupted by an additive noise $b(n)$. The recorded signal is expressed by:

$$x(n) = s(n) + b(n) \tag{14.6}$$

Using a M Gaussian mixture model $s(n)$ can be expressed as follows [4, 10]:

$$\tilde{s}(n) = \sum_{i=1}^{M} A_i \exp\left[-(n-\mu_i)^2/\sigma_i^2\right] \; n = 0, \ldots N-1 \tag{14.7}$$

where, A_i, μ_i and σ_i^2 stand for the amplitude, the mean and the variance of the ith Gaussian function, respectively.

Therefore, optimal model parameters can be found by minimizing a distance between the observed signal $x(n)$ and the model $\tilde{s}(n)$. For instance, one can minimize the following distance (14.8) with respect to $\Theta = \{A_i, \mu_i, \sigma_i;\; i = 1, 2, \ldots, M\}$:

$$J = \left\| x(n) - \sum_{i=1}^{M} A_i \exp\left[-(n-\mu_i)^2/\sigma_i^2\right] \right\|_2^2 \tag{14.8}$$

J is considered as an objective function J [11]: the standard process of setting the partial derivatives to zero results in a set of non-linear coupled equations. Thereafter, the system is solved through numerical techniques. As previously mentioned, the continuous nature of the problem does not allow for an exhaustive enumerating of every possible combination in order to identify the optimal solution. Although for simple cost functions several very effective greedy algorithms can be used [12, 13], arbitrary functions show unpredictable surfaces with numerous local minima and require a stochastic approach which is in principle suitable for identifying global minima. For this purpose, the PSO Algorithm is applied in the next section.

14.3.2 Optimization Using PSO

14.3.2.1 What is PSO?

The notion of employing many autonomous agents (particles) that act together in simple ways to produce seemingly complex emergent behaviors was initially considered to solve the problem of rendering images in computer animations which attempt to resemble natural phenomena. In 1993, Reeves implemented a particle system that used many individuals working together to form what appeared to be a fuzzy objects (e.g. a cloud or an explosion). A particle system stochastically generates a series of moving points which are typically being initialized to predefined locations. Each particle is assigned an initial velocity vector. In graphical simulations, these points may also have additional characteristics such as color, texture and limited lifetime. Iteratively, velocity vectors are adjusted by some random factors. Each particle is then moved by taking its velocity vector from its current position and, where necessary, by a constrained angle, to make the movement appear realistically interactive graphical environments. Such systems are widely deployed in the generation of special effects and realistic interactive environment [6].

PSO was introduced by Kennedy and Eberhart in 1995 [7] as an alternative to Genetic Algorithms. Kennedy and Eberhart tried to reflect social behaviors of searching food. The researchers used non trivial mathematical problems such as fitness function for flock members (called agents, since the model is more general than the bird model.) The PSO technique has since become a competitor in the field on numerical optimization and is only briefly described in this paper (for further details see [8, 6, 9]) Although PSO share many similarities with the Genetic Algorithms, there is no evolution operator and a population of potential solutions is used in the search.

The technique starts with a random initialization of a swarm of particles in the search space (Table 14.3). Each particle is modeled by its position in the search space and its velocity. At each time step, all particles adjust their positions and velocities hence their trajectories according to their best locations and the location of the best particle of the swarm in the global version of the algorithm, or of the neighbors, in the local version. Indeed, each individual is influenced not only by its own experience, but also by the experience of other particles.

Table 14.3 PSO Algorithm

1. **Initialization** of a population of particles with random positions and velocities.
2. **Evaluate** the objective function for each particle and compute g.
3. **Repeat** while the stopping criterion is not met
    3.1. For each individual i, L_i is initialized at P_i
    3.2. Update the velocities and the positions of the particles.
    3.3. Evaluate the objective function for each individual.
    3.4. Compute the new L_i and g.
4. Show the best solution.

In a K-dimensional search space, the position and the velocity of the ith particle are defined by: $P_i = (p_{i,1}, \ldots, p_{i,K})$ and $V_i = (v_{i,1}, \ldots, v_{i,K})$, respectively. Each particle is characterized by its best location: $L_i = (l_{i,1}, \ldots, l_{i,K})$, that corresponds to the best location reached at the iteration t. The best location reached by the entire swarm is saved in the vector $G = (g_1, \ldots, g_K)$. The velocity of each particle is updated using the following equation:

$$v_{ij}(k+1) = w.v_{ij}(k) + c_1.r_1.\left(l_{ij} - v_{ij}(k)\right) + c_2.r_2.\left(g_j - v_{ij}(k)\right) \tag{14.9}$$

where, $j = 1, \ldots, K$, w is a constant called inertia factor, c_1 and c_2 are constants called acceleration coefficients, r_1 and r_2 are two independent random numbers uniformly distributed in [0,1]. If the computed velocity leads one particle out of the search space, the particle goes out of the search space and its fitness is not computed. The computation of the position at iteration k+1 is derived using:

$$p_{ij}(k+1) = p_{ij}(k) + v_{ij}(k+1) \tag{14.10}$$

for $j = 1, \ldots, K$. the inertia weight w controls the impact of the previous velocity on the current one, which allows the particles to avoid the local optima. Similarly, c_1 controls the attitude of the particle searching around its best location, and c_2 controls the influence of the swarm on the particle's behavior.

The parameters of PSO must be tuned in order to work well. This is also the case with the other metaheuristics (e.g. for the tuning of GA mutation rates). However, changing a PSO parameter can have a proportionally large effect. For instance, leaving out the constriction coefficient and not constricting particle speed by maximum velocity will result in an increase of speed, also partially due to the particle inertia. Particles with such speeds might explore the search space, but loose the ability to fine-tune a result. Thus, tuning the constriction coefficient or settings for maximum velocity is non-trivial and neither is the tuning of the settings for the inertia weight or the random values of c_1 r_1 and c_2 r_2. These parameters also control the abilities of the PSO. The higher the inertia weight, the higher the particle speed. As with the constriction coefficient, the setting of the inertia must balance between having good exploration of the search space and good fine-tuning abilities. As previously described, the setting of these parameters also determines the exploratory vs. the fine-tuning abilities of PSO. With the tradeoffs just described, the performance of PSO is problem-dependent. As described above, some general rules of thumb can be used, but they do not guarantee optimal PSO performance.

In the last standard 2006, the population size is given by $S = 10 + 2\sqrt{K}$, where K is the dimension of the problem. The inertia factor is given by $w = \frac{1}{2 \cdot \log(2)}$ and the coefficients are $c_1 = c_2 = 0.5 + \log(2.0)$.

14.3.2.2 An Example on Event-Related Potential (ERP)

As mentioned in Chap. 9 the classic method used to extract Event Related Potential (ERP) from brain signals recorded with EEG, consists in averaging over many trials to remove the noise, namely the EEG. Such technique bears the following drawbacks:

- Generally, tens to hundreds of evoked brain responses are required,
- The method is time consuming,
- A long experimental session can easily be strenuous on the participants/patients and significantly alter the signal (e.g. a lack of attention can attenuate cortical responses).
- The repetition of the same stimuli to obtain a large EEG response can be counter-intuitive to our current understanding of neural population responses (e.g. neurons adapt and lessen their responses to repetitive stimulations).

In some cases, ERPs are estimated from a single trial or by averaging just a few responses. This can be achieved because the SNR is not as low as in the BAEP case. Consequently, by processing separately each ERP, one can analyze both ERP waveforms as well as their dynamics over time. In our example, we try to highlight the way one can model or fit an ERP signal from a single trial, or from a signal obtained after a few averaged responses. For this purpose, we consider the Gaussian model described in Eq. (14.7). In this example, the parameter M is fixed to 15 namely, 15 Gaussians are used to model the ERP (i.e. 45 parameters to estimate). The setting parameters c and w required by PSO are empirically set to 0.5 and 2.1 as shown in Table 14.4. As a result, the optimal parameters are estimated following the convergence phase (cf. Fig. 14.6). Based on Eq. (14.7), these parameters are then used to reconstruct an ERP model, which is compared to the observed signal in Fig. 14.7.

Table 14.4 Modeling condition of the ERP using PSO

Number of classes (Gaussians)	Final value of fitness	PSO setting
15	30.402	Swarm size: 23, $w = 0.5$, $c = 2.1$

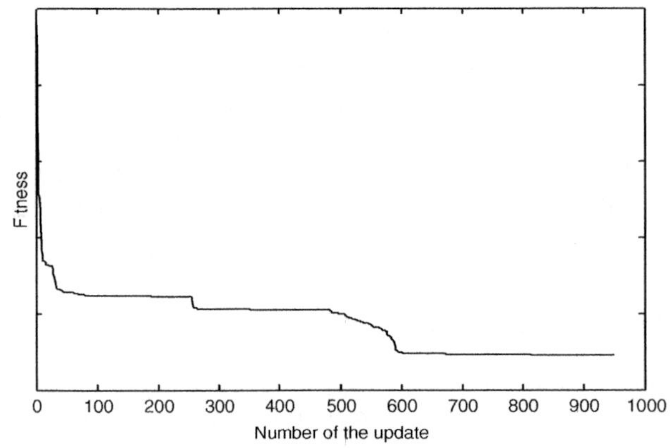

Fig. 14.6 Evolution of the best position of the swarm over iterations

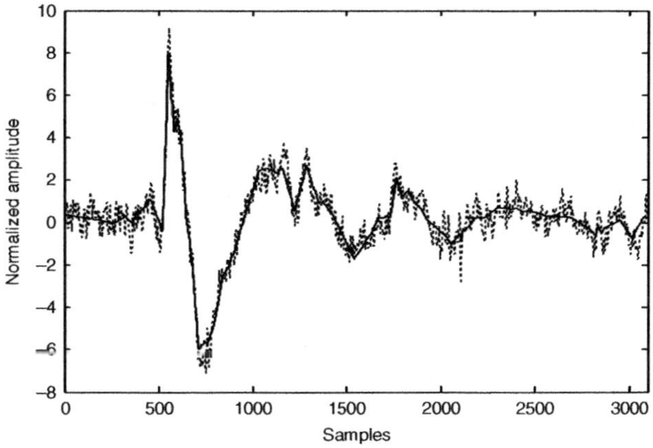

Fig. 14.7 ERP modeling. The observed signal is the dashed trace, the estimated model using 15 Gaussians is the solid trace

Except for the fact that the model can be used to extract some ERP characteristics, such as wave latencies, modeling ERPs over time allows to analyze their dynamics based on the variation of their corresponding parameters. Even if this aspect is not presented in this paper, we recommend this type of analysis to the reader. In addition, we should mention that an improvement of PSO can be performed if setting parameters are automatically determined. This aspect is currently under consideration.

14.3.2.3 An Example on ECG Modeling (Fitting)

Let's consider using a second example ECG beat fitting using the same Gaussian model optimized by PSO. Algorithm parameters are set empirically, based on some experiments achieved on numerous signals available in the database MeDEISA (www.medeisa.net).

Since the PSO doesn't require any prior knowledge of the starting point, the initialization is then defined randomly. Figures 14.8a and 14.8b illustrate the obtained models using 5 and 7 Gaussians, respectively. Optimal parameters found after the convergence of PSO are gathered in Table 14.5 corresponding to the selected setting. Note that an existing variant of this technique consists of preprocessing the ECG signal by detecting automatically its peaks; the purpose of the process presented here is to accelerate PSO convergence [11].

From the literature [12, 13], another idea consists of optimizing a criterion under a constraint, such as:

$$\sum_{i=1}^{M} A_i = 1 \tag{14.11}$$

where, A_i stands for peak amplitude.

In such a situation, the number of parameters is reduced by one. However, this constraint produces some errors in the fitting due to the integral:

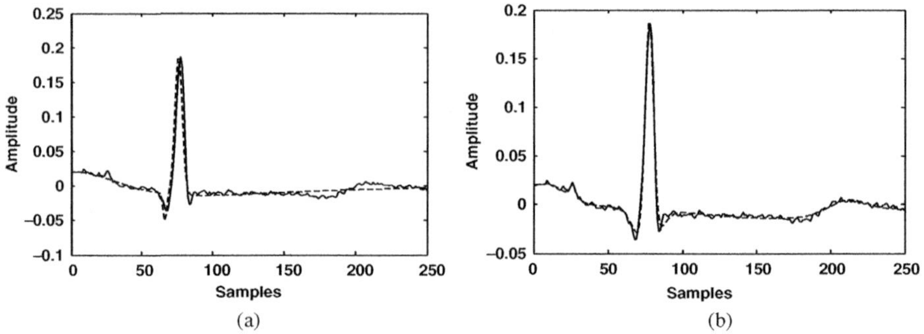

Fig. 14.8 ECG beat fitting. The real ECG beat is the solid trace, the ECG beat fitting is the dotted trace. (**a**) using 5 Gaussians, (**b**) using 7 Gaussians

Table 14.5 Modeling condition of the ERP using the PSO

Number of classes	Parameters of Gaussian curves	Final value of fitness	PSO setting
5	A:(0.199; –0.019; 0.200; –0.040; 0.040) μ:(202.3599; 0.000; 21.3140; 37.9300; 0.0000) σ:(249.9900; 179.8900; 3.5900; 1.7700; 47.1500)	0.0021	Swarm size: 17 $w = 0.72$ $c = 1.19$
7	A(0.237; –0.043; 0.000; 0.013; – 0.014; 0.022; 0.014) μ:(77.51; 76.21; 6.958; 26.27; 161.5; 7.739; 211.7) σ:(4.04; 10.86; 0.001241; 2.779; 81.13; 18.96; 18.20)	0.0037	Swarm size: 19 $w = 0.72$ $c = 1.19$

$$1 = \int x(t)\, dt \neq \sum_{i=0}^{N-1} x(i) \tag{14.12}$$

In the current chapter, this constraint is not taken into consideration.

Since the number of Gaussians (M) is supposed to be *a priori* known, the stopping criterion is fixed empirically to $10000\times M$. Additionally, a second stopping criterion is used based on the variations of the fitness function. Consequently, if the value does not decrease significantly after $100\times M$ evaluations of the objective function, convergence is considered to have been reached. Thus, the optimization process is stopped.

14.4 Conclusion

Two efficient tools have been presented, namely, GAs and PSO, to solve "hard" optimization problems. Throughout the chapter, these techniques have been used to estimate optimal parameters of parametric non-linear models commonly used in biosignal processing, namely BAEPs, ERPs and ECG beats. These signals have been specifically taken into consideration here and illustrated our use of the described tools.

Overall, the most important benefit of dealing with metaheuristics is that a variety of criteria can be estimated irrespective of the non-linearity of the objective function. Their efficiency is high mainly when classical methods fail. Additionally, it is convenient to identify automatically the setting parameters in order to avoid substantial manipulations. Finally, given the high-performance calculation platforms available nowadays, we believe that this class of optimization would be suited for real-time implementations.

References

1. Cherrid N, Naït-Ali A and Siarry P (2005) Fast Simulated Annealing Algorithm for BAEP Time Delay Estimation using a reduced order dynamics model. Med Eng Phys 27(8):705–711
2. Naït-Ali A and Siarry P (2002) Application of simulated annealing for estimating BAEPs in some pathological cases. Med Eng Phys 24:385–392
3. Naït-Ali A and Siarry P (2006) A new vision on the averaging technique for the estimation of non-stationary Brainstem Auditory Evoked Potentials: Application of a Metaheuristic method. Comp Biol Med 36(6):574–584
4. El Khansa L and Naït-Ali A (2007) Parametrical modeling of a premature ventricular contraction ECG beat: comparison with the normal case. Comp Biol Med 37(1):1–7
5. Haas P, Naughton J, Seshadri S et al. (1996) Selectivity and cost estimation for joins based on random sampling. J Comput Syst Sci 52(3):550–569
6. Banks A, Vincent J and Anyakoha C (2007) A review of particle swarm optimization. Part I: background and development. Natural Comput DOI 10.1007/s11047-007-9049-5, 1–18
7. Kennedy J and Eberhart R C (1995) Particle swarm optimization. IEEE Int Conf Neuronal Netw 4:1942–1948
8. Clerc M and Kennedy J (2002) The particle swarm: explosion, stability, and convergence in multi-dimensional complex space. IEEE Trans Evol Comput 6:58–73
9. Banks A, Vincent J and Anyakoha C (2007) A review of particle swarm optimization. Part II: hybridisation, combinatorial, multicriteria and constrained optimization, and indicative applications. Natural Comput DOI 10.1007/s11047-007-9050-z, 1–16
10. Flexer A et al. (2001) Single Trial estimation of Evoked potentials Using Gaussian mixture models with integrated noise component. ICANN 2001: International conference on artificial neural networks, Vienna, 2130:609–616
11. Nakib A, Oulhadj H and Siarry P (2008) Non-supervised image segmentation based on multiobjective optimization. Pattern Recognit Lett 29(2):161–172
12. Collette Y and Siarry P (2003) Multiobjective Optimization. s.l.: Eyrolles 2002
13. Dréo J, Pétrowski A, Siarry P and Taillard E (2005) Metaheuristics for hard optimization, Springer

Chapter 15
Nonlinear Analysis of Physiological Time Series

Anisoara Paraschiv-Ionescu and Kamiar Aminian

Abstract Biological systems and processes are inherently complex, nonlinear and nonstationary, and that is why nonlinear time series analysis has emerged as a novel methodology over the past few decades. The aim of this chapter is to provide a review of main approaches of nonlinear analysis (fractal analysis, chaos theory, complexity measures) in physiological research, from system modeling to methodological analysis and clinical applications.

15.1 Introduction

During the last two decades, theoretical and experimental studies of various biomedical systems, including heart, brain, respiratory system, immune system, human movement etc., have shown that these systems are best characterized as complex dynamical processes subjected and updated by nonlinear feed-forward and feedback inputs. The essential nonlinearities and the complexity of physiological interaction limit to the ability of linear analysis to provide full description of the underlying dynamics. In addition to classical linear analysis tools, more sophisticated nonlinear analysis methods are needed to quantify physiological dynamics and get insight into the understanding of underlying system/function.

Nonlinear dynamical analysis of biological/physiological signals lies at crossroads of frontier research in physics, engineering, medicine and biology, since current medical research studies of biochemical and biophysical mechanisms must deal with mathematical modeling. The basic goals of mathematical modeling and analysis are to: *understand* how a system works, *predict* the future behavior of a system and, *control* the system in order to guide it to a preferred state or keep it away from undesired states. These basic goals are closely related to the practical goal

A. Paraschiv-Ionescu (✉)
Laboratory of Movement Analysis and Measurement, Ecole Polytechnique Federale de Lausanne (EPFL), Switzerland
e-mail: anisoara.ionescu@epfl.ch

A. Naït-Ali (ed.), *Advanced Biosignal Processing*,
DOI 10.1007/978-3-540-89506-0_15,

of *diagnosis and treatment*, which underlie much of the rationale for research into physiological systems.

One of the aims of nonlinear time series analysis is the quantification of complexity of a system. 'Complexity' is related to intrinsic patterns hidden in the dynamics of the system (if however, there is no recognizable structure in the system, it is considered to be stochastic). In physiological research, a growing attention was devoted to the quantification of the complexity level of heath vs. disease through the analysis of relevant time series. Modern hypothesis postulate that when physiological systems became less complex, their information content is reduced [69, 38]. As a result, they are less adaptable and less able to cope with the exigencies of a continuous changing environment, this 'decomplexification' of systems appearing to be a common feature of many diseases.

Different metrics can be used to measure complexity of physiologically derived time series. Depending on the formulated medical/biological problem, different analysis methods can be used to quantify various aspects of physiological complexity related to long range correlations, regularity *vs.* randomness, chaotic *vs.* stochastic behavior, signal roughness. The analysis methods allowing the calculation of quantitative measures of complex temporal behavior of biosignals as well as classification of different physiological and pathological states are mainly derived from statistical physics: chaos theory, fractal analysis, information theory and, symbolic dynamics.

This chapter is organized as follows. Sect. 15.2 briefly defines the basic concepts related to *nonlinear dynamics theory* and *complex systems.* Sect. 15.3 describes the mathematical background of several nonlinear time series analysis methods derived from chaos theory, fractal analysis and symbolic dynamics. Sect. 15.4 reviews the relevance of these methods to quantify health and disease states of various physiological systems (e.g. heart rate, brain, respiration) with special emphasis on nonlinear analysis of *human physical activity/movement patterns* in health and disease. Finally, conclusions and research perspectives related to complexity analysis of physiological systems are drawn in Sect. 15.5.

15.2 Background Concepts

Time series – A time series is a sequence of data points, measured typically at successive uniform time intervals. Time series analysis accounts for the fact that points taken over time may have an internal structure (such as correlations, trend, etc.) that should be accounted for.

Dynamical systems – A dynamical system is any deterministic system whose state is defined at any time by the values of a set of variables and its evolution in time is determined by a set of rules employing time-delayed or differential equations. The term 'dynamic' introduces time and patterns as major factors in understanding the behavior of the system [14].

Linear vs. nonlinear systems – Dynamical systems (engineering, chemical, biological, etc.) are either linear or nonlinear. The fundamental distinction between them is made according to how the inputs to the system interact with each other to produce outputs. *Linear systems* satisfy the properties of superposition and proportionality; the magnitude of a response may be expressed as a sum of multiple, mutually independent variables, each with its own independent linear coefficient. The overall system behavior can be fully understood and predicted by analyzing their components individually using an analytical/*reductionist approach*.

Nonlinear systems contravene the principle of superposition and proportionality involving components/variables that interact in a complex manner: small changes can have striking and unanticipated effects, and strong stimuli will not always lead to drastic changes in the behavior of the system. Therefore, in order to understand the behavior of a nonlinear system, it is obligatory to perceive not only the behavior of its components, using the reductionist approach, but also the logic of the interconnections between components.

Deterministic vs. stochastic (random) systems – A *deterministic system* is a system in which the later states of the system follow from, or are determined by, the earlier ones (however, this does not necessary imply that later states of the system are predictable from the knowledge of earlier ones). Such a system contrasts with a *stochastic* or *random system* in which futures states are not determined from previous ones.

Complex (nonlinear) systems – A *complex system* is defined as a set of simple (and often nonlinearly) interacting units characterized by a high degree of connectivity or interdependence. Complex nonlinear systems are dynamic in the sense that interactions of their components are cooperative, self-enhancing and have feedback properties. This is known as *emergence* and gives systems the flexibility to adapt and *self-organize* in response to external challenge. The properties of the system are distinct from the properties of the parts, and they depend on the properties of the whole; the systemic properties vanish when the system break apart, whereas the properties of the parts are maintaining. The understanding of actual mechanism involved in complex nonlinear systems is usually attempted using two combined approaches: (1) the *reductionist approach* which identifies the elements of the system and attempts to determine the nature of their interactions; (2) the *holistic approach* which looks at detailed records of variations of system parameters and seek a consistent pattern indicative of the presence of a control scheme [15, 13].

Complex nonlinear systems are ubiquitous throughout the natural world. Nonlinear behavior is present at many structural levels in living organism from cellular to organism and behavioral level. The nonlinear properties of complex physiological signals are the result of a myriad of interactions between subcellular, cellular, organ and systemic components of that system. Information is continually transferred between these components to modify the functionality of the system, giving rise to a highly adaptive physiological system capable of responding to internal and external perturbations [68, 69].

The recognition of dynamic nature of regulatory physiological processes challenges the classical view of *homeostasis,* which asserts that systems normally operate to reduce variability and fluctuations in a very limited range. The more realistic *homeodynamic* concept conveys the idea that physiological systems in health and disease display an extraordinary range of temporal and structural patterns that cannot be explained by the classical theories based on linear construct, reductionistic strategies and homeostasis [113].

Recently, there has been growing interest in the development of new measures to describe the dynamics of physiological systems, and the use of these measures to distinguish healthy function from disease, or to predict the on-set of adverse health-related events [35]. A variety of measures have been derived from *chaos* theory and the field of nonlinear dynamics and statistical physics. Many of these are based on the concept of *fractals*.

Chaos and Fractal Physiology – Chaotic behavior is a feature of nonlinear dynamical systems that give rise to a number of important characteristics that can be identified and characterized using mathematical techniques. The paradox with the term *chaos* is the contradiction between its meaning in colloquial use and its mathematical sense. In normal usage chaos means disordered, without form, while in mathematics chaos is defined as stochastic behavior occurring in a deterministic system. Chaos can be understood by comparing it with two other types of behavior – randomness and periodicity: although it looks random, it is generated from deterministic rules.

The specific features of chaotic systems are sensitivity to initial conditions and the presence of multiple equilibriums (chaotic attractor). Sensitivity to initial conditions means that small differences in the initial state of the system can lead to dramatic differences in its long-term dynamics (known as the 'butterfly effect'). In practice, this extreme sensitivity to initial conditions makes chaotic systems unpredictable. Multiple equilibriums means that in chaotic dynamical systems the trajectory (i.e. the state of a system as a function of time) will never repeat itself but forms a unique pattern as it is attracted to a particular area of phase space – *a chaotic attractor*. Chaotic/*strange attractors* demonstrate *fractal* properties characterized by similar features at different levels of scales or magnification.

Fractals – The term fractal is used to characterize *objects in space* or *sequences of events in time* that possess a form of *self-similarity* or *scale-invariance*: fragments of the object or sequence can be made to fit to the whole object or sequence by shifting and stretching. The concept of fractals may be used for modeling certain aspects of temporal evolution (e.g. breathing pattern) of spatially extended dynamical systems (e.g. branching pattern of lungs). The temporal fluctuations exhibit structure over multiple orders of temporal magnitude in the same way that fractal forms exhibit details over several orders of spatial magnitude. Self-similar variations on different time scales will produce a frequency spectrum having an *inverse power law* distribution or *1/f* – like distribution and imply long-range temporal correlations signifying persistence or 'memory'.

Fractals can be *exact* or *statistical* copies of the whole. Mathematically, fractals can exactly represent the concept of self-similarity, whereas the fragments of natural fractals are only statistically related to the whole. Fractal structures have been noticed in many biological phenomena, including complex anatomical structures (e.g. bronchial tree, His-Purkinje conduction system, human retinal vessels, blood vessel system, etc.) and the *fluctuation pattern* of physiological signals (e.g. heart rate, breathing, ion-channel kinetics etc) [39] (Fig. 15.1).

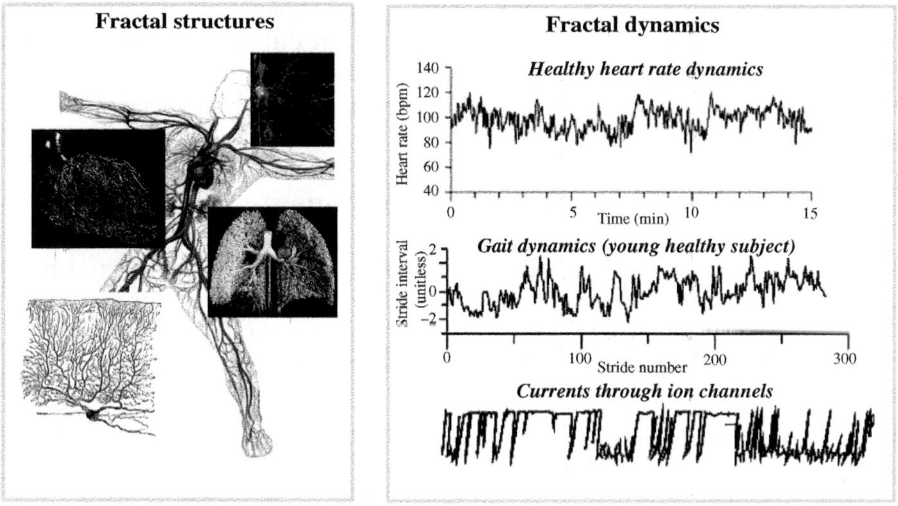

Fig. 15.1 Fractal physiology as structure of organs and temporal variation of time series. Reprint from Physionet (http://www.physionet.org/tutorials/fmnc [37])

15.3 Nonlinear Time Series Analysis

A time series is not a very compact representation of a time evolving phenomenon therefore it is necessary to quantify the information that contains the most relevant features of the underlying system into *metrics*.

Traditional analysis of physiological time series often tends to focus on time domain statistics with comparison of means and variances. Additional analyses based on frequency/frequency-scale domain techniques involving spectral and wavelet analysis are also frequently applied. If the data have some very specific mathematical properties (e.g. linear, stationary, Gaussian distributed) these linear time- and frequency-domain methods are useful and can be adopted in studies of health and disease because the resulted metrics are easy to interpret in physiological terms. However, the analysis of physiological signals is more challenging because the different time scales involved in dynamical processes give rise to non-Gaussian, non-stationary, and nonlinear behavior. It has been suggested that important information can be hidden within complex data fluctuations which need appropriate analysis tools to be quantified [35, 39, 46].

The new modern theories of complex systems and nonlinear dynamics have suggested strategies where the focus has shifted from the traditional study of averages, histograms and simple power spectra of physiological variables to the study of the pattern in the fluctuations of variables using nonlinear analysis methods [39]. They revealed that (i) physiological processes operate far from equilibrium, (ii) their fluctuations exhibit long-range correlations that extend across many time scales, (iii) and underlying dynamics can be nonlinear driven by deterministic chaos.

The analysis of a defined physiological signal raise the important question of what property one may wish to characterize from the data and, equivalently, what is the appropriate technique for extracting this information. The nonlinear analysis methods described in the following are aimed to capture different features or metrics of time series data: (i) long-range power low correlations or fractal organization of fluctuations (ii) degree of determinism/chaos, (iii) regularity or predictability.

15.3.1 Quantification of Fractal Fluctuations

The dynamics of a time series can be explored through its correlation properties, or in other words, the time ordering of the series. Fractal analysis is an appropriate method to characterize complex time series by focusing on the time-evolutionary properties on the data series and on their correlation properties.

15.3.1.1 Detrended Fluctuation Analysis

The *detrended fluctuation analysis* (*DFA*) method was developed specifically to distinguish between intrinsic fluctuations generated by complex systems and those caused by external or environmental stimuli acting on the system [82, 83]. The *DFA* method can quantify the temporal organization of the fluctuations in a given non-stationary time series by a single scaling exponent α – a self-similarity parameter that represents the long-range power-law correlation properties of the signal. The scaling exponent α is obtained by computing the root-mean-square fluctuation $F(n)$ of integrated and detrended time series at different observation windows of size n and plotting $F(n)$ against n on a log-log scale. Fractal signals are characterized by a power law relation between the average magnitudes of the fluctuations $F(n)$ and the number of points n, $F(n) \sim n^{\alpha}$. The slope of the regression line relating $log(F(n))$ to $log(n)$ determines the scaling exponent α. The time series of fractal signals can therefore be indexed by the deviation of α from 0.5 to 1. For a value of $\alpha = 0.5$ the signal is random; increasing values of α ($0.5 < \alpha \leq 1$) indicate rising power-law scaling behavior of the signal and the presence of long-range (fractal like) correlations (Fig. 15.2).

Practical considerations related to the effect of the trends, noise and nonstationarities on the *DFA* were studied in detail by Hu et al. [52], and by Chen et al. [17], respectively. An important issue of fluctuations analysis in practical applications is the influence of the time series length upon the reliability of the estimated scaling behavior on short scales n. Solutions to improve the performances of *DFA* method for short recordings were suggested by Kantelhardt et al. [57] and Govindan et al. [40].

15.3.1.2 Fano/Allan Factor Analysis

Some physiological phenomena, such as heart rate, neural spike trains, biological ion-channel opening, occur at some random location in time [70]. A stochastic *point process* is a mathematical description which represents these events as random points on the time axis [21, 22]. A useful representation of a point process is

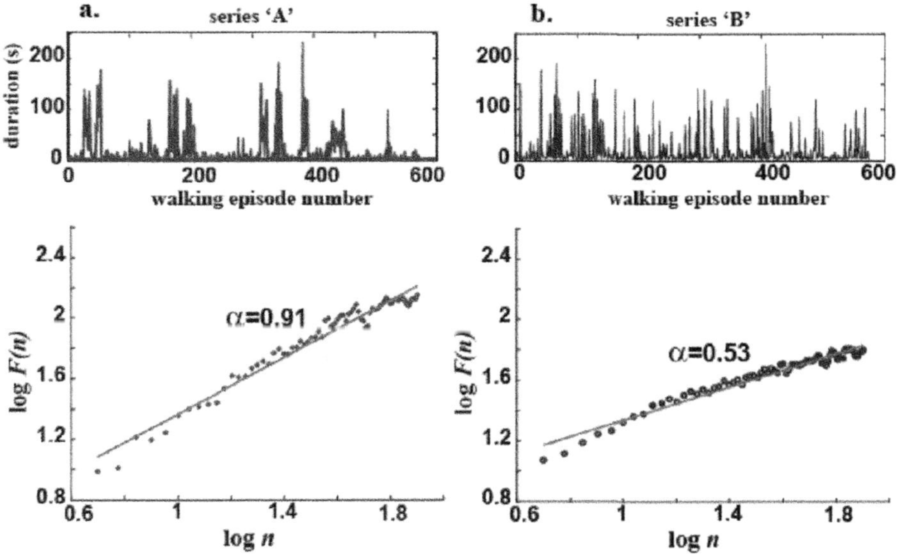

Fig. 15.2 Detrended fluctuation analysis of two time series representing sequences of walking episodes recorded during five consecutive days. Even if the two data series are characterized by similar mean value, they look different: the 'difference' resides in the temporal structure of the data and can be quantified by the DFA-fractal scaling exponent [81]

given by dividing the time axis into equally spaced contiguous counting windows of duration T, and producing a sequence of counts $\{N_i(T)\}$, with $N_i(T)$ denoting the number of events in the ith window, (Fig. 15.3) [71].

Such a process is fractal if some relevant statistics display scaling, characterized by cluster of points/events over a relatively large set of time scales [70]. The presence of fractality can be revealed using the Fano factor (FF) or the Allan factor (AF) [108, 106]. These methods involve the calculation of the $FF(T)$ or $AF(T)$ for window sizes of different lengths. The $FF(T)$ is defined as the ratio of the variance of the number of events to the mean number of events in window sizes of specified length T. The Fano factor curve is constructed by plotting $FF(T)$ as a function of the window size on a log-log scale. For a random process in which the fluctuation in the number of events are uncorrelated, $FF(T) = 1$ for all window sizes. For a fractal process, $FF(T) = 1 + (T/T_0)^{\alpha_F}$, where $0 < \alpha_F < 1$ is the fractal scaling exponents and T_0 is the fractal onset time that marks the lower limit for significant scaling behavior in FF [70, 101]. Fractal-rate stochastic point processes generate a hierarchy of clusters of different durations, which lead to an $FF(T)$ plot that continues to rise as each cluster time scale is incorporated in turn.

The $AF(T)$ is defined as the ratio of the event-number Allan variance to twice the mean number of events in windows of specified length T. The Allan variance is expressed in terms of the variability of the difference in the number of events in successive windows, rather than in terms of the variability of the number of events

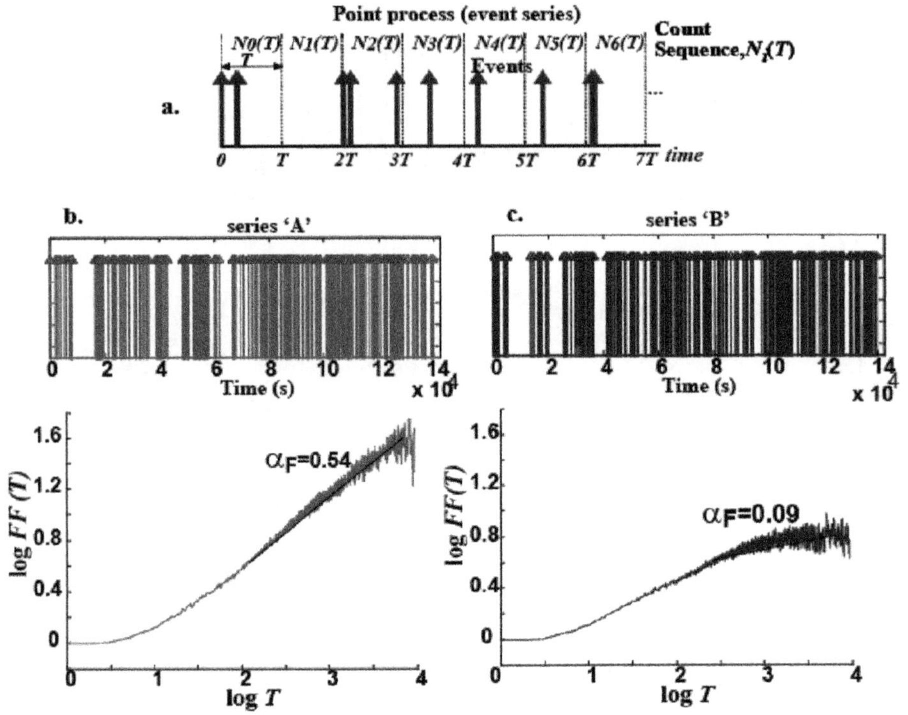

Fig. 15.3 Fractal analysis of point processes. (**a**) Representation of a point process: physiological events such as R-wave occurrence, neural spike trains, human activity-rest postural transitions, etc., can be represented as point processes on the real time axis; (**b**), (**c**) the Fano factor analysis reveals fractal-rate behavior for series 'A' and random Poisson-like behavior for series 'B'

in individual windows. The Allan factor curve is constructed similarly to the Fano factor curve. An advantage offered by Fano factor analysis is that the window size at which the power law begins is usually much smaller than for Allan factor analysis [102, 106]. Thus Fano factor analysis may reveal a power law relationship extending over more than one time scale (indicative of fractal behavior) when the data block is too short to show this with Allan factor analysis.

15.3.1.3 Multifractal Analysis

Many physiological time series do not exhibit a simple monofractal scaling behavior, which can be accounted for by a single scaling exponent. In some cases, the need for more than one scaling exponent to characterize complex inhomogeneous fluctuations can derive from the existence of a crossover timescale, which separate regimes with different scaling behaviors [52]. Different scaling exponents could be required for different segments of the same time series, indicating a time variation of the scaling behavior. In other cases, different scaling exponents can be revealed for many interwoven fractal subsets of the time series [89]; in this case the process is *multifractal*. Thus, multifractals are intrinsically more complex and inhomogeneous that monofractals and characterize systems featured by very irregular dynamics, with sudden and intense bursts of high frequency fluctuations [96].

There are two main techniques for the multifractal characterization of time series: the wavelet-transform modulus maxima method (*WTMM*) and the *multifractal DFA* (*MF-DFA*). The *WTMM* method is based on wavelet analysis and involves tracing the maxima lines in the continuous wavelet transform over all scales [55, 9, 11]. The *MF-DFA* is based on the ientification of scaling of qth-order moments and is a generalization of the standard *DFA* using only the second moment ($q = 2$). The *MF-DFA* allows a global detection of multifractal behavior, while the *WTMM* method is suited for the local characterization of the scaling properties of signals [58, 59, 80].

15.3.2 Quantification of Degree of Chaos/Determinism

The analysis of a degree of chaos in a dynamical system is a procedure that consists of three distinct steps: (i) reconstruction of the system dynamics in phase/state space, (ii) characterization of the reconstructed attractor, (iii) validation of the procedure with 'surrogate data testing'. We attempt to give an intuitive explanation of what is involved in each of the three steps and focus on practical methodological issues (surrogate data analysis will be discussed in Sect. 15.3.4) without an extensive discussion of mathematical details which can be found in a few review papers and textbooks (Grassberger et al. 1991) [44, 92, 33, 62, 63, 59].

15.3.2.1 Phase Space Reconstruction

The principle of chaos analysis is to transform the properties of a time series into the topological properties of a geometrical object (attractor) constructed out of a time series, which is embedded in a *state/phase space*. The concept of state space reconstruction is central to the analysis of nonlinear dynamics. A valid state space is any vector space in which the state of the dynamical system can be unequivocally defined at any point [62]. The most used way of reconstructing the full dynamics of the system from scalar time measurements is based on the embedding theorem [99] that states that one can 'reconstruct' the attractor of the system from a the original time series and its time-delayed copies:

$$\vec{x} = [x(t), x(t+\tau), x(t+2\tau), \ldots, x(t+(d_E-1)\tau]$$

where $\vec{x}$ is the d_E–dimensional state vector, $x(t)$, $t = 1,\ldots, N$ is the original 1-D data, τ is the time delay, and d_E is the *embedding dimension*. Every instantaneous state of the system is therefore represented by the vector $\vec{x}$, which defines a point in the phase space (Fig. 15.4a,b). Appropriate values for both τ and d_E can be obtained in a number of ways described in detail in a series of publications [77, 1, 62].

When selecting a time delay, τ, the goal is to find a delay large enough so that the resulting individual coordinates are relatively independent, but not so large that they are completely independent statistically [1]. One could, for example, choose τ as: (1) the position of the first local minimum of the autocorrelation function of the data [1, 77], (2) the first minimum of the average mutual information function, which evaluate the amount of information shared between two data sets over a range of time delays [32]. While the first approach evaluates only linear relations among the data, the mutual information method examines also the nonlinear signal structure providing adjacent delay coordinates with a minimum of redundancy.

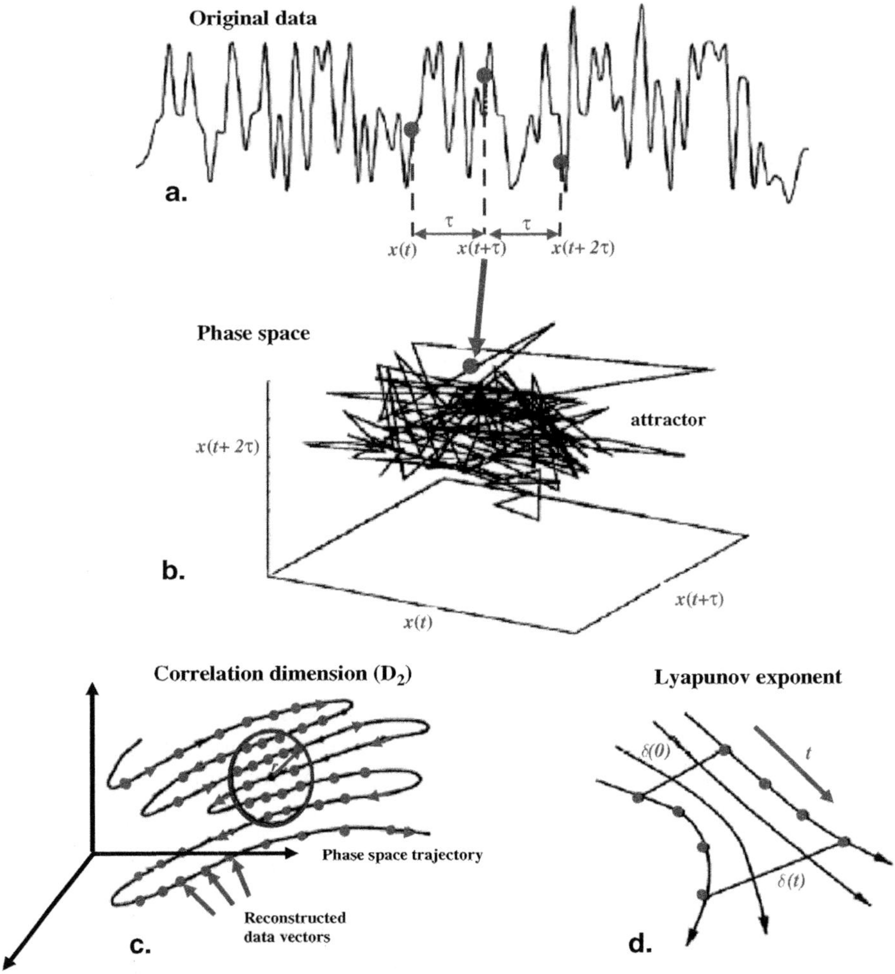

Fig. 15.4 Schematic representation of dynamical nonlinear analysis: (**a**) original time series data, (**b**) reconstruction of 3-dimensional attractor in the phase space based on based on Taken's 'time shift method', (**c**) illustration of correlation dimension which is measure of the 'complexity' of the system, (**d**) illustration of Lyapunov exponent which is a measure of 'sensitivity to initial conditions'

A valid state space must include a sufficient number of coordinates (d_E) to unequivocally define the state of the system at all times (i.e. there must be no intersecting or overlapping of trajectories from different regions of the attractor). An interesting method to estimate an acceptable minimum embedding dimension is the method of false neighbor [65] which compares distances between neighboring trajectories at successively higher dimensions. 'False neighbors' occur when trajectories that overlap in dimension d_i are distinguished in dimension $d_{i+1.}$ As i increases the total percentage of false neighbors (%FN) across the entire attractor declines and d_E is chosen where %FN $\rightarrow$ 0 [1, 3, 62]

15.3.2.2 Characterization of the Reconstructed Attractor

Several methods and algorithms are available to characterize a reconstructed attractor in a quantitative way. The most basic measures are the *correlation dimension*, the *Lyapunov exponents* and the *entropy*. The correlation dimension emphasizes the geometrical properties (complexity) of the attractor while Lyapunov exponents and entropy focus on the dynamics of trajectories in the phase space.

- Correlation dimension

The *correlation dimension* or D_2 is a measure of the complexity of the process/time series being investigated, which characterize the distribution of the points in the phase space. The most frequently used algorithm to calculate D_2 was introduced by Grassberger and Procaccia [42]. In this algorithm the computation of D_2 is based upon the correlation integral C_r which is the likelihood that any two randomly chosen points on the attractor $\vec{x}_i, \vec{x}_j$ will be closer than a given distance r:

$$C_r = \lim_{N\to\infty} \frac{1}{N^2} \sum_{i\neq j} \Theta(r - |\vec{x}_i - \vec{x}_j|) \tag{15.2}$$

where N is the number of data points and Θ is the Heavyside function.

In the case the attractor is a simple curve in the phase space, the number of pair of vectors whose distance is less than a certain radius r will be proportional to r^1. In the case the attractor is a two dimensional surface, Cr $\sim$ r2; and for a fixed point Cr $\sim$ r0 (Figs. 15.4c, 15.5) Generalizing, Cr can be expressed as Cr $\sim$ rD2 then, if the number of data and the embedding dimension are sufficiently large we obtain:

$$D_2 = \lim_{r\to 0} \frac{\log C_r}{\log r} \tag{15.3}$$

The main point is that C_r behaves as a power of r for small r. By plotting log C_r versus log r, D_2 can be calculated from the slope of the curve. The correlation dimension gives also a measure of the 'fractal dimension' of the attractor: it is in the phase space where chaos meets fractals, since strange attractors have fractal dimension.

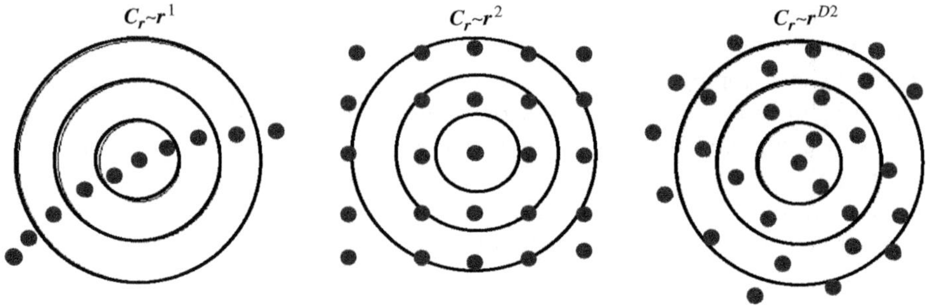

Fig. 15.5 Representation of correlation integral for different topologies of the attractor in the phase space

It was shown that the original algorithm has several limitations related to an inaccurate estimation of D_2 for short and noisy data sets [27] therefore many modifications and improvements have been proposed during the last two decades [41, 49, 78, 103].

- Lyapunov exponents

For a dynamical system, the hallmark of deterministic chaos is represented by sensitivity to initial conditions [62] quantified by the Lyapunov exponents (*LE*). Lyapunov exponents quantify the exponential divergence or convergence of initially close phase space trajectories. *LE* quantifies also the amount of instability or predictability of the process. An d_E-dimensional dynamical system has d_E exponents but in most applications it is sufficient to compute only the largest Lyapunov exponent (*LLE*) instead of all *LE*. If the *LLE* is positive then the nonlinear deterministic system is chaotic and consequently the divergence among initially close phase-space trajectories grows exponentially in time (it is unpredictable) [2, 62, 112].

Robust approaches to estimate *LLE* are based on the idea that the maximal Lyapunov exponent (λ_1) for a dynamical system can be defined from [61, 88]:

$$\delta(t) = \delta(0)e^{\lambda_1 t} \tag{15.4}$$

where $\delta(t)$ is the mean Euclidian distance between neighboring trajectories in state space after some evolution time t and $\delta(0)$ is the initial separation between neighboring points (Fig. 15.4d). Taking the natural logarithm of the both sides of the Eq. (15.4), one obtains:

$$\ln \delta_j(i) \cong \lambda_1(i\Delta t) + \ln \delta_j(0) \tag{15.5}$$

where $\delta_j(0)$ is the Euclidian distance between the *j*th pair of initially nearest neighbors after *i* time steps. For a sufficiently large embedding dimension d_E Eq. (15.5) yields a set of curves (one for each *j*) of approximately parallel lines. If these lines are approximately linear, their slope approximates λ_1 which can then be robustly estimated from the slope of a linear fit to the 'average' log divergence curve defined by:

$$S(d_E, i) = \frac{1}{\Delta t} \langle \ln \delta_j(i) \rangle_j \qquad (15.6)$$

where $\langle . \rangle$ denotes the average over the all values of j [88]. More details related to estimation of *LE* can be found in [112, 61, 88, 24, 28].

The concern for practical applications is that Lyapunov exponents are very sensible to the election of the time lag τ, the embedding dimension d_E, and especially to the election of the evolution Δt. If the evolution time is too short, neighbor vectors will not evolve enough in order to obtain relevant information. If the evolution time is too large, vectors will jump to other trajectories thus giving unreliable results.

- Entropy

Another important quantity of the characterization of deterministic chaos is the entropy. A wide variety of algorithms for the computation of entropy measures have been introduced based on correlation integral [42, 43], approximate entropy [86], multi resolution entropy [107], etc. One interesting entropy measure is the Kolmogorov entropy [62] which quantifies the average rate at which the information about the state of the system is lost over time. The Kolmogorov entropy is determined from the embedded time series data by finding points on the trajectory that are close together in the phase space (i.e., have a small separation) but which occurred at different times (i.e. are not time correlated). These two points are then followed into the future to observe how rapidly they move apart from one another. The time it takes for points pairs to move apart is related to the so-called Kolmogorov entropy, K, by $\langle t_{div} \rangle = 2^{-Kt}$, where $\langle t_{div} \rangle$ is the average time for the pair to diverge apart and K is expressed in bits per second. Entropy reflects how well one can predict the behavior of each respective part of the trajectory from the other. Higher entropy indicates less predictability and a closer approach to stochastic behavior.

15.3.3 Quantification of Roughness, Irregularity, Information Content

15.3.3.1 Fractal Dimension

The term 'fractal dimension' (*FD*) refers to a non-integer or fractional dimension of an object. Application of fractal dimension for analysis of data series include two types of approaches, those in the phase space domain (like correlation dimension D_2) and ones in the time domain where the signal waveform is considered a geometric figure. Waveform *FD* values indicates the 'roughness' of a pattern (Fig. 15.6), [34] or the quantity of information embodied in a waveform pattern in terms of

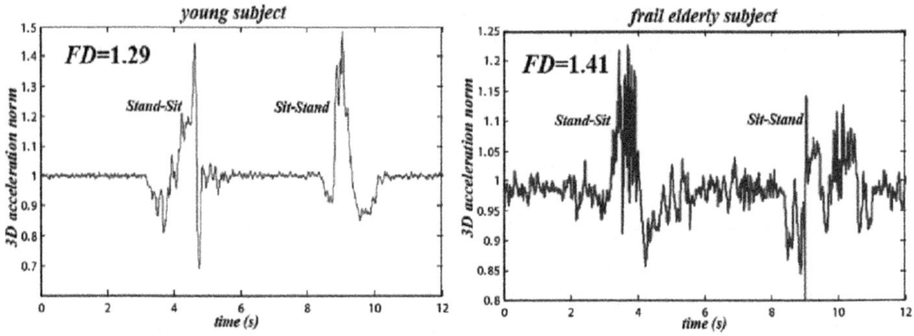

Fig. 15.6 Fractal dimension quantifying 'movement roughness' during postural changes (Stand-Sit/Sit-Stand) recorded in a healthy young subject and a frail elderly subject

morphology, entropy, spectra or variance [50, 60, 95]. The most common methods of estimating the *FD* directly in the time domain were analyzed and compared by Esteller et al. [31].

15.3.3.2 Approximate Entropy

Approximate entropy (*ApEn*) provides a measure of the degree of irregularity or randomness within a series of data. *ApEn* assigns a non-negative number to a sequence or time series, with larger values corresponding to greater process randomness or serial irregularity, and smaller values corresponding to more instances of recognizable features or patterns in the data [86]. *ApEn* measures the logarithmic likelihood that runs of patterns that are close (within a tolerance window 'r') for length m continuous observations remain close (within the same tolerance r) on next incremental comparison. The input variables m and r must be fixed to calculate *ApEn*. The method can be applied to relatively short time series, but the amounts of data points has an influence on the value of *ApEn*. This is due to the fact that the algorithm counts each sequence as matching itself to avoid the occurrence of ln(0) in the calculations [87]. The *sample entropy* (*SampEn*) algorithm excludes self-matches in the analysis and is less dependent of the length of data series.

15.3.3.3 Multiscale Entropy

The *multiscale entropy* (*MSE*) was developed as a more robust measure of complexity of physiological time series which typically exhibit structure over multiple time scales [20]. Given a time series $x_1, \ldots, x_N$ the first step is to construct multiple 'coarse-grained' time series by averaging τ consecutive data points:

$$y_j(\tau) = 1/\tau \sum_{i=(j-1)\tau+1}^{j\tau} x_i, \ 1 \le j \le N/\tau \tag{15.7}$$

where τ is the scale factor. For each coarse grained data $y_j(\tau)$ an entropy measure is calculated (e.g. *SampEn*) and then plotted as a function of the scale factor τ.

As a guideline, a *MSE* profile of an uncorrelated sequence (e.g. white noise) monotonically decreases with the scale factor, whereas the profile for a fractal or long-range correlated time series is steady across scales [20].

From practical point of view it is important to note that approximate, sample and multiscale entropies require evaluations of vectors representing consecutive data points, and thus the order of the data is essential for their calculations. In addition, significant noise and nonstationary data compromise meaningful interpretation of estimated values.

15.3.3.4 Symbolic Dynamics

Symbolic time series analysis (symbolic dynamics) involves the transformation of the original time series into a series of discrete symbols that are processed to extract useful information about the state of the system generating the process.

The first step of symbolic time series analysis is the transformation of the time series into a symbolic/binary sequence using a *context-dependent* symbolization procedure. After symbolization, the next step is the construction of words from the symbol series by collecting groups of symbols together in temporal order. This process typically involves definition of a finite word-length template that can be moved along the symbol series one step at a time, each step revealing a new sequence [81].

Quantitative measures of word sequence frequencies include statistics of words (word frequency, transition probabilities between words) and information theoretic based on entropy measures. In addition to approximate/sample entropy which can also be applied to binary sequences or other symbolic dynamics, other entropy measures (e.g. Shannon entropy, Rényi entropy, conditional entropy) can be used to evaluate the relative complexity of the word sequence frequency.

Shannon entropy gives a number that characterize the probability that different words of length L occur. First, the probabilities of each word of length L are estimated from the whole binary sequence [93]:

$$p(w_1, w_2, \ldots, w_N) = \frac{n_{w_1 \ldots w_N}}{n_{tot}} \tag{15.8}$$

where $n_{w_1 \ldots w_N}$ is the number of occurrences of the words $w_1, w_2, \ldots, w_N$, $(N = 2^L)$ and n_{tot} is the total number of words. Next the entropy estimation *SE*(*N*) is defined as:

$$SE(N) = -\frac{1}{N} \sum_{w_1, \ldots, w_N} p(w_1, w_2, \ldots, w_N) \log_2 p(w_1, w_2, \ldots, w_N) \tag{15.9}$$

For a very regular binary sequence, only a few distinct words occur. Thus Shannon entropy would be small because the probability for these patterns is high and only little information is contained in the whole sequence. For a random binary sequence, all possible L-length words occur with the same probability and the Shannon entropy is maximal.

Other complexity measures of symbolic sequences were proposed recently by Zebrowski et al. [115] and Aboy et al. [4].

15.3.4 Methodological Issues

15.3.4.1 Surrogate Data Analysis

The method of surrogate data has become a central tool for validating the results of nonlinear dynamics analysis. The surrogate data method tests for a statistical difference between a test statistics (e.g. complexity/fractal metrics) computed for the original time series and for an ensemble of test statistics computed on linearised version of the data, the so-called 'surrogate data'. The major aspects of the surrogate data that need to be considered are: (1) the definition of the *null hypothesis* about the nonlinear dynamics underlying a given time series, (2) the realization of the null hypothesis, i.e., the generation method for the surrogate data and (3) the test statistic. There are three main methods to generate surrogate data [64, 91]: (1) *phase randomized surrogates* which preserve the linear correlation of data (i.e. power spectrum) but destroy the nonlinear structures by randomizing the phases using discrete Fourier transform, (2) *amplitude-adjusted surrogate* which preserve the probability distribution but destroy any linear correlation by random shuffling the samples and (3) *polished surrogates* which preserve amplitude distribution and power spectrum but destroy correlations by random shuffling the samples.

Generically stated, the procedure of surrogate data can be reduced to the following steps: (1) a nonlinear dynamical measure (e.g. fractal analysis) is applied to the original time series obtaining the result, M_{orig}, (2) surrogate data sets are constructed using the original time series (3) the nonlinear dynamical measure that was applied to the original time series is applied to the surrogate sets and the average and standard deviation values are denoted $\langle M_{surr} \rangle$ and σ_{surr}, respectively, (4) a statistical criterion is used to determine if M_{orig} and $\langle M_{surr} \rangle$ are sufficiently different. If they are, the surrogate null hypothesis (the original and the surrogate data come from the same population) is rejected. An estimation of the difference between M_{orig} and $\langle M_{surr} \rangle$ may be obtained by means of the estimate S_M defined as [104]:

$$S_M = \left| M_{orig} - \langle M_{surr} \rangle \right| / \sigma_{surr} \qquad (15.10)$$

The larger the S_M the larger is the separation between the nonlinear measures derived from surrogate data and the nonlinear measure derived from original data. The probability (p-value) of observing a significant S_M, given that the null surrogate hypothesis is true, is specified by the complementary error function $p = erfc(S_M/\sqrt{2})$.

15.3.4.2 Practical Limitations

Generally, practical problems with nonlinear analysis methods arise because of shortness of the time series, lack of stationary, and the influence of noise on measurement. Stationary represent a fundamental requirement in the application of nonlinear analysis to biological systems. The problem lies in the fact that the physiological time series are not stationary over periods of sufficient length to permit a reliable estimation of these nonlinear quantities. One solution is to use short epochs of stationary data and to assume that the error in the nonlinear estimates that arises because of the small data samples will be systematic, thus enabling control and experimental conditions to be made. Nevertheless, it is difficult to determine to what extent this assumption affects the result. Another solution would be using mathematical transformation (e.g. differentiation) to obtain stationary data while assuring that the dynamical properties of transformed data are equivalent to the original data [23].

The noise and the artifacts, frequently present in physiological recordings, have a pronounced adverse effect on the estimation of the nonlinear measures.

15.4 Quantifying Health and Disease with Nonlinear Metrics

15.4.1 Heart Rate Analysis

The heart rate (HR) analysis was one of the first physiological time series studied with the tools of nonlinear dynamics. Analysis of short-term fractal properties of HR fluctuations by *DFA* method has provided prognostic power among patients with acute *myocardial infarction*, *depressed left ventricular function* [53, 54, 74, 72, 73], and *chronic congestive heart failure* [51, 75]. Changes in regularity (*ApEn*) and fractal properties (*DFA*) have been reported to precede the spontaneous onset of *atrial fibrillation* in patients with no structural heart disease [111, 8]. The *LLE* was shown to be lower in *old myocardial infarction* and *diabetic* patients than in normal subjects and to decreases with aging indicating that HR variability becomes less chaotic as the healthy subject grows old [6].

15.4.2 Respiration Pattern Analysis

The *DFA* analysis of inter-breath interval time series suggested fractal organization in physiologic human breathing cycle dynamics, which degrades in elderly men [84]. Identification of such fractal correlation properties in breathing rate has open exciting new possibilities that use external fractal fluctuations in life-support systems to improve lung function [79]. Several studies described respiratory irregularity in patients with panic disorder and illustrated the utility of nonlinear measures such as *ApEn* and *LLE* as additional measures toward a better understanding of the abnormalities of respiratory physiology [16, 114].

15.4.3 EEG Analysis

The nonlinear analysis methods have effectively applied to the EEG, to study the dynamics of its complex underlying behavior. In *sleep* studies, the general idea that emerged was that deeper sleep stages are associated with a lower 'complexity' as exemplified by lower values of D_2 and *LLE* [97, 98]. The useful of nonlinear analysis tools (D_2, *ApEn*, *DFA fractal scaling exponent*) to monitor *anesthetic depth* was also suggested by a series of studies [97, 98, 5]. Probably the most important application of nonlinear EEG analysis is the study of *epilepsy* [30]. This is because epileptic seizures, in contrast to normal background activity, are highly nonlinear phenomena. This important fact has driven a significant amount of studies and has open the way for localization of the epileptogenic zone, detection and prediction of epileptic seizure [66, 105]. Nonlinear EEG analysis has been extensively applied to quantify the *effect of pharmacological agents* and to characterize mental and psychiatric diseases (*cognition, emotion, depression, schizophrenia* [97, 67, 56, 85].

15.4.4 Human Movement Analysis

15.4.4.1 Gait Pattern

Stride interval *variability* is a quantifiable feature of walking that is altered both in terms of magnitude and dynamics in clinically relevant syndromes such as *falling, frailty* and *neurodegenerative diseases* (e.g. Huntington's disease, Parkinson's disease) [45, 46, 47, 48]. Several investigations have examined long-range stability of walking patterns, by studying the fractal properties of stride interval variability over long walking durations. The *DFA* analysis showed that over extended periods of walking, long-range self-similar correlations of stride-to-stride temporal pattern exist in healthy adults. In contrast, ageing and disease decreases long-range stability (stride duration is more random) of the walking patterns, presumably due to a degradation of central nervous system locomotor generators [47] (Fig. 15.7). *Fractal dimension* (*FD*) analysis has been applied to gait-related trunk accelerations, with higher *FD* associated with post-stroke hemiplegia [7] and Parkinson's disease [94] compared to healthy elderly individuals.

The approximate entropy (*ApEn*) of the lateral trunk acceleration during walking was suggested by Arif et al. [12] as a metric for gait stability in elderly subjects.

15.4.4.2 Postural Control

Human posture is a prototypical example of a complex control system. Upright posture of human body is maintained by a complex processing of sensory signals originating from the vestibular, visual and somatosensory system, which govern a cascade of muscular correctional movements.

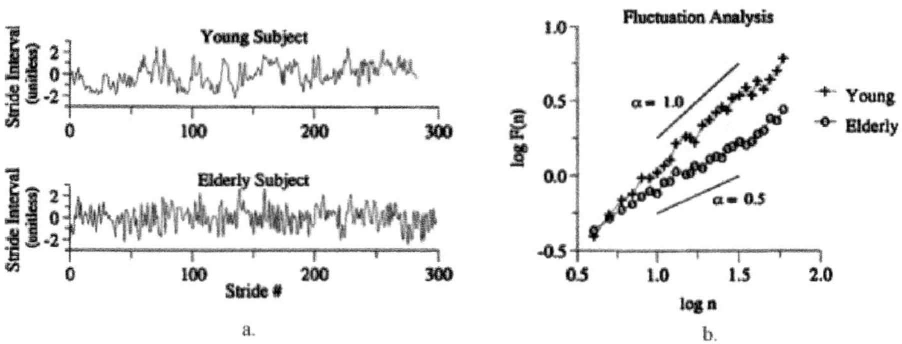

Fig. 15.7 Example of effects of aging on fluctuation analysis of stride-interval dynamics. Stride interval time series (**a**) and fluctuation analysis (**b**) for a 71-yr-old elderly subject and a 23-yr-old young subject. For the elderly subject, fluctuation analysis shows that stride-interval fluctuations, F(n), increase more slowly with time scale n. This indicates a more random and less correlated time series. Indeed, scaling index (α) is 0.56 for the elderly subject and 1.04 for the young subject. Reprint with the permission from J Appl Physio, [47]

In posturography, the main parameter used to estimated balance during standing posture is the center of pressure (COP) location using a force plate. The COP represents the trajectory of the application's point of the resultant of vertical forces acting on the surface of support. COP characteristics are used for outcome evaluation of the postural control system. Concepts that emerged from the fractal analysis nonlinear dynamical system perspective were discussed in the context of coordination and functional aspects of variability in postural control. The fractal nature of COP movement during prolonged unconstraint standing (30 min) was shown in normal subjects (Fig. 15.8) [26]. The implication of these concepts was studied in order to understand postural instability due to aging and movement disorders, with special emphasis on aging and Parkinson's disease [29, 19, 18, 25].

15.4.4.3 Daily-Life Physical Activity Pattern

Like other physiological systems, control of human activity is also complex being influenced by many factors both extrinsic and intrinsic to the body. The most obvious extrinsic factors that affect activity are daily schedule of planned events (work, recreation) as well as reaction to unforeseen random events. The most obvious intrinsic factors are the circadian and ultradian rhythms, the homeostatic control of body weight, and the (chronic) disease state. During the last years an important research challenge was to try to quantify the intrinsic feature of real life daily activity pattern in order to provide objective outcomes/metrics related to chronic disease severity and treatment efficacy.

In [10] long-duration time series of human physical activity were investigated under three different conditions: healthy individuals in (i) a constant routine protocol and (ii) in regular daily routine, and (iii) individuals diagnosed with multiple chemical sensitivities. Time series of human physical activity were obtained by inte-

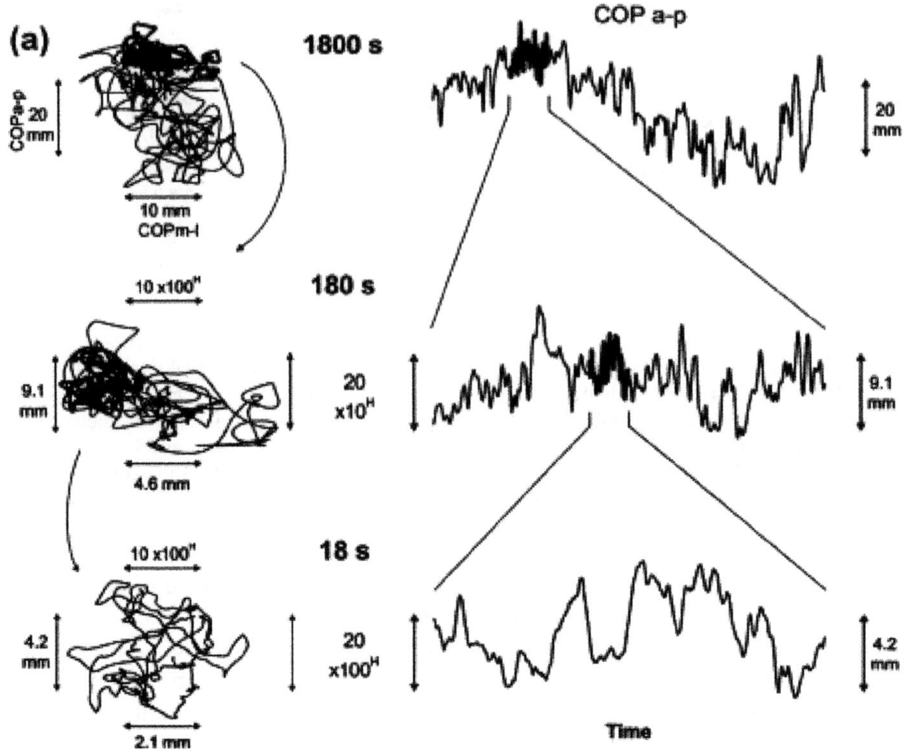

Fig. 15.8 COP trajectory in horizontal plane and anterior-posterior (a-p) COP time series (*right*) for the entire data during natural standing (1800 s, *first row*), for 1/10 of the data (180 s, *second row*), and for 1/100 of the data (18 s, *third row*). Notice that after each scaling (that is related to the fractal exponent and to the periods of time), the three COP trajectories and the time series present roughly the same amplitudes in space. Reprint with the permission from Neuroscience Letter- [26]

gration of vertical acceleration signal (raw data) at the waist over 8 s interval. The *DFA* showed that the time series of integrated acceleration signal display power law decaying temporal auto-correlations. It was found that under regular daily routine, time correlations of physical activity are significantly different during diurnal and nocturnal periods but that no difference exists under constant routine conditions. Finally, it was found significantly different auto-correlations for diurnal records of patients with multiple chemical sensitivities.

In [79] the objective was to study the temporal correlation of physical activity time series in patients with chronic fatigue syndrome during normal daily life and to examine if it could identify the altered physical activity in these patients. Physical activity was monitored with Actilog V3.0 device and time series were obtained by integration of acceleration counts above a threshold level (integration for every 5 min). Using *DFA* it was shown that time series of acceleration counts display fractal time structure for both control fatigue syndrome patients and healthy control subjects. Moreover, chronic fatigue syndrome patients have significantly smaller fractal scaling exponents than control healthy subjects.

A more recent study [81] has investigated patterns of daily-life physical activity of chronic pain patients and healthy individuals. After estimations of body postures and activities (e.g. sitting, standing, lying, walking) using body fixed kinematical sensors, physical activity time series were defined as: (i) the sequence of posture allocation, i.e. lying=1, sitting=2, standing=3, walking=4, (ii) the sequence of daily walking episodes characterized by their duration, (iii) the occurrence time of activity-rest postural transitions (activity = walking and standing, rest = lying and sitting) treated as a point process, i.e., as a sequence of events distributed on the time axis. The dynamics (temporal structure) of defined physical activity patterns were analyzed using *DFA*, *FFA* and symbolic dynamics statistics. For both groups the *DFA* showed fractal time structure in the daily posture allocation pattern; however the average scaling exponent was significantly smaller in chronic pain patients that in healthy control. Similarly, the *DFA* analysis of the sequence of daily walking episodes showed a fractal scaling exponent smaller in the chronic pain group. The *FFA* revealed that under healthy conditions the timing of activity-to-rest transitions follows a power law distribution, suggesting time clustering of activities at different time scales.

Symbolic dynamics approach revealed that under healthy conditions activity periods preceded and followed by shorter rest were more likely than under chronic pain conditions.

The conclusion that emerges from this study is that parameters quantifying the temporal structure of physical activity pattern locate a significant difference between healthy and chronic pain conditions.

15.4.4.4 Movement Irregularity/Roughness

The 'geometrically' fractal analysis method (fractal dimension) was used to determine irregularities/complexity of human body raw acceleration data. Fractal dimension analysis was applied to gait-related trunk accelerations, with higher *FD* associated with post-stroke hemiplegia [7] and Parkinson's disease [94] compared to healthy elderly individuals. *FD* was also used to analyze movement smoothness during sit-to-stand/stand-to-sit postural transitions in frail elderly subjects; *FD* of body kinematical signals recorded during postural transitions task was significantly lower after a rehabilitation program as compared to baseline and was associated with an improvement of the functional state of the subjects [34].

15.5 Discussion and Conclusion

Application of nonlinear analysis approaches in biomedical research leads to new insight into the complex pattern of signals emerging from physiological systems. This chapter highlights the basic modern concepts and methodologies developed to characterize complexity of physiological signals. Each with distinct theoretical background and significance, they contribute various information regarding signal characteristics, complementary to the linear approach (time or frequency domain).

Fractal analysis shows that (i) behind the seemingly random fluctuations of a number of physiological parameters, a special order – the structure of fractal geometry can be found, (ii) a single measure – the fractal scaling exponent can describe the complexity of fluctuations and (iii) this parameter does change in response to disturbance of the system, such as disease and aging.

Concepts derived from chaos theory allowed to show that many systems are not normally in a stable state of homeostatic balance, but in a dynamically stable state characterized by chaotic fluctuations within a certain area of phase space.

As the term 'complexity' has a broad range of significations in physiology, when the complexity measures are to be used for diagnosis purposes, it is of great importance to correctly interpret the deviation of the estimated metric from a 'normal' value characterizing the healthy state. For example, in this chapter we have seen that while the complexity in heart rate variability measured with *DFA* is associated with the better health status, in cerebral disease (epileptic seizure) EEG signals are more nonlinear and complex. Long-range correlations of physiological fluctuations appear to decrease with disease in some situations, while irregularity and roughness of body kinematics increase in pathological conditions, suggesting that the lack of complexity in the control of movement leads to irregular and erratic motion. Therefore, the fundamental premise in physiological research is that increase and decrease complexity may occur in disease and aging [36, 38, 109, 110]. The expected trend depends on the formulated problem including the medical hypothesis, the definition of physiological time series and the mathematical tool for analysis.

Finally, we conclude that like reductionism and synthesis (holism), the linear and nonlinear approaches are the essential dual aspects of any system, which strongly suggest that both are needed to better understand any complex process (physiological, physical, behavioral/social etc.).

References

1. Abarbanel, H (1996) Analysis of Observed Chaotic Data. Springer -Verlag, New York
2. Abarbanel H, Brown R and Kennel M B (1991) Variation of Lyapunov exponent on a strange attractor. J Nonlinear Sci 1:175–199
3. Abarbanel H and Kennel M B (1993) Local false nearest neighbors and dynamical dimensions from observed chaotic data. Phys Rev E 47:3057–3068
4. Aboy M, Hornero R, Abásolo D and Álvarez D (2006) Interpretation of the lempel-ziv complexity measure in the context of biomedical signal analysis. IEEE Trans. Biomed Eng 53(11):2282–2288
5. Accardo A, Affinito M, Carrozzi M and Bouquet F (1997) Use of the fractal dimension for the analysis of electroencephalographic time series. Biol Cybern 77:339–350
6. Acharya U R, Kannathal N, Sing O W, Ping L Y and Chua T (2004) Heart rate analysis in normal subjects of various age groups. Biomed Eng Online 3:24
7. Akay M, Sekine M, Tamura T, Higashi Y and Fujimoto T (2004) Fractal dynamics of body motion in post-stroke hemiplegic patients during walking. J Neural Eng, 1:111–116

8. Al-Angari H M and Sahakian A V (2007) Use of sample entropy approach to study heart rate variability in obstructive sleep apnea syndrome. IEEE Trans Biomed Eng 54: 1900–1904
9. Amaral L A N, Ivanov P Ch, Aoyagi N, Hidaka I, Tomono S, Goldberger A L, Stanley H E and Yamamoto Y (2001). Behavioral-independent features of complex heartbeat dynamics. Phys Rev Lett 86:6026–6029
10. Amaral L A N, Bezerra Soares D J, da Silva L R et al. (2004) Power law temporal auto-correlations in day-long records of human physical activity and their alteration with disease. Europhys Lett 66(3):448
11. Arneodo A, Grasseau G and Holschneider M (1988) Wavelet transform of multi-fractals. Phys Rev Lett 61:2281–2284
12. Arif M, Ohtaki Y, Nagatomi R and Inooka H (2004) Estimation of the Effect of Cadence on Gait Stability in Young and Elderly People Using Approximate Entropy Technique. Meas Sci Rev 4:29–40
13. Bell I and Koithan M (2006) Models for the study of whole systems. Integrat Cancer Ther 293–307
14. Beuter A, Glass L, Mackey M C and Titcombe M S (2003). Nonlinear Dynamics in Physiology and Medicine. Interdisciplinary Applied Mathematics, Vol. 25, Springer, New York, 2003, xxvi+434
15. Brandon R (1996) Reductionism versus holism versus mechanism. Concepts and Methods in Evolutionary Biology, Cambridge: Cambridge University Press, 179–204
16. Caldirola D, Bellodi L, Caumo A, Migliarese G, Perna G (2004) Approximate entropy of respiratory patterns in panic disorder. Am J Psychiatry 161:79–87
17. Chen Z, Ivanov P Ch, Hu K and Stanley H E (2002) Effect of nonstationarities on detrended fluctuation analysis. Phys Rev E 65: 041197
18. Collins J J, De Luca C J, Burrows A and Lipsitz L A (1995) Age-related changes in open-loop and closed-loop postural control mechanisms. Exp Brain Res 104:480–492
19. Collins J J and De Luca C J (1994) Random walking during quiet standing. Phys Rev Lett 73(5):764–767
20. Costa M, Goldberger A L and Peng C K (2002) Multiscale entropy analysis of complex physioogic time series. Phys Rev Lett 89:068102
21. Cox D R and Isham V (1980) Point Processes. London, U.K.: Chapman and Hall
22. Cox D R and Lewis P A W (1966) The Statistical Analysis of Series of Events. New York, Wiley
23. Dingwell J B and Marin L C (2006) Kinematic variability and local dynamic stability of upper body motions when walking at different speeds. J Biomech 39:444–452
24. Dingwell J B and Cusumano J P (2000) Nonlinear Time Series Analysis of Normal and Pathological Human Walking. Chaos 10(4):848–863
25. Doyle T L A, Dugan E L, Humphries B and Newton R U (2004) Discriminating between elderly and young using a fractal dimension analysis of centre of pressure. Int J Med Sci 1(1):11–20
26. Duarte M and Zatsiorsky V M (2000) On the fractal properties of natural human standing. Neurosci Lett 283:173–176
27. Eckmann J-P and Ruelle D (1992) Fundamental limitations for estimating dimensions and Lyapunov exponents in dynamical systems. Physica D 6:185–187
28. Eckmann J-P, Kamphorst S O, Ruelle D and Ciliberto D (1992) Lyapunov exponents from time series. Phys Rev A 34, 4971
29. van Emmerick R E A and van Wegen E E H (2002) On the functionalaspects of variability in postural control. Exerc Sport Sci Rev 30:177–183
30. Elger C E, Widman G, Andrzejak R et al. (2000) Nonlinear EEG analysis and its potential role in epileptology. Epilepsia 41 Suppl 3:S34–38
31. Esteller R, Vachtsevanos G, Echauz J and Litt B (2001) A comparison of waveform fractal dimension algorithms. IEEE Trans Circuits Syst I: Fundam. Theory Appl 48:177–183

32. Fraser A M and Swinney H L (1986) Independent coordinates for strange attractors from mutual information. Phys Rev A 33:1134–1140
33. Galka A (2000) Topics in Nonlinear Time Series Analysis – With Implications for EEG Analysis (Advanced Series in Nonlinear Dynamics, edited by R.S. MacKay, Vol. 14), 342 pages, World Scientific Publishing Company, Singapore; ISBN 981-02-4148-8
34. Ganea R, Paraschiv-Ionescu A, Salarian A et al. (2007) Kinematics and dynamic complexity of postural transitions in frail elderly subjects. Conf Proc IEEE Eng Med Biol Soc 2007, 1:6118–6121
35. Goldberger A L (1996) Non-linear dynamics for clinicians: chaos theory, fractals, and complexity at the bedside. Lancet 347:1312–1314
36. Goldberger A L (1997) Fractal variability versus pathologic periodicity: complexity loss and stereotypy in disease. Perspect Biol Med 40:543–561
37. Goldberger A L, Amaral L A N, Glass L, Hausdorff J M, Ivanov P Ch, Mark R G, Mietus J E, Moody G B, Peng C K and Stanley H E (2000) PhysioBank, PhysioToolkit, and PhysioNet: Components of a New Research Resource for Complex Physiologic Signals. Circulation 101(23):e215–e220 [Circulation Electronic Pages http://circ.ahajournals.org/cgi/content/full/101/23/e215]
38. Goldberger A L, Peng C K and Lipsitz L A (2002) What is physiologic complexity and how does it change with aging and disease? Neurobiol Aging 23:23–26
39. Goldberger A L (2006) Giles f. Filley lecture. Complex systems. Proc Am Thorac Soc 3:467–471
40. Govindan R B et al. (2007) Detrended fluctuation analysis of short datasets: An application to fetal cardiac data. Physica D 226:23–31
41. Grassberger P (1990) An optimal box-assisted algorithm for fractal dimensions. Phys Lett A 148:63–68
42. Grassberger P and Procaccia I (1983a) Measuring the strangeness of a strange attractors, Physica D 9:189–208
43. Grassberger P and Procaccia I (1983b) Estimation of the Kolmogorov entropy from a chaotic signal. Phys Rev A 28:2591
44. Grassberger P, Schreiber T and Schaffrath C (1991) Non-linear time sequence analysis. Internat J Bifurcation and Chaos 1:521–547
45. Hausdorff J M, Peng C K, Ladin Z et al. (1995) Is walking a random walk? Evidence for long-range correlations in stride interval of human gait. J Appl Physiol 78:349–358
46. Hausdorff J M, Purdon P L, Peng C K et al. (1996) Fractal dynamics of human gait: stability of long-range correlations in stride interval fluctuations. J Appl Physiol 80:1448–1457
47. Hausdorff J M, Mitchell S L, Firtion R, Peng C K et al. (1997) Altered fractal dynamics of gait: reduced stride-interval correlations with aging and Huntington's disease. J Appl Physiol 82:262–269
48. Hausdorff J M, Lertratanakul A, Cudkowicz M E et al. (2000) Dynamic markers of altered gait rhythm in amyotrophic lateral sclerosis. J Appl Physiol 88:2045–2053
49. Havstad J W and Ehlers C L (1989) Attractor dimension of nonstationary dynamical systems from small data sets. Phys Rev A 39(2):845–853
50. Higuchi T (1988) Approach to an irregular time series on the basis of the fractal theory. Physica D 31:277–283
51. Ho K K, Moody G B, Peng C K et al. (1997) Predicting survival in heart failure case and control subjects by use of fully automated methods for deriving nonlinear and conventional indices of heart rate dynamics. Circulation 96:842–848
52. Hu K, Ivanov P Ch, Zhi C et al. (2001) Effects of trends on detrended fluctuation analysis. Phys Rev E 64:011114
53. Huikuri H V, Makikallio T H, Airaksinen K E et al. (1998) Power-law relationship of heart rate variability as a predictor of mortality in the elderly. Circulation 97:2031–2036
54. Huikuri H V, Makikallio T H, Peng C K et al. (2000) Fractal correlation properties of R-R interval dynamics and mortality in patients with depressed left ventricular function after an acute myocardial infarction. Circulation 101:47–53

55. Ivanov P Ch, Amaral L A N, Goldberger A L et al. (1999) Multifractal in human heartbeat dynamics. Nature 399:461–465
56. Jospin M et al. (2007) Detrended fluctuation analysis of EEG as a measure of depth of anesthesia. IEEE Trans Biomed Eng 54:840–846
57. Kantelhardt J W, Koscielny-Bunde E, Rego H H A et al. (2001) Detecting long-range correlations with detrended fluctuation analysis. Physica A 295, 441–454
58. Kantelhardt J W, Zschiegner S A, Koscielny-Bunde E, Havlin H, Bunde A and Stanley H E (2002) Multifractal detrended fluctuation analysis of nonstationary time series. Physica A 316:87–114
59. Kantelhardt J W, Rybski D, Zschiegner S A et al. (2003) Multifractality of river runoff and precipitation: comparison of fluctuation analysis and wavelet methods. Physica A 330: 240–245
60. Kantz H (1988) Fractals and the analysis of waveforms. Comput Biol Med 18(3):145–156
61. Kantz H (1994) A robust method to estimate the maximal Lyapunov exponent of a time series. Phys Lett A 185:77
62. Kantz H and Schreiber T (1997) Nonlinear Time Series Analysis, Cambridge
63. Kaplan D T and Glass L (1995) Understanding Nonlinear Dynamics. New York: Springer Verlag
64. Kaplan D T (1997) Nonlinearity and nonstationarity: the use of surrogate data in interpreting fluctuations. In: Frontiers of Blood Pressure and Heart Rate Analysis, edited by M. Di Rienzo, G. Mancia, G. Parati, A. Pedotti, and A. Zanchetti. Amsterdam: IOS
65. Kennel M B, Brown R, and Abarbanel H (1992) Determining embedding dimension for phase-space reconstruction using a geometrical construction. Phys Rev A 45:3403–3411
66. Lehnertz K (1999) Non-linear time series analysis of intracranial EEG recordings in patient with epilepsy – an overview. Int J Psychophysiol 34:45–52
67. Leistedt S et al. (2007) Characterization of the sleep EEG in acutely depressed men using detrended fluctuation analysis. Clinical Neurophysiol 118:940–950
68. Lipsitz L A (2002) The dynamics of stability: the physiologic basis of functional health and frailty. J Gerontol A Biol Sci Med Sci 57:B115–B125
69. Lipsitz L A (2004) Physiological Complexity, Aging, and the Path to Frailty. Sci Aging Knowl Environ 16.pe16
70. Lowen S B and Teich M C (2005) Fractal-Based Point Processes. Hoboken, NJ: Wiley
71. Lowen S B and Teich M C (1991) Doubly stochastic point process driven by fractal shot noise. Phys Rev A 43:4192–4215
72. Makikallio T H, Seppanen T, Airaksinen K E et al. (1997) Dynamic analysis of heart rate may predict subsequent ventricular tachycardia after myocardial infarction. Am J Cardiol 80: 779–783
73. Makikallio T H, Ristimae T, Airaksinen K E et al. (1998) Heart rate dynamics in patients with stable angina pectoris and utility of fractal and complexity measures. Am J Cardiol 81:27–31
74. Makikallio T H, Seppanen T, Niemela M et al. (1996) Abnormalities in beat-to-beat complexity of heart rate dynamics in patients with a previous myocardial infarction. J Am Coll Cardiol 28:1005–1011
75. Makikallio T H, Huikuri H V, Hintze U et al. (2001) Fractal analysis and time and frequency-domain measures of heart rate variability as predictors of mortality in patients with heart failure. Am J Cardiol 87:178–182
76. Mutch W A C, Graham M R, Girling L G and Brewster J F (2005) Fractal ventilation enhances respiratory sinus arrhythmia. Respir Res 6:41
77. Nayfeh A H and Balachandran B (1995) Applied Nonlinear Dynamics: Analytical, Computational, and Experimental Methods. New York: Wiley-Interscience
78. Nolte G, Ziehe A and Muller K R (2001) Noisy robust estimates of correlation dimension and K2 entropy. Phys Rev E 64:016112
79. Ohashi K, Bleijenberg G, van der Werf S et al. (2004) Decreased fractal correlation in diurnal physical activity in chronic fatigue syndrome. Methods Inf Med 43:26–29

80. Oswiecimka P, Kwapien J and Drozdz S (2006) Wavelet versus detrended fluctuation analysis of multifractal structures. Phys Rew E 74(2): 016103–016117
81. Paraschiv-Ionescu A, Buchser E, Rutschmann B et al. (2008). Nonlinear analysis of the human physical activity patterns in health and disease. Phys Rev E 77:021913
82. Peng C K, Buldyrev S V, Havlin S et al. (1994) Mosaic organization of DNA nucleotides. Phys Rev E 49:1685
83. Peng C K, Havlin S, Stanley H E et al. (1995) Quantification of scaling exponents and crossover phenomena in nonstationary heartbeat time-series. Chaos 5:82–87
84. Peng C K, Mietus J E, Liu Y, Lee C, Hausdorff J M, Stanley H E, Goldberger A L and Lipsitz L A (2002) Quantifying fractal dynamics of human respiration:age and gender effects. Ann Biomed Eng 30(5):683–692
85. Petrosian A (1995). Kolmogorov complexity of finite sequences and recognition of different preictal EEG patterns. Proc IEEE Symp Computer- Based Medical Syst 212–217
86. Pincus S M (1991). Approximate entropy as a measure of system complexity. Proc Natl Acad Sci USA 88(6):2297–2301
87. Richman J S and Moorman J R (2000) Physiological time-series analysis using approximate entropy and sample entropy. Am J Physiol Heart Circ Physiol 278:H2039–H2049
88. Rosenstein M T, Collins J J and De Luca C J (1993) Reconstruction expansion as a geometry-based framework for choosing proper delay times Physica D 65:117
89. Sachs D, Lovejoy S and Schertzer D (2002) The multifractal scaling of cloud radiances from 1 M to 1 KM Fractals 10(3):253–264
90. Schreiber T and Schmitz A (1996) Improved surrogate data for nonlinearity tests. Phys Rev Lett 77:635–638
91. Schreiber T and Schmitz A (2000) Surrogate time series. Physica D 142:346–382
92. Schreiber T (1999). Is nonlinearity evident in time series of brain electrical activity?. In: Lehnertz K et al. (Ed), Chaos in Brain? Interdisc. Workshop, World Scientific, Singapore 13–22
93. Schurmann T and Grassberger P (1996) Entropy estimation of symbol sequences. Chaos 6:414–427
94. Sekine M, Akay M, Tamura T et al. (2004). Fractal dynamics of body motion in patients with Parkinson's disease. J Neural Eng 1:8–15
95. Sevcik C (1998). A Procedure to Estimate the Fractal Dimension of Waveforms. Appeared in Complexity International 5, the article is also available at the URL: http://www.csu.edu.au/ci/vol05/sevcik/sevcik.htm
96. Stanley H E, Amaral L A N, Goldberger A L, Halvin S, Ivanov P Ch and Peng C.-K. (1999). Statistical physics and physiology: Monofractal and multifractal approaches. Physica A 270:309–324
97. Stam C J (2005) Nonlinear dynamical analysis of EEG and MEG: review of an emerging field. Clin Neurophysiol 116:2266–2301. doi: 10.1016/j.clinph.2005.06.011
98. Stam C J (2006) Nonlinear brain dynamics. New York: Nova Science Publishers
99. Takens F (1981) Detecting Strange Attractors in Turbulence. Warwick, Lecture notes in Mathematics, v.898, Ed. D. Rand & L.-S Young, Springer, 366–381
100. Teich M C, Turcott R G and Lowen S B (1990) The fractal doubly stochastic Poisson point process as a model for the cochlear neural spike train. In: The mechanics and biophysics of hearing (Dallos P, Geisler CD, Matthews JW, Ruggero MA, Steele CR, eds), 354–361. New York: Springer
101. Teich M C (1992) Fractal neuronal firing patterns. In: Single Neurons Computation. edited by McKenna T, Davis J, and Zormetzer SF. Boston, MA: Academic, 589–625
102. Teich M C, Heneghan C, Lowen S B et al. (1997). Fractal character of the neuronal spike train in the visual system of the cat. J Opt Soc Am A 14:529–546
103. Theiler J (1986) Spurious dimension from correlation algorithms applied to limited time-series data. Phys Rev A 34(3):2427–2432
104. Theiler J, Eubank S, Longtin A et al. (1992) Testing for nonlinearity in time series: the method of surrogate data. Physica D 58, 77–94

105. Theiler J (1995) On the evidence for low-dimensional chaos in an epileptic electroencephalogram. Phys Lett A 196:335–341
106. Thurner S, Lowen S B, Feurstein M C et al. (1997). Analysis, synthesis and estimation of fractal-rate stochastic point processes. Fractals 5:565–595
107. Torres M, Añino M, Gamero L and Gemignani M (2001) Automatic detection of slight changes in nonlinear dynamical systems using multiresolution entropy tools, Int J Bifurc Chaos 11:967–981
108. Turcott R G and Teich M C (1996) Fractal character of the electrocardiogram: distinguishing heart-failure and normal patients. Ann Biomed Eng 24:269–293
109. Vaillancourt D E and Newell K M (2002a) Changing complexity in human behavior and hysiology through aging and disease. Neurobiol Aging 23:1–11
110. Vaillancourt D E and Newell K M (2002b) Complexity in aging and disease: response to commentaries. Neurobiol Aging 23:27–29
111. Vikman S, Mäkikallio TH, Yli-Mäyry S, Pikkujämsä S et al. (1999). Altered complexity and correlation properties of R-R interval dynamics before the spontaneous onset of paroxysmal atrial fibrillation Circulation 100, 2079–2084
112. Wolf A, Swift J B, Swinney H L et al. (1985) Determining Lyapunov exponents from a time series. Physica D 16:285–317
113. Yates F E (1994) Order and complexity in dynamical systems: homeodynamics as a generalized mechanics for biology. Math Comp Model 1:49–74
114. Yeragani V K, Radhakrishna R K, Tancer M et al. (2002) Non-linear measures of respiration: respiratory irregularity and increased chaos of respiration in patients with panic disorder. Neuropsychobiology 46:111–120
115. Zebrowski J J, Poplawska W, Baranowski R and Buchner T (2000) Symbolic dynamics and complexity in a physiological time series. Chaos Solitons & Fractals 11:1061–1075

Chapter 16
Biomedical Data Processing Using HHT: A Review

Ming-Chya Wu and Norden E. Huang

Abstract Living organisms adapt and function in an ever changing environment. Even under basal conditions they are constantly perturbed by external stimuli. Therefore, biological processes are all non-stationary and highly nonlinear. Thus the study of biomedical processes, which are heavily depending on observations, is crucially dependent on the data analysis. The newly developed method, the Hilbert-Huang Transform (HHT), is ideally suited for nonlinear and non-stationary data analysis such as appeared in the biomedical processes. Different from all other data analysis existing methods, this method is totally adaptive: It derives the basis from the data and based on the data. As a result, it is highly efficient in expanding any time series in their intrinsic modes, which reveal their full physical meaning. In this article, we review biomedical data processing by using HHT. We introduce two exemplary studies: cardiorespiratory synchronization and human ventricular fibrillation. The power and advantages of HHT are apparent from the achievements of these studies.

16.1 Introduction

Physiological systems are complex and their dynamical properties and the underlying biomedical processes can only be studied through some physicological and pathological data. The adaptation and the interactions and feedbacks amongst our body systems, however, make the physiological and pathological signals highly nonlinear and nonstationary [1]; consequently, the resultant biomedical signals are among the most complicated data there is. Since the underlying dynamics can only be studied through limited data, data analysis methods play a crucial role in the outcome. An essential task to analyze biomedical data is to extract essential component(s) that will be fully representative of the underlying biological

M.-C. Wu (✉)
Research Center for Adaptive Data Analysis, Department of Physics and National Central University, Chungli 32001, Taiwan; Institute of Physics, Academia Sinica, Nankang Taipei 11529, Taiwan
e-mail: mcwu@ncu.edu.tw

A. Nait-Ali (ed.), *Advanced Biosignal Processing*,
DOI 10.1007/978-3-540-89506-0_16,

processes. For this purpose, there should be criteria derived from the data itself to judge what is the inherent dynamics and what are the contributions of external factors and noises in the data. To accommodating the variety of complicated data, analysis method would then have to be adaptive. Here, adaptivity means that the definition of the basis has to base on and be derived from the data. Unfortunately, most currently available data analysis methods are based an a priori basis (such as trigonometric functions in Fourier analysis, for example); they are not adaptive [2]. From the viewpoint of data analysis, the ultimate goal is not to find the mathematical properties of data, but is to uncover the physical insights and implications hidden in the data. There are no a priori reasons to believe that a basis functional, however cleverly designed, is capable of representing the variety of the underlying physical processes. An a posteriori adaptive basis provides a totally different approach from the established mathematical paradigm; though it may also present a great challenge to the mathematical community.

The recently developed Empirical Mode Decomposition (EMD) and the associated Hilbert Spectral Analysis (HSA), together designated as the Hilbert-Huang Transform (HHT) [2], represents such a paradigm shift of data analysis methodology. The HHT is designed specifically for analyzing data from nonlinear and nonstationary processes. From the very beginning, HHT had been proved to be a powerful tool for biomedical data analysis in research [3–6].

The EMD uses the sifting process to extract monocomponent signal by eliminating riding waves and making the wave-profiles more symmetric. The expansion of any data set via the EMD method has only a finite-number of locally non-overlapping time scale components, known as Intrinsic Mode Functions (IMFs) [2]. Each intrinsic mode, linear or nonlinear, represents a simple oscillation, which has the same number of extrema and zero-crossing. In comparison with simple har-

Table 16.1 Comparison between Fourier, wavelet, and HHT analysis. Adapted from Ref. [7]

	Fourier	Wavelet	HHT
Basis	a priori	a priori	a posteriori, Adaptive
Frequency	Integral transform over global domain, Uncertainty	Integral transform over global domain, Uncertainty	Differentiation over local domain, Certainty
Presentation	Energy in frequency space	Energy in time-frequency space	Energy in time-frequency space
Nonlinearity	No	no	Yes
Nonstationarity	No	yes	Yes
Feature Extraction	No	Discrete: no Continuous: yes	Yes
Theoretical base	Complete mathematical theory	Complete mathematical theory	Empirical

monic functions, the IMF can have a variable amplitude and frequency as functions of time. Furthermore, these IMFs as bases are complete and orthogonal to each other. All IMFs enjoy good Hilbert transform such that they are suitable for spectral analysis. The adaptive properties of HHT to empirical data further make it easy to give physical significations to IMFs. Table 16.1 summarizes comparisons between Fourier, wavelet, and HHT analysis [7].

The power and effectiveness of HHT in data analysis have been demonstrated by its successful applications to many important problems covering engineering, biomedical, financial and geophysical data. Recently, 2 dimensional version of HHT [8–12] has also been developed and applied to image processing. Readers interested in completed details can consult Refs. [2, 7] and Refs. [9–12]. In this article, we review biomedical data processing by using 1D HHT. Due to the limit of space, here we focus on two exemplary studies: cardiorespiratory synchronization (CS) [13–15], and human ventricular fibrillation (VF) [16, 17]. From the outcome of these studies, the advantages and power of HHT are apparent.

16.2 Empirical Mode Decomposition

The EMD in HHT is developed on the assumption that any time series consists of simple intrinsic modes of oscillations, and the essence of the method is to identify the intrinsic oscillatory modes by their characteristic time scales in the data empirically and then decompose the data accordingly [2]. The resultant components decomposed from the EMD are IMFs, which are symmetric with respect to the local mean and have the same numbers of zero crossings and extremes. This is achieved by sifting data to generate IMFs. The algorithm to create IMFs in the EMD consists of two main steps [2]:

Step-1: Identify local extrema in the experimental data $x(t)$. All the local maxima are connected by a cubic spline line $U(t)$, which forms the upper envelope of the data. Repeat the same procedure for the local minima to produce the lower envelope $L(t)$. Both envelopes will cover all the data between them. The mean of upper envelope and lower envelope $m_1(t)$ is given by:

$$m_1(t) = \frac{U(t) + L(t)}{2} \tag{16.1}$$

Subtracting the running mean $m_1(t)$ from the original time series $x(t)$, we get the first component $h_1(t)$,

$$h_1(t) = x(t) - m_1(t) \tag{16.2}$$

The resulting component $h_1(t)$ is an IMF if it is symmetric and has all maxima positive and all minima negative. An additional condition of intermittence can be imposed here to sift out waveforms with certain range of intermittence for physical consideration. If $h_1(t)$ is not an IMF, the sifting process has to be repeated as many times as it is required to reduce the extracted signal to an IMF. In the subsequent sifting process steps, $h_1(t)$ is treated as the data to repeat above steps mentioned above,

$$h_{11}(t) = h_1(t) - m_{11}(t) \tag{16.3}$$

Again, if the function $h_{11}(t)$ does not yet satisfy criteria for IMF, the sifting process continues up to k times until some acceptable tolerance is reached:

$$h_{1k}(t) = h_{1(k-1)}(t) - m_{1k}(t) \tag{16.4}$$

Step-2: If the resulting time series is an IMF, it is designated as $c_1 = h_{1k}(t)$. The first IMF is then subtracted from the original data, and the difference r_1 given by

$$r_1(t) = x(t) - c_1(t) \tag{16.5}$$

is the residue. The residue $r_1(t)$ is taken as the original data, and we apply to it again the sifting process of *Step-1*.

Following the procedures of *Step-1* and *Step-2*, we continue the process to find more intrinsic modes c_i until the last one. The final residue will be a constant or a monotonic function which represents the general trend of the time series. Finally, we obtain

$$x(t) = \sum_{i=1}^{n} c_i(t) + r_n \tag{16.6}$$

$$r_{i-1}(t) - c_i(t) = r_i(t) \tag{16.7}$$

Here we remark that an extended version of EMD, named Ensemble EMD (EEMD) [18], has recently been developed to improve the mode-mixing problem which may occur in EMD and makes each component lack of full physical meaning. The EEMD is implemented by constructing a sufficiently large amount of realizations each combines $x(t)$ and a white noise, and taking average for the IMFs decomposed by EMD from these realizations. By using the fact that average of large amount of white noise converges to null, the noise has zero net effect on the data but is beneficial to effective sifting in decomposition. It has been shown that EEMD indeed performs better than the primary version of EMD from avoiding the mode mixing problem. Since EMD and EEMD are essentially in the same framework, here we only discuss EMD. Details of EEMD can be found in Ref. [18].

The instantaneous phase of IMF can be calculated by applying the Hilbert transform to each IMF, say the rth component $c_r(t)$. The procedures of the Hilbert transform consist of calculation of the conjugate pair of $c_r(t)$, i.e.,

$$y_r(t) = \frac{1}{\pi} P \int_{-\infty}^{\infty} \frac{c_r(t')}{t - t'} dt' \tag{16.8}$$

where P indicates the Cauchy principal value. With this definition, two functions $c_r(t)$ and $y_r(t)$ forming a complex conjugate pair, define an analytic signal $z_r(t)$:

$$z_r(t) = c_r(t) + i y_r(t) \equiv A_r(t) e^{i\phi_r(t)} \tag{16.9}$$

with amplitude $A_r(t)$ and instantaneous phase $\phi_r(t)$ defined by

$$A_r(t) = [c_r^2(t) + y_r^2(t)]^{1/2} \tag{16.10}$$

$$\phi_r(t) = \tan^{-1}\left(\frac{y_r(t)}{c_r(t)}\right) \tag{16.11}$$

Then, one can calculate the instantaneous phase from Eqs. (16.8) and (16.11).

16.3 Cardiorespiratory Synchronization

First, we present an application of HHT to the study of CS [13, 14, 15]. CS is a phenomenon originating from the interactions between cardiovascular and the respiratory systems. The interactions can lead to a perfect locking of their phases whereas their amplitudes remain chaotic and non-correlated [19]. The nature of the interactions has been extensively studied in recent years [20–42]. Recently, Schäfer et al. [32, 33] and Rosenblum et al. [34] applied the concept of phase synchronization of chaotic oscillators [43] to analyze irregular non-stationary bivariate data from cardiovascular and respiratory systems, and introduced the cardiorespiratory synchrogram (CRS) to detect different synchronous states and transitions between them. They found that there were sufficiently long periods of hidden synchronization and concluded that the CS and respiratory sinus arrhythmia (RSA) are two competing factors in cardiorespiratory interactions. Since then, CS has been reported in young healthy athletes [32, 33], healthy adults [34–36], heart transplant patients [34], infants [38], and anesthetized rats [39]. More recently, Kotani et al. [40] further developed a physiologically model from these observations to study the phenomena, and showed that both the influence of respiration on heartbeat and the influence of heartbeat on respiration are important for CS. Since aforementioned studies are mostly based on measured data, the data processing method plays a crucial role in the outcome. The essential part of such investigations is the extraction of respiratory rhythms from noisy respiratory signals. A technical problem in the analysis of the respiratory signal then arises: insufficiently filtered signals may still have too many noises, and over-filtered signal may be too regular to lose the characteristics of respiratory rhythms: Improper analysis methods may lead to misleading results.

To overcome these difficulties, Wu and Hu [13] proposed using the EMD for such studies and got significantly reasonable results. Unlike conventional filters, the EMD provides an effective way to extract respiratory rhythms from experimental respiratory signals. The adaptive properties of EMD to empirical data make it easy to give physical significations to IMFs, and allow one to choose a certain IMF as a respiratory rhythm [13]. In the implement of EMD, respiratory rhythms are extracted from empirical data by using the number of respiratory cycles per minute for human beings as a criterion in the sifting process of EMD [13]. They considered empirical data consisting of 20 data sets which were collected by the Harvard medical school in 1994 [44]. Ten young (21–34 years old) and ten elderly (68–81 years old) rigorously-screened healthy subjects underwent 120 minutes of continuous supine resting while continuous electrocardiogram (ECG) and respiration signals were collected. The continuous ECG and respiration data were digitized at 250 Hz (respiratory signals were later preprocessed to be at 5 Hz). Each heartbeat was annotated using an automated arrhythmia detection algorithm, and each beat annotation was verified by visual inspection. Each group of subjects includes equal numbers of men and women. In the following, we will review the scheme proposed by Wu and Hu [13], and focus on the application of HHT to CS. Details of the study and extended investigations which are not included herein will be referred to the original paper [13].

The respiratory signals represent measures of the volume of expansion of ribcage, so the corresponding data are all positive numbers and there are no zero crossings. In addition to respiratory rhythms, the data also contain noises originating from measurements, external disturbances and other factors. From the EMD decomposition, one can select one component as the respiratory rhythm according to the criteria of intermittencies of IMFs imposed in *Step-1* as an additional sifting condition [13]. Among IMFs, the first IMF has the highest oscillatory frequency, and the relation of intermittence between different modes is roughly $\tau_n = 2^{n-1}\tau_1$ with τ_n the intermittence of the n th mode. The reason for such a dyadic intermittence criterion selection is that the EMD indeed represents a dyadic filter bank as suggested by Flandrin et al. [45] and Wu and Huang [46].

More explicitly, the procedures for the analysis are as follows [13]: (i) Apply the EMD to decompose the recorded data into a number of IMFs. Since the respiratory signal was preprocessed to a sampling rate of 5 Hz, there should be (10–30) data points in one respiratory cycle[1]. Thus, for example, one can use c_1: (3–6), c_2: (6–12), c_3: (12–24), ..., etc. After the sifting processes of the EMD, the original respiratory data are decomposed into n IMFs $c_1, c_2, \ldots, c_n$, and a residue r_n. (ii) Visually inspect the resulting IMFs. If the amplitude of a certain mode is dominant and the waveform is well distributed, then the data are said to be well decomposed, and the decomposition is successfully completed. Otherwise, the decomposition may be inappropriate, and one has to repeat step (i) with different parameters.

Figure 16.1 shows the decomposition of an empirical signal with a criterion of the intermittence being (3–6) data points for c_1, and $(3 \times 2^{n-1} - 3 \times 2^n)$ data points for c_n's with $n > 1$. Comparing $x(t)$ with c_i's, it is obvious that c_3 preserves the

[1] The number of breathing per minute is about 18 for adults, and about 26 for children. For different healthy states, the number of respiratory cycles may vary case by case. To include most of these possibilities, respiratory cycles ranging from 10 to 30 times per minute were taken. Each respiratory cycle then roughly takes 2–6 s, i.e., (10–30) data points.

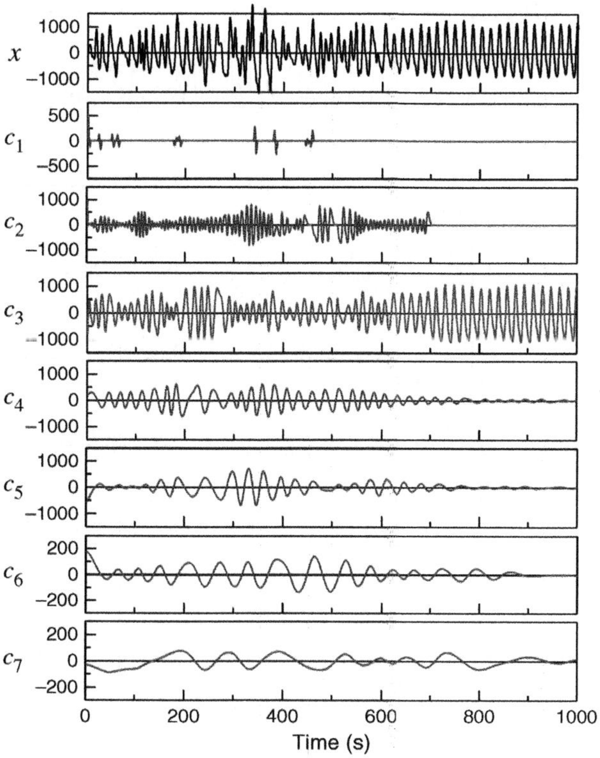

Fig. 16.1 Example of EMD for a typical respiratory time series (code f1o01 in the database [44]). The criterion for intermittence in the sifting process is (3–6) data points per cycle for c_1. Signal $x(t)$ is decomposed into 14 components including 13 IMFs and 1 residue; here, only the first 7 components are shown. After Ref. [13]

main structure of the signal and is also dominant in the decomposition. One can see that component c_3,with (12–24) data points per respiratory cycle, corresponds to the respiratory rhythm. Figure 16.2 compares respiratory signal in various stages. In Fig. 16.2a, a typical respiratory time series data $x(t)$ is shown. The preprocessed signal $x'(t)$ by a proper Fourier band filter is shown in Fig. 16.2b, in which only fast oscillatory noises are filtered out, and main structures of the signal are preserved. Figure 16.2c shows the IMF $c_3(t)$ obtained by performing EMD on $x'(t)$. The process is similar to that used to obtain $c_3(t)$ in Fig. 16.1. Obviously, IMF $c_3(t)$ of Fig. 16.2c still preserves characteristic structure of $x(t)$ shown in Fig. 16.2a. Here we should emphasize that the preprocessing to have $x'(t)$ is not necessary in the framework of EMD. We show $x'(t)$ only for comparison.

After selecting one IMF as the respiratory rhythm, one can proceed in the calculation of the instantaneous phase by the Hilbert transform and incorporating with heartbeat signals to construct CRS, which is a visual tool for inspecting synchronization. Let us denote the phase of the respiratory signal calculated by using Eq. (16.11) as ϕ_r and the heartbeat as ϕ_c. If the phases of respiratory signal $\phi_r$and heartbeat ϕ_c are coupled in a fashion that a cardiovascular system completes n heartbeats in m respiratory cycles, then a roughly fixed relation can be proposed. In general, there is a phase and frequency locking condition [13, 19, 32, 33].

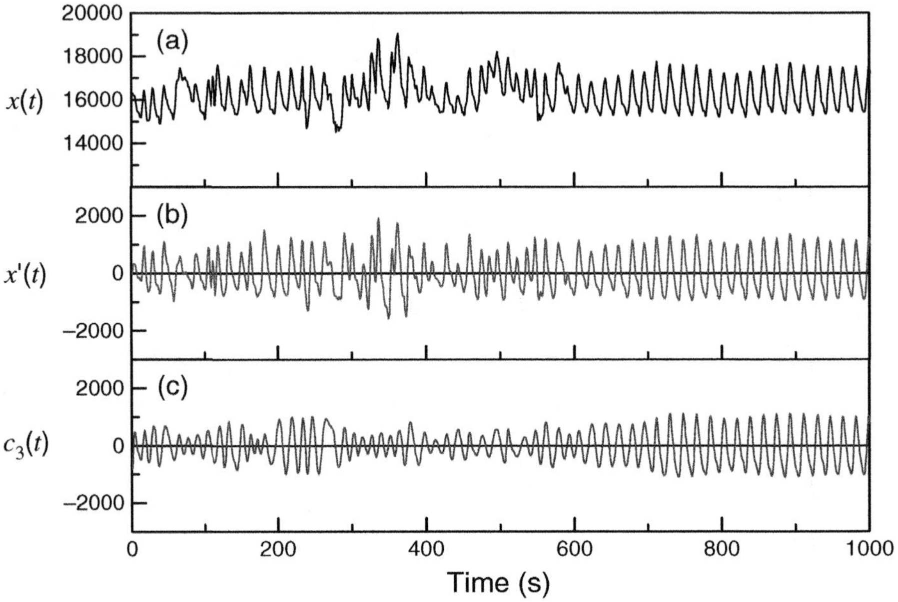

Fig. 16.2 Comparison of respiratory signals for a typical subject (code f1o01) in different data processing stages: (**a**) original experimental time series $x(t)$, (**b**) after performing low pass filtering, $x'(t)$, and (c) the third IMF $c_3(t)$ in Fig. 16.1, after performing EMD on $x'(t)$. Adapted from Ref. [13]

$$|m\phi_r - n\phi_c| \leq const \tag{16.12}$$

with m, n integer. According to Eq. (16.12), for the case that ECG completes n cycles while the respiration completes m cycles, it is said to be synchronization of n cardiac cycles with m respiratory cycles. Using the heartbeat event time t_k as the time frame, Eq. (16.12) implies the relation

$$\phi_r(t_{k+m}) - \phi_r(t_k) = 2\pi m \tag{16.13}$$

Furthermore, by defining

$$\Psi_m(t_k) = \frac{1}{2\pi}\left[\phi_r(t_k) \mod 2\pi m\right] \tag{16.14}$$

and plotting $\Psi_m(t_k)$ versus t_k, synchronization will result in n horizontal lines in case of $n : m$ synchronization. By choosing n adequately, a CRS can be developed for detecting CS [32, 33]. Example of 3:1 synchronization with $n = 6$ and $m = 2$ is shown in Fig. 16.3, where phase locking appear in several epochs, e.g., at 2800–3600 s, and there are also frequency locking, e.g., at 400 s, near which there are n parallel lines with the same positive slope.

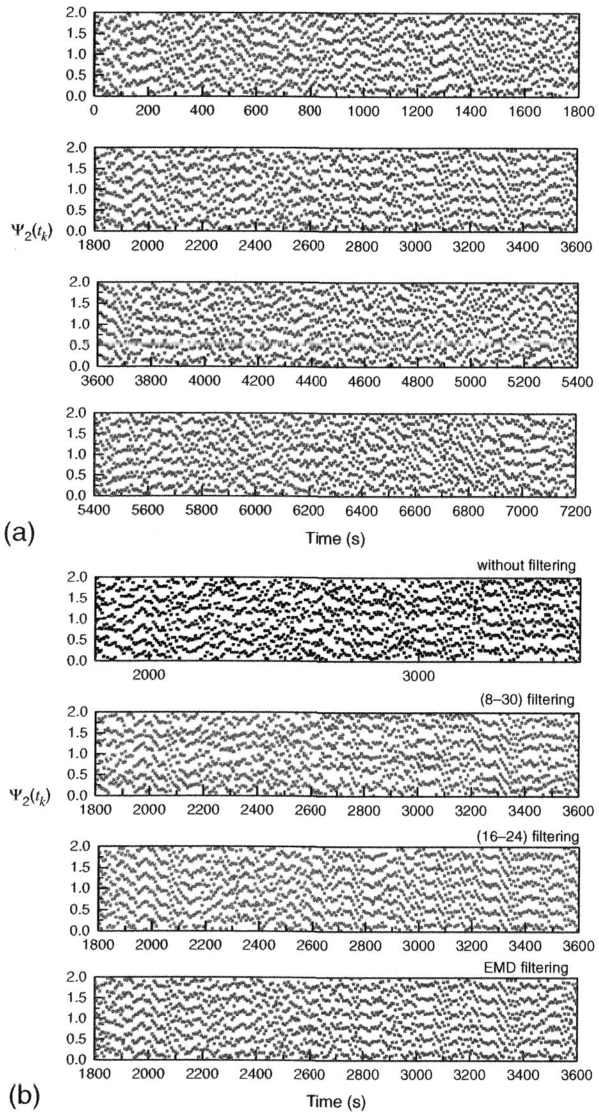

Fig. 16.3 CRS for a typical subject (code f1o06). (**a**) Empirical data are preprocessed by the EMD method. There are about 800 sec synchronization at 2800–3600 sec, and several spells of 50–300 sec at other time intervals. (**b**) Comparison of the results without filtering (top one), preprocessed by the standard filters with windows of (8–30) and (16–24) cycles per minute (the second and the third ones), and the EMD method (bottom one). After Ref. [13]

For comparison, the results of the same subject in 1800–3600s, but with respiratory signals without filtering, preprocessed by the standard filters and the EMD are shown in Fig. 16.3b. The windows of the standard filters are (8–30) and (16–24) cycles per minute. In general, some noise dressed signals can still show synchronization in some epochs but the HSA failed at some time intervals (e.g., around 3400–3600 s of the case without filtering), and over-filtered signals reveal too strong synchronization (filter with window of 16–24). In other words, global

Fig. 16.4 Histogram of phase for the phase locking period from 2800 sec to 3600 sec for a typical subject (code f1o06) shown in Fig. 16.3a. After Ref. [13]

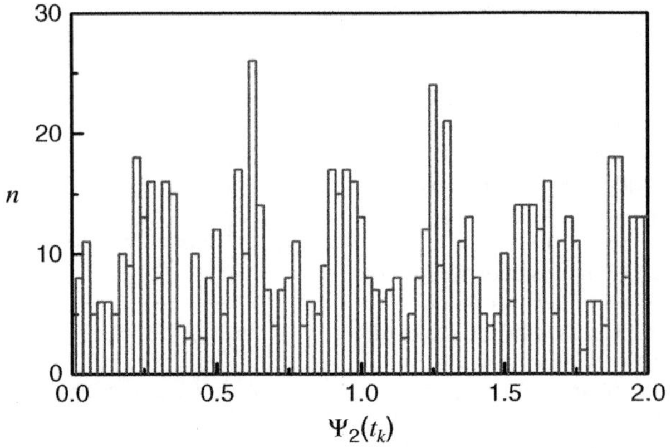

frequency used in standard filters may dissolve local structures of the empirical data. This does not happen in the EMD filtering.

Figure 16.4 shows the histogram of phases for the phase locking period from 2800 to 3600 s in Fig. 16.2a. Significant higher distribution can be found at $\Psi_2 \approx$ 0.25, 0.6, 0.9, 1.25, 1.6, 1.9 in the unit of 2π, indicating heartbeat events occur roughly at these respiratory phase during this period. Following above procedures, we analyze data of 20 subjects, and the results are summarized in Table 16.2. The results are ordered by the strength of CS. From our results, we do not find specific relations between the occurrence of synchronization and sex of the subjects as in Refs. [32, 33]. Here we note that if we use other filters to the same empirical data, we will have different results depending on the strength of synchronization. As noted above, data processing method plays a crucial role in the analysis of real data. Over-filtered respiratory signals may lose detailed structures and become too regular. It follows that final conclusions are methodological dependent.

To compare EMD and Fourier-based filter, we use the intermittency in the EMD analysis as the bandwidth of a generic Fourier-based filter to filter the same empirical data. We usually have different results depending on the strength of synchronization. For example, for the f1o06 subject, the intermittency of the third IMF is (12–24). Using (12–24) as a bandwidth of the generic Fourier-based filter, we have similar epochs of synchronization. However, for the f1y02 subject with intermittency of the second IMF (16–32) being selected to optimize the decomposition, we have more epochs of 3:1 synchronization lasting 50 s and new few 7:2 synchronization lasting 50 s and 80 s when the bandwidth of (16–32) is used. For the f1o05 subject in which the second IMF with intermittency (10–20) being selected, epochs of 5:2 synchronization lasting 50 s are found when the same bandwidth (10–20) is used. In comparison with the results presented in Table 16.2, the Fourier-based filter with a bandwidth of the same intermittency appears to smooth the data to have more regular wave form, and in turn usually have stronger synchronization. For a time series with variable intermittencies, the smoothing of data may introduce

Table 16.2 Summary of our results. 20 subjects are ordered by the strength (total time length) of CS. After Ref. [13]

Code	Sex	Age	Synchronization
flo06	F	74	3:1(800 s, 300 s, 250 s, 150 s, 100 s, 50 s)
fly05	M	23	3:1(350 s, 300 s, 200 s, 100 s)
flo03	M	73	3:1(200 s, 50 s, 30 s)
fly10	F	21	7:2(200 s, 50 s), 4:1(50 s)
flo07	M	68	7:2(120 s, 100 s, 80 s)
flo02	F	73	3:1(100 s, several spells of 50 s)
fly01	F	23	7:2(several spells of 30 s)
fly04	M	31	5:2(80 s, 50 s, 30 s)
flo08	F	73	3:1(50 s, 30 s)
fly06	M	30	4:1(50 s, 30 s)
flo01	F	77	7:2(several spells of 50 s)
fly02	F	28	3:1(50 s)
fly08	F	30	3:1(50 s)
flo10	F	71	3:1(30 s)
flo05	M	76	No synchronization detectable
fly07	M	21	No synchronization detectable
fly09	F	32	No synchronization detectable
fly03	M	34	No synchronization detectable
flo09	M	71	No synchronization detectable
flo04	M	81	No synchronization detectable

additional modes which do not exist in some segments of the primary data and thus leads to misleading results. For example, in Fig. 16.5, comparisons for the results of the fly02 subject obtained by using the Fourier-based filter and the EMD approach are shown. The original time series $x(t)$ is dressed with noises such that the signal almost disappears at $t = 2320 - 2380$ sec. The Fourier-based filter introduces a new waveform at this epoch, but the new waveform having a local minimum larger than 0 can not be processed directly by the Hilbert transform. This is not the case for the waveform obtained from EMD method. Furthermore, at $t = 2000 - 2100$ sec, the Fourier-based filter does not preserve the structure of the original time series, and the waveform obtained from EMD is similar to the original time series. Therefore, from the aspect of data processing that could preserve the essential features of original empirical data, the EMD approach is better than Fourier based filtering.

From above investigation, we conclude that: (i) In most cases, cardiac oscillations are more regular than respiratory oscillations and the respiratory signal is the key factor for the strength of CS. (ii) Cardiorespiratory phase locking and frequency locking take place when respiratory oscillations become regular enough and have a particular frequency relation coupling with cardiac oscillations. Therefore, CS and RSA are competing factors [32, 33]. We observed the intermittence of respiratory oscillation varies with time but synchronization persists in some subjects. This confirms correlations in CS. (iii) Over-filtered respiratory signals may be too regular, and in turn, appear to have stronger synchronization than they shall have. As a result,

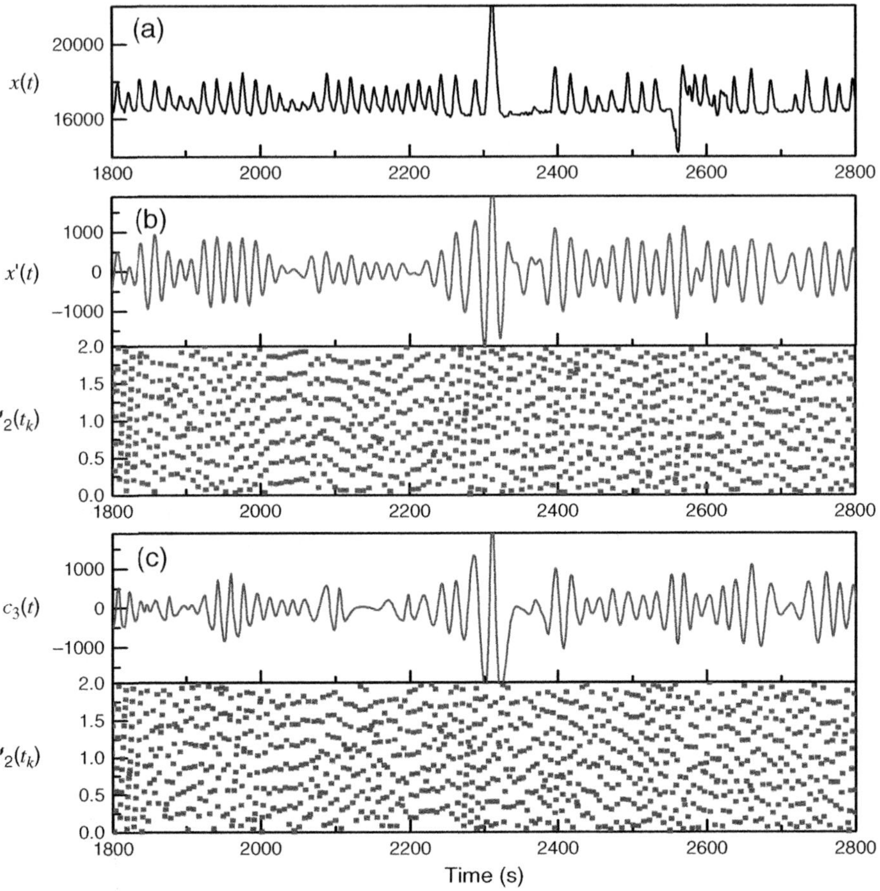

Fig. 16.5 Comparison of the data processing for a typical subject (code fly02). (**a**) The empirical time series, (**b**) the time series filtered by the Fourier-based filter with bandwidth (16–32) and the corresponding synchrogram, and (**c**) the time series of the third IMF decomposed by the EMD method with intermittency (16–32) and the corresponding synchrogram. After Ref. [13]

if the Fourier based approach with narrow band filtration is used, some epochs of phase locking or frequency locking should be considered as being originated from these effects.

16.4 Human Ventricular Fibrillation

Cardiac arrhythmias are disturbances in the normal rhythm, and fibrillation is manifested as irregular electrical activity of the heart. During fibrillation the coordinated contraction of the cardiac muscle is lost and the mechanical pumping effectiveness of the heart fails. Among these, ventricular fibrillation (VF) is known as the most dangerous cardiac arrhythmia frequently leading to sudden cardiac death (SCD) [47]. Prediction of VF is thus an important issue in cardiology, yet to date; there exists no effective measure capable of predicting fatal VF. Since short-term VF can also occurs in the ECG of healthy people, the first task to this issue is to distinguish fatal VF from non-fatal VF.

Recently, Wu et al. [16, 17] investigated the empirical data of VF patients by using the approach of phase statistics to estimate the correlation between characteristic properties of VF ECG and the corresponding consequences, i.e., dead/alive outcome. They found that there is an explicit correlation which can be used as a predictor for fatal VF. The phase statistics approach was first introduced by Wu et al. for the study of financial time series [48, 49]. The authors found that the approach can catch structure information of time series and is suitable for the analysis of wave profiles of VF ECG. The phase statistics analysis is in principle an extension of HHT, and is capable of describing morphologies of a wave in a statistical sense [16, 17]. The study of Wu et al. includes collections of ECG for SCD and VF from patients, and the signal analysis of the resultant VF ECG. In this section, we will shortly review their analysis by HHT. Again, details of the study and extended investigations which are not included herein will be referred to the original paper [16, 17].

ECG records the electric potential of myocardial cells at the body surface as a function of time, and the occurrence of VF signals implies the heart does not work (to pump blood) normally. More precisely, normal ECG explicitly shows P wave, QRS complexes, and T wave, while there is lack of QRS complexes wave form in VF ECG. Figure 16.6 shows the comparison between a typical normal ECG and VF ECG signal used in the study. Apparently, it is possible to extract the intervals of VF from ECG chart by technician by directly visual inspection. In this study, the ECG for SCD and VF by using a portable 24-hour Holter has been collected. The

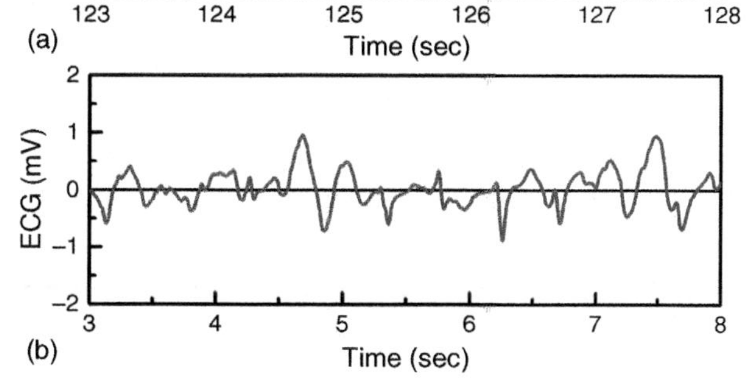

Fig. 16.6 (**a**) A typical normal ECG, and (**b**) VF ECG signal used in the study

data were recorded by the CM5 lead (the bipolar V5 lead) at a sampling frequency of 125 Hz. There were totally 24 patients involved in the study, but 7 of the patients did not suffer from the VF problem and data for one patient was not recorded. Among the remaining 16 subjects, there were 6 survivors and 10 non-survivors. The VF ECG segments were picked up by a medical doctor. Some patients have more than one VF ECG segment, and finally 27 VF ECG data were available for the analysis.

From the viewpoint of cellular eletrophysiology, the appearance of ventricular tachycardia is due to the formation of a reentrant wave in cardiac tissue, driving the ventricle at a rate much faster than the normal sinus rhythm. VF is a spatially and temporally disorganized state arising from the subsequent breakdown of the reentrant wave into multiple drifting and meandering spiral waves [50, 51]. Therefore, the detection of the characteristic features corresponding to the disordered state in ECG is likely a path toward the early evaluation of VF. In normal ECG, there are sharp P waves which are not suitable for direct analysis, while waveforms of VF ECG behave better and can be used for morphology analysis. Therefore, the analysis is only applied to the ECG data during VF.

The timing characteristics of transient features of nonstationary time series like VF ECG are best revealed using the concept of the instantaneous phase. The analysis is thus carried out by phase statistics. The phase statistics approach consists of the calculation of the instantaneous phase of a time series and the statistics of the calculated phases. In order faithfully to calculate the phase, we decompose empirical data into a number of well-defined IMFs by EMD, and calculate the instantaneous phase of resultant IMFs directly by the Hilbert transform. The phase statistics is achieved by defining the histogram of instantaneous phase satisfying the following normalization condition

$$\int P(\rho)d\rho = 1 \tag{16.15}$$

where $P(\cdots)$ stands for the probability density function (PDF). Direct calculations show that the PDF of instantaneous phase of the first IMF can be classified into three types of patterns, CV (convex), UF (uniform), and CC (concave), according to the morphologies of the histograms [17]. Furthermore, the statistics of 27 VF intervals and the best fit in the logistic regression concluded that the CV type VF is likely to be the fatal VF [17].

To quantify the phase distribution patterns, we define a measure χ,

$$\chi = \langle P_1(\phi_1)\rangle - \langle P_2(\phi_1)\rangle \tag{16.16}$$

where $\langle\cdots\rangle$ denotes an average, P_1 is the PDF of the instantaneous phase ϕ_1 of the first IMF in the range $-0.5\pi \leq \phi_1 \leq 0.5\pi$, and P_2 is for the phase in the ranges $-\pi \leq \phi_1 < -0.5\pi$ and $0.5\pi < \phi_1 \leq \pi$. More specifically, χ is a measure of the difference between the average of the PDFs of the phases located in the range $-0.5\pi \leq \phi_1 \leq 0.5\pi$ and the average of those not in this range. According to this definition, we have $\chi > \varepsilon$ for the CV type, $|\chi| \leq \varepsilon$ for the UF type, and $\chi < -\varepsilon$

Fig. 16.7 χ as a function of time for typical VFs of survivals (*dashed line/blue color*) and non-survivals (*thick solid line/red color*). *Dotted line* is the threshold which separates regimes of survival and non-survival. After Ref. [16]

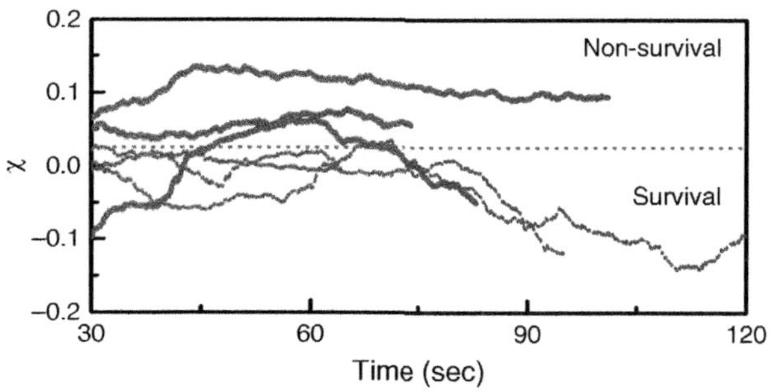

for the CC type. The value of ε is determined by the properties of statistics. It is better to establish a proper threshold of χ such that there is a clear separation for the CV type pattern from the UF and CC types. From the analysis of Holter data from 16 individuals, it was found that taking $\varepsilon = 0.025$ gives reasonable results which are consistent with direct visual inspection. Note that $\varepsilon = 0.025$ corresponds to a tolerance of 5% from the probability $P = 0.5$. Hence, one can describe the temporal evolution of the phase histogram by measuring $\chi(t)$ with a fixed window. As $\chi(t)$ enters into the regime of the CV type pattern $\chi(t) > \varepsilon$, it is considered as an indication of the occurrence of fatal VF. For practical purposes, here we take a window of 30 sec. Figure 16.7 shows χ as a function of time for typical VFs of three survivals and three non-survivals. The threshold ε of $\chi(t)$ can substantially separate the survival and non-survival groups into survival and non-survival regimes, and $\chi(t)$ for survivors rarely enter into the non-survival regime. As a result, the technique offers a new possibility to improve the effectiveness of intervention in defibrillation treatment and limit the negative side effects of unnecessary interventions. It can also be implemented in real time and should provide a useful method for early evaluation of fatal VF [16].

16.5 Conclusions

We have briefly explored the applications of HHT to biomedical data processing. The remarkable advantage of HHT in these applications is that it can catch primary structures of intrinsic rhythms from empirical data based on its adaptive feature [13, 17]. It should be pointed out that although intermittence test was used in this study, a more general method of EEMD [18] should be tested in the future. The EEMD enjoys an advantage of not setting the intermittence criterion subjectively.

In the study of CS, we found that from a physiological viewpoint, it is difficult to precisely identify the mechanisms responsible for the observed non-linear interactions in CS. However, cardiac oscillations are more regular than respiratory oscillations, and CS occurs at the period when respiratory signals become regular enough. Therefore, the regularity of respiratory signals contributes dominantly to the synchronization. Consequently, over-filtered signals may cause a misleading conclusion that there is CS. The adaptivity of HHT allows us to effectively keep the signal structures and avoid the introduction of artificial periodicity easily appear in the Fourier-based filters with priori bases [13] and lead to a conclusion of too strong CS. From this aspect, HHT is better than other methods.

In the study of VF, we have used the phase statistics approach [48] to investigate ECG during VF in humans. In this approach, the HHT was used to calculate the instantaneous phase of IMFs decomposed from VF ECG, and the corresponding momentary phase histograms was then construct to inspect the evolution of the waveform of the time series. The HHT's capability of handling nonstationary and nonlinear time series allows us to define a measure as a monitor to inspect temporal evolution of the phase histogram of ECG during VF. The classification of VF ECG from the phase histograms further provides a possible route for early evaluation of fatal VF. Since to date there is no predictor available for fatal VF, this breakthrough may indicate the power and promise of HHT.

From the impressive achievements of the applications of HHT to CS and VF ECG time series analysis as presented in this article, we expect that HHT can also be applied to other biomedical data. Among others, the importance of biomedical image has been emphasized for a long time, and 2D HHT to biomedical imaging applications is promising. We are working this direction, and results will be reported in the near future.

Acknowledgments This work was supported by the National Science Council of the Republic of China (Taiwan) under Grant Nos. NSC 96-2112-M-008-021-MY3 (M.-C. Wu) and NSC 95-2119-M-008-031-MY3 (N. E. Huang).

References

1. Peng CK, Costa M, Goldbergr AL (2009) Adaptive Data Analysis of Complex Fluctuations in Physiologic Time Series. Adv Adapt Data Analy 1: 61–70
2. Huang NE, Shen Z, Long SR, Wu MC, Shin HH, Zheng Q, Yen NC, Tung CC, Liu HH (1998) The Empirical Mode Decomposition Method and the Hilbert Spectrum for Non-stationary Time Series Analysis. Proc Roy Soc London A 454: 903–995
3. Huang W, Shen Z, Huang NE, Fung YC (1998) Engineering analysis of biological variables: An example of blood pressure over 1 day. Proc Natl Acad Sci USA 95: 4816–4821
4. Huang W, Shen Z, Huang NE, Fung YC (1998) Engineering Analysis of Intrinsic Mode and Indicial Response in Biology: the Transient response of Pulmonary Blood Pressure to Step Hypoxia and Step Recovery. Proc Natl Acad Sci USA 95: 12766–12771
5. Huang W, Shen Z, Huang NE, Fung YC (1999) Nonlinear indicial response of complex nonstationary oscillations as pulmonary hypertension responding to step hypoxia. Proc Natl Acad Sci USA 96: 1834–1839
6. Cummings D, Irizarry R, Huang E, Endy T, Nisakak A, Ungchusak K, Burke D (2004) Travelling waves in dengue hemorrhagic fever incidence in Tailand. Nature 427: 344–347
7. Huang NE, Shen SSP (2005) edited. Hilbert-Huang Transform and Its Applications, World Scientific
8. Wu Z, Huang NE, Chen XY (2008) The 2 Dimensional Ensemble Empirical Mode Decomposition method, Patent (submitted)
9. Nunes JC, Bouaoune Y, Deléchelle E, Niang O, Bunel P (2003) Image analysis by bidimensional empirical mode decomposition. Image Vis Comput 21: 1019–1026

10. Nunes JC, Guyot S, Delećhelle E (2005) Texture analysis based on local analysis of the bidimensional empirical mode decomposition. J Machine Vision Apply (MVA) 16: 177–188
11. Nunes JC, Deléchelle E (2009) Empirical Mode Decomposition: Applications on Signal and Image Processing. Adv Adapt Data Analy 1: 125–175
12. Yuan Y, Jin M, Song PJ, Zhang J (2009) Two dimensional empirical mode decomposition and dynamical detection of bottom topography with SAR picture. Adv Adapt Data Analy (in press)
13. Wu MC, Hu CK (2006) Empirical mode decomposition and synchrogram approach to cardiorespiratory synchronization. Phys Rev E 73: 051917
14. Wu MC (2007) Phase statistics approach to time series analysis. J Korean Phys Soc 50: 304–312
15. Wu MC (2007) Phase statistics approach to physiological and financial time series. AAPPS Bulletin 17: 21–26
16. Wu MC, Struzik ZR, Watanabe E, Yamamoto Y, Hu CK (2007) Temporal evolution for the phase histogram of ECG during human ventricular fibrillation. AIP Conf Proc 922: 573–576
17. Wu MC, Watanabe E, Struzki ZR, Hu CK, Yamamoto Y (2008) Fatal ventricular fibrillation can be identified by using phase statistics of human heart beat signals, submitted for publication
18. Wu Z, Huang NE (2009) Ensemble Empirical Mode Decomposition: A Noise Assisted Data Analysis Method. Adv Adapt Data Analy 1: 1–42
19. Tass P, Rosenblum MG, Weule J, Kurths J, Pikovsky A, Volkmann J, Schnitzler A, Freund HJ (1998) Detection of *n:m* Phase Locking from Noisy Data: Application to Magnetoencephalography. Phys Rev Lett 31: 3291
20. Guyton AC (1991) Text book of medical physiology. 8th ed (Saunders, Philadelphia)
21. Bernardi L, Salvucci F, Suardi R, Solda PL, Calciati A, Perlini S, Falcone C, Ricciardi L (1990) Evidence for an intrinsic mechanism regulating heart-rate-variability in the transplanted and the intact heart during submaximal dynamic exercise. Cardiovasc Res 24: 969–981
22. Almasi J, Schmitt OH (1974) Basic technology of voluntary cardiorespiratory synchronization in electrocardiology. IEEE Trans Biomed Eng 21: 264–273
23. Rosenblum MG, Pikovsky AS, Kurths J (2004) Synchronization approach to analysis of biological systems. Fluctuation and Noise Lett 4: L53–L62
24. Rosenblum MG, Pikovsky AS (2004) Controlling Synchronization in an Ensemble of Globally Coupled Oscillators. Phys Rev Lett 92: 114102
25. Schreiber T (2000) Measuring Information Transfer. Phys Rev Lett 85: 461
26. Paluš M, Stefanovska A (2003) Direction of coupling from phases of interacting oscillators: An information-theoretic approach. Phys Rev E 67: 055201(R)
27. Jamsek J, Stefanovska A, McClintock PVE (2004) Nonlinear cardio-respiratory interactions revealed by time-phase bispectral analysis. Phys Med Bio 49: 4407–4425
28. Richter M, Schreiber T, Kaplan DT (1998) Fetal ECG extraction with nonlinear state-space projections. IEEE Eng Med Biol Mag 45: 133–137
29. Hegger R, Kantz H, Schreiber T (1999) Practical implementation of nonlinear time series methods: The TISEAN package. Chaos 9: 413–435
30. Kantz H, Schreiber T (1998) Human EGG: nonlinear deterministic versus stochastic aspects. IEE Proc Sci Meas Technol 145: 279–284
31. Toledo E, Roseblum MG, Schäfer C, Kurhts J, Akselrod S (1998) Quantification of cardiorespiratory synchronization in normal and heart transplant subjects. In: Proc. of Int. Symposium on Nonlinear Theory and its Applications, vol.1. Lausanne: Presses polytechniques et universitaries romandes, 171–174
32. Schäfer C, Rosenblum MG, Kurths J, Abel HH (1998) Heartbeat synchronized with ventilation. Nature (London) 392: 239–240
33. Schäfer C, Rosenblum MG, Abel HH, Kurths J (1999) Synchronization in the human cardiorespiratory system. Phys Rev E 60: 857

34. Rosenblum MG, Kurths J, Pikovsky A, Schäfer C, Tass P, Abel HH (1998) Synchronization in noisy systems and cardiorespiratory interaction. IEEE Eng Med Biol Mag 17: 46–53
35. Toledo E, Rosenblum MG, Kurths J, Akselrod S (1999) Cardiorespiratory synchronization: is it a real phenomenon? In: Computers in Cardiology, vol. 26, Los Alamitos (CA), IEEE Computer Society, 237–240
36. Lotric MB, Stefanovska A (2000) Synchronization and modulation in the human cardiorespiratory system. Physica A 283: 451–461
37. Toledo E, Akselrod S, Pinhas I, Aravot D (2002) Does synchronization reflect a true interaction in the cardiorespiratory system? Med Eng Phys 24: 45–52
38. Mrowka R, Patzak A (2000) Quantitative analysis of cardiorespiratory synchronization in infants. Int J Bifur Chaos 10: 2479–2488
39. Stefanovska A, Haken H, McClintock PVE, Hozic M, Bajrovic F, Ribaric S (2000) Reversible Transitions between Synchronization States of the Cardiorespiratory System. Phys Rev Lett 85: 4831
40. Quiroga RQ, Arnhold J, Grassberger P (2000) Learning driver-response relationships from synchronization patterns. Phys Rev E 61: 5142
41. Quiroga RQ, Kraskov A, Kreuz T, Grassberger PP (2002) Performance of different synchronization measures in real data: A case study on electroencephalographic signals. Phys Rev E 65: 041903
42. Kotani K, Takamasu K, Ashkenazy Y, Stanley HE, Yamamoto Y (2002) Model for cardiorespiratory synchronization in humans. Phys Rev E 65: 051923
43. Rosenblum MG, Pikovsky AS, Kurths J (1996) Phase Synchronization of Chaotic Oscillators. Phys Rev Lett 76: 1804
44. Iyengar N, Peng CK, Morin R, Goldberger AL, Lipsitz LA (1996) Age-related alterations in the fractal scaling of cardiac interbeat interval dynamics. Am J Physiol 271: 1078–1084, Data sets are available from http://physionet.org/physiobank/database/fantasia/. The subject codes used in the article follow those in the database
45. Wu Z, Huang NE (2004) A study of the characteristics of white noise using the empirical mode decomposition method. Proc R Soc London A 460: 1597–1611
46. Flandrin P, Rilling G, Gonçalvès P (2004) Empirical mode decomposition as a filter bank. IEEE Signal Proc Lett 11: 112–114
47. Zipes DP, et al (2006) ACC/AHA/ESC 2006 guidelines for management of patients with ventricular arrhythmias and the prevention of sudden cardiac death: a report of the American College of Cardiology/American Heart Association Task Force and the European Society of Cardiology Committee for Practice Guidelines (Writing Committee to Develop Guidelines for Management of Patients with Ventricular Arrhythmias and the Prevention of Sudden Cardiac Death). Europace 8: 746
48. Wu MC, Huang MC, Yu HC, Chiang TC (2006) Phase distribution and phase correlation of financial time series. Phys Rev E 73: 016118
49. Wu MC (2007) Phase correlation of foreign exchange time series. Physica A 375: 633–642
50. Christini DJ, Glass L (2002) Introduction: Mapping and control of complex cardiac arrhythmias. Chaos 12: 732–739
51. Bursac N, Aguel F, Tung L (2004) Multiarm spirals in a two-dimensional cardiac substrate. Proc Natl Acad Sci USA 101: 15530–15534

Chapter 17
Introduction to Multimodal Compression of Biomedical Data

Amine Naït-Ali, Emre Zeybek and Xavier Drouot

Abstract The aim of this chapter is to provide the reader with a new vision of compressing jointly medical images/videos and signals. This type of compression is called "multimodal compression". The basic idea is that only one codec can be used to compress, at the same time, a combination of medical data (i.e. images, videos and signals). For instance, instead of using one codec for each signal or image/video which might complicate the software implementation, one should proceed as follows: for the encoding process, data are merged using various specific functions before being encoded using a given codec (e.g. JPEG, JPEG 2000 or H.264). For the decoding phase, data are first decoded and afterwards separated using an inverse merging-function. The performance of this approach in terms of compression-distortion appears promising.

17.1 Introduction

Nowadays compression of biosignals and biomedical images for the purpose of storage and transmission is becoming more and more fundamental. This can be explained by an important increase in the number of clinical applications such as long duration biosignal recording or even their real-time transmission. In this context, data compression is particularly useful in telemedicine applications including data sharing, monitoring, medical system control and so on. Consequently, reducing the size of data without losing valuable clinical information becomes crucial when acquisition systems provide huge amounts of data under tough real-time constraints and limited bandwidths.

As an example, the next section will present a clinical application for which multimodal compression might be useful. This application concerns the

---

A. Naït-Ali (✉)
Université Paris 12, Laboratoire, Image, Signaux et Systèmes Intelligents, LiSSi, EA. 3956. 61, avenue du Général de Gaulle, 94010, Créteil, France
e-mail: naitali@univ-paris12.fr

A. Naït-Ali (ed.), *Advanced Biosignal Processing*,
DOI 10.1007/978-3-540-89506-0_17,

polysomnography (PSG) which requires recording, during the analysis process, various biosignals, in parallel with a video recording the patient.

Before describing the idea of multimodal compression, Sect. 17.2 will address the benefits of multimodal recording analysis. Sect. 17.3 will then present a global vision on biomedical data compression, including signals and medical images. An initial joint compression "*image-signal*" scheme is presented first. Afterwards, an extension of this technique to the video is evoked in Sect. 17.4.

Since the objective of this chapter is to introduce this new approach of compression, only the basics are presented.

17.2 Benefits of Multimodal Recording Analysis: An Example from Sleep Medicine

The most notable development in medicine at the end of the 20th century is indisputably the progress of biomedical engineering that made an extensive variety of techniques available for physicians with which to examine their patients.

While the increase in the power and speed of microprocessors has benefitted medical imagery, ambulatory medicine took advantage of the reduction in size of amplifiers, analog-digital convertors and of the increase in memory capacity; These advances allowed the production of handheld devices and permitted the recording of various physiological biosignals such as blood pressure for several hours at a time. Nowadays, handheld devices can simultaneously record and memorize up to 25 different signals during periods as long as 24 h.

17.2.1 Sleep Recordings

Sleep medicine is a specialty that has benefitted considerably from such technological progress. Sleep medicine was kept confidential until the late 70s and was really born during the late 80s when physicians had the possibility of recording biosignals over the space of at least one day and one night successively. Sleep is a vital physiological state, during which the entire organism enters into a specific functional regime. All physiological functions, including heart rate, breathing, blood pressure, muscle activity, body movements and brain activity slow down or operate in a specific manner during normal sleep. Thus, the precise analysis of sleep requires multimodal recordings, that is, simultaneous recordings of various biosignals (electrophysiological as well as infrared video images), captured using various sensors and captors. Nevertheless, sleep is a highly complex phenomenon, regulated by sophisticated mechanisms that organize the succession of the different phases of sleep; its main aim is to restore the energy that has been consumed during waking hours.

Sleep is mainly considered in terms of brain behaviour. Thus, analysis of sleep requires the recording of neuronal activity with an electroencephalogram (EEG). Neuronal activity is recorded through electrodes (usually five) pasted with biological glue onto the scalp. Monopolar EEG signals are amplified, A/D converted and finally stored with 200 Hz sampling (with a 12 bit resolution). Recordings of eyeball movements are also required by the study of sleep, since they are specific to paradoxical sleep (cf. infra). These recordings (electro-oculograms) are performed using electrodes placed around each eye. Muscular activity (electromyogramm, EMG) is recorded with bipolar electrodes disposed on the chin. EMG must be sampled at higher rates (more than 200 Hz). These are the minimum requirements needed and an example of the equipment is illustrated in Fig. 17.1

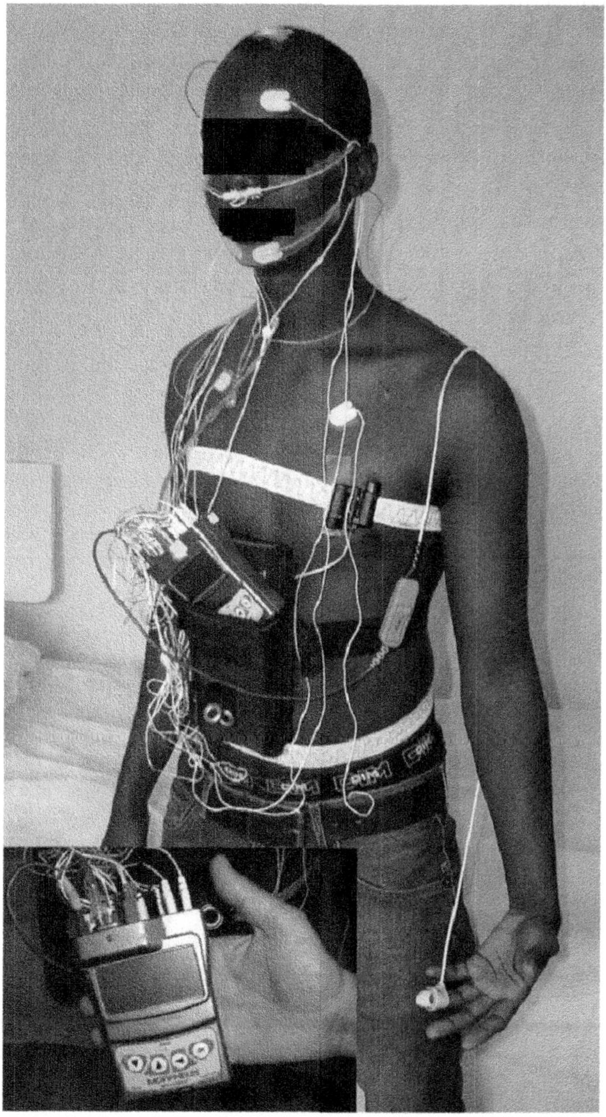

Fig. 17.1 Handheld device used to carry out ambulatory sleep recordings. This device can record up to 40 different biosignals over a period of 8 h

It is also common for other sensors and captors to be added especially to assess respiratory functions: pressure sensors for oro-nasal airflow, belts with inductance plethysmography for thoracic and abdominal respiratory movements, pulse oxymetry, electrocardiogram. The main biological signals acquired for sleep studies are indicated in Table 17.1. An example of a handheld device is presented in Fig. 17.1.

The aspects of all the signals recorded are modified during sleep. For example, the low amplitude (10–20 microvolt) and high frequency (20–40 Hz) oscillations observed on EEG during wakefulness are replaced by slow frequency (1–2 Hz) high amplitude (50–100 microvolt) oscillations during slow wave sleep. EMG amplitude and pulse rate also decrease, and virtually all signals are affected during sleep. These modifications, especially those of EEG, EOG and EMG, are specific to the different sleep stages that compose nocturnal sleep. Sleep studies are composed of "scoring", that is to attribute to each 30s period, a sleep stage (deep sleep, light sleep or paradoxical sleep) depending of EEG, EOG and EMG signals aspects. Sleep disorders are characterized by the occurrence and the repetition during sleep of some events that fragment sleep, alter sleep continuity and impede the restorative properties of sleep. Patients awake in the morning tired and could be sleepy during the daytime. The most frequent events are sleep apneas: these are brief and repetitive interruptions of nasal airflow that are diagnosed using sleep recordings.

Table 17.1 Main biological signals required for sleep analysis. Physiological function measured and techniques used are indicated. Note that each signal requires a specific sampling rate. Frequency bands of interest (low frequency and high frequency filters) also differ for each signal

Physiological parameter	Technique/sensor	Biological signal	Sampling rate (Hz)
Neuronal activity	Electrical field	Electroencephalography	200
Muscular activity	Electrical field	Electromyography	500
Eyeball movements	Electrical field	Electro-oculagraphy	200
Nasal airflow	Pressure captor	Respiratory airflow	50
Respiratory noise	Sound	Snoring	200
Thoraco-abdominal volume	Inductance plethysmography	Respiratory movements	50
Cardiac electrical activity	Electrical field	Electrocardiography	100
haemoglobin saturation	Infrared spectroscopy	Oxymetry	20
Finger volume	Infrared plethysmography	Cardiac pulse	20
Body movement, position	Images	Video	15

17.2.2 Different Levels of "Multimodality" in Sleep Recordings

From a signal processing point of view, we can distinguish several levels of "multimodality" in sleep recording. Regarding the various physiological functions recorded during sleep (Table 17.1), sleep recordings can be considered as multimodal. These signals are acquired using various specific sensors and digitalized through A/D convertors working with a specific frequency band, depending on the speed of variation of the parameter. Because sampling frequency can vary from 10 to 500 Hz, specific algorithms have been designed to store signals in a unique file type. This has constituted one of the main limitations for the diffusion of knowledge and scientific production since each manufacturer has developed its own algorithm and file format. Such difficulties recently improved with the generalization of a new data format dedicated to data exchange. However, the need for efficient algorithms is still on the agenda.

Once digitalized, and excepting the sampling frequencies, these biological signals these are somewhat quite similar.

Additional degrees of complexity of multimodality arise when video images are also acquired during sleep recordings. In this case, sleep recordings are truly multimodal, combining images and biological signals.

17.2.3 Utility of Multimodal Recording in Sleep Medicine

First, the recording of video images simultaneously with other biosignals allows the detection of artifacts and acquisition problems such as the displacement or disconnection of electrodes, which are two of the main constraints which can arise. A precise evaluation of the quality of the signal is the preliminary condition before sleep analysis.

Since body movements are frequently associated with artifacts or saturation of the EMG signal, video images are crucial to determine whether these movements are preceded by cerebral abnormalities on EEG, such as epilepsy; these combined recordings (video + biosignals) are essential in order to establish the link between movements observed on the video and brain dysfunctions (Fig. 17.2).

Finally, some sleep events could occur when the body is lying in a specific position. For instance, sleep apnea are commonly seen when patients are lying in dorsal decubitus. In this case, the combined analysis of video and biosignals is also crucial to establish an appropriate diagnosis.

17.2.4 Need for Signal Compression

Storage of sleep recordings is mandatory and represents a major problem for sleep laboratories. Nowadays, a conventional sleep recording is a 300 Mb file, with another 300 Mb video file.

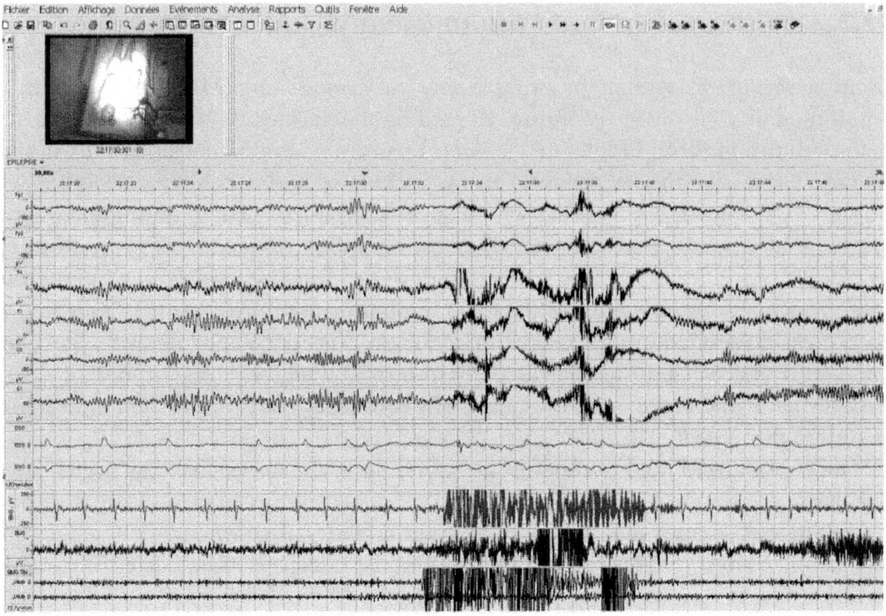

Fig. 17.2 Screen capture of a multimodal sleep recording. The video is on the *upper left*. Biosignals are from *top* to *bottom*: EEG (*6 traces*), EOG (*2 traces*), ECG (*1 trace*) chin EMG (*1 trace*) and limbs EMG (*2 traces*), and the screen a 30s segment. Note the artifacts on the left half of the screen, which affect all the signals and are due to body movements

Beside the storage, the exchange between sleep laboratories of these recordings files will be extended over the few next years. Small files in a readable format will then be required.

17.3 Biomedical Data Compression

From a clinical point of view, we have seen in the previous section that a multimodal analysis of biomedical data is particularly interesting for diagnostic purposes. Nowadays, sharing, storing and transmitting medical information through a given network are useful functionalities which clinicians and physicians have become accustomed to using. Of course, these functionalities concern, on the one hand, medical images (e.g. ultrasound, X ray), and on the other hand, biosignals. This section is important in the sense that general aspects of compression will be presented for both medical images and biosignals. It might be considered, to a certain extent, as a survey. The reader will notice that the common techniques and the current available codecs have been designed or evaluated, separately for some specific data.

17.3.1 Generalities on Data Compression

As is well known, data can be compressed according to two schemes, namely lossless compression or lossy compression. When lossless compression is used, the reconstructed information is exactly identical to the original one, whereas when using a lossy compression, the quality of reconstruction depends on the compression ratio (CR), or the bit-rate. In other words, when the CR is low, the loss of visual quality is not perceptible. However, the distortion is more significant for higher CRs as we will demonstrate later. So choosing between lossless compression and lossy compression is highly dependent on the application and the expected reconstruction quality. Therefore, one has to point out that when using lossless compression of signals and images, the CR that can be reached in this case is very limited (i.e., generally about 2 or 3). The only solution to overcome this limitation is to deal with lossy compression for which the CR can easily attain 20, 30, 40 or more, depending, of course, on the quality of the reconstruction.

Moreover, nowadays we tend to design progressive codecs (i.e. encoding/decoding data according to a progressive way, from low frequencies to high frequencies, up to a given requested bit-rate). In addition, codec scalability is a functionality that allows the user to adjust the bit-rate/quality of the decoded data according to the specific reception system being employed (i.e. network type, permitted quality of service and so on).

17.3.1.1 Lossless Compression

Generally speaking, lossless compression of images and signals are achieved according to a two stage scheme. In the first stage, a temporal (spatial in case of an image) or a frequency reversible transformation is performed in order to reduce the entropy of the input data. For example, in the time-domain (or spatial domain), one can use the predictive based techniques whereas transforms such as Discrete Fourier Transform (DFT), Discrete Cosine Transform (DCT) or Discrete Wavelet Transform (DWT) can be employed as a frequency representation. In the second stage, the redundancy of the information is reduced by means of an entropic coding such as Huffman or Run Length Encoding (RLE). The lossless compression generic scheme is shown in Fig. 17.3a.

17.3.1.2 Lossy Compression

Lossy compression schemes generally use three stages as shown in Fig. 17.3b. The loss of information occurs at the quantization level which can be either scalar or vectorial. As we have already mentioned, this technique enables the achievement of high CRs in comparison to the performances allowed by lossless compression techniques. Many standards can handle this functionality. For instance, JPEG or JPEG 2000.

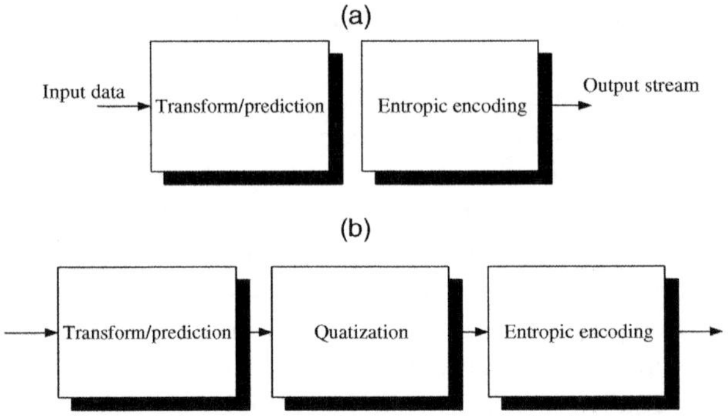

Fig. 17.3 Generic compression scheme. (**a**) Lossless compression (**b**) Lossy compression

17.3.2 Medical Image Compression

Should one compress medical images using lossless or lossy techniques? Looking back several years, the answer would have frequently been "*lossless compression*", (e.g. using LZW, FELICS, CLIC, JPEG-LS, CALIC, SPHIT (lossless), JPEG 2000 (lossless), LOCO, etc.). Maybe this seems to be logical because the priority here is obviously the quality of the diagnostic. In fact, one can imagine that if a physician provides a wrong diagnosis because of a badly compressed medical image, this could end in disaster! However, if the loss is well controlled, no risk of erroneous interpretation would occur. Fortunately, mindsets have changed over the last few years, and physicians, now commonly agree, on the analysis, under certain conditions of lossy compressed images. Actually, accepting lossy compression methods is becoming increasingly widespread thanks to the numerous publications available in this field. Moreover, as pointed out by the American College of Radiology, compression should only be carried out if it results in no loss of diagnostic information.

As is well known, medical images are frequently stored using the DICOM format. This standard can encapsulate other general purpose standards such as JPEG and JPEG 2000, either in lossless or lossy mode. Therefore, even if the lossy mode is also supported, the DICOM committee doesn't provide directives on how to choose the parameters of compression (i.e. compression ratio).

As is also well known, the JPEG standard is based on the DCT whereas JPEG 2000 uses the DWT which seems to provide better performances in terms of bit-rate vs. distortion. From the images shown in Fig. 17.4, we can highlight a well known block effect that occurs when JPEG compression is used at 0.2 bpp. For the same bit-rate, one can notice that a better visual quality is achieved using JPEG 2000, which is basically a DWT based standard.

According to recent works, as has been reported in Table 17.2, some CRs, for which acceptable image qualities (in terms of clinical criteria), are obtained using JPEG 2000. For more details, the reader can refer to [45].

From the literature, it is clear that various approaches dedicated to medical images compression have been published to date. As stated in Table 17.3,

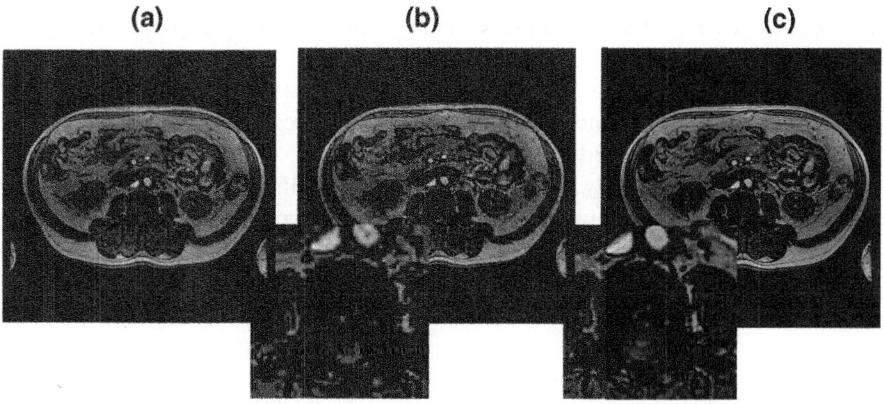

Fig. 17.4 JPEG and JPEG 2000 comparison: (**a**) original image from MeDEISA "medeisa.net", (**b**) JPEG compressed image at 0.2 bpp, (**c**) JPEG 2000 compressed image at 0.2 bpp. In the image (**b**), one can emphasize the block effect due to the DCT transform. For the same bite-rate, JPEG 2000 "(**c**)" doesn't present this drawback

Table 17.2 JPEG 2000 evaluated on various medical images [45]

Image Type	Acceptable compression ratio	Reference
Digital chest radiograph	20:1 (so that lesions can still be detected)	[57, 13]
Mammography	20:1 (detecting lesions)	[58]
Lung CT image	10:1 (so that the volume of nodules can still be measured)	[31]
Ultrasound	12:1	[16]
Coronary angiogram	30:1 (after optimizing JPEG 2000 options)	[64]

Table 17.3 Some recent works related to medical image compression

References	Image type	Approaches
[40, 54]	All; X ray.	Wavelets; Fractal-Wavelets
[17, 19, 27, 41, 42, 52]	All; All; All; All; Angio.	Autoregressive model; wavelets; JPEG; Wavelets; Wavelets; wavelets ROI
[7, 11, 25, 28, 59, 55]	X ray; All; All; Tomography; All; All	JPEG; ROI; Polynomial decomposition; Wavelets; 3D DCT; AR models
[10, 18, 62]	Angio.; echo.; all	Wavelets; wavelets; DCT
[33, 61]	Chromosome images; All	ROI/ Wavelets; DCT
[1, 56, 30, 50]	All; 3D; Tomography; X ray.	Fast JPEG 2000; Wavelets; Wavelets; Wavelets
[3, 23, 51, 37]	Echo.; all; all; all; all	ROI Quantization; Optimal quantization; wavelets; wavelets
[29]	Echo	Quantization

wavelet based compression techniques have been intensively explored during the last decade. Moreover, object oriented compression has been often included in the compression process. In this context, a Region of Interest (ROI), defined either manually or automatically, is encoded using lossless compression, whereas the remaining region is encoded according to a lossy mode.

As mentioned previously, the progressivity in encoding/decoding as well as the scalability are regarded, nowadays, as important functionalities.

17.3.3 Biosignal Compression

From the literature, we have gathered in this sub-section, the main compression techniques developed especially for some biosignals such as EEG, ECG and EMG. Of course, one can easily observe that most of the publications are primarily concerned with the ECG and subsequently the EEG. This seems to be obvious regarding the various applications dedicated to monitoring using biosignals.

17.3.3.1 EEG Compression

Based on the publications related to the EEG compression, four main classes of techniques can be highlighted, namely time-domain compression, frequency domain compression, time-frequency compression and finally, spatio-temporal compression. Generally, most of the proposed approaches in the literature devoted to the EEG compression in time-domain are mainly prediction based. This can be explained by the fact that the EEG is a low-frequency signal, which is characterized by a high temporal correlation. Some of these techniques are, in fact, a direct application of classical digital signal processing methods such as the Linear Prediction Coding (LPC), the Markovien Prediction, the Adaptive Linear Prediction and Neural Network Prediction based methods. Moreover, some approaches include the information related to the long-term temporal correlation of the samples due to the fact that spaced samples are also correlated.

By considering the frequency-domain, it is well known that the main energy of the EEG signal is concentrated essentially in low frequencies (i.e. lower than 20 Hz for the α rhythm). Consequently, a frequency transform of this particular signal makes it suited to compression. For this purpose, many techniques such as Karhunen-Loève Transform (KLT) and Discrete Cosine Transform (DCT) have been evaluated.

As we have pointed out above, EEG can also be compressed using the time-frequency approach, in particular, using wavelets. For instance, in [12], the signal is segmented and decomposed using Wavelet Packets. The coefficients are afterwards encoded. Other algorithms such as the well known EZW (Embedded Zerotree Wavelet) have also been successfully applied to compress the EEG signal [34].

Finally, one can evoke the fact that the EEG can be compressed by taking into account the spatio-temporal characteristics. For this purpose, the reader can refer to [5].

17.3.3.2 ECG Compression

In this section, ECG compression techniques will be examined. As one will notice, some of the presented approaches are appropriate for real-time transmissions, whereas other techniques are more suitable for storage purpose (e.g. Holters).

For clarity reasons, Table 17.4 gathers together the most recent research works in this field. In fact, if one considers the number of articles published over the past six years, no less than 20 papers have been published in international journals. This indicator clearly emphasizes the importance of this field of application.

ECG compression techniques can be classified into four broad categories, namely, time domain compression methods, transform-based methods, parameter extraction methods (i.e. modeling) and bi-dimensional processing. When time domain methods (also called direct techniques) are considered, ECG samples are directly encoded. On the other hand, in transform-based methods category, the original samples are transformed and the compression is achieved in the new domain. For this purpose, several approaches have been handled such as DFT, DCT and DWT.

When dealing with model-based techniques, the features (i.e. parameters) of the processed signal are extracted then used *a posteriori* for the purpose of reconstruction.

Finally, after a bi-dimensional transformation, the ECG can be regarded as an image. In such case, standards or other algorithms dedicated to image processing can be adapted to the ECG context.

Table 17.4 Some recent research works related to the ECG compression

References	Approaches	References	Approaches
[26]	Wavelet packet	[53]	Wavelet transform
[2]	Wavelet transform of the prediction error	[38]	Vectorial quantization
[4]	Wavelet transform	[8]	JPEG 2000
[6]	Optimal quantization of the DCT	[15]	Shape adaptation
[47]	Minimization of the distortion rate	[32]	"Review"
[39]	Vectorial quantization	[36]	Vectorial quantization-wavelets
[60]	SVD	[14, 9, 46, 48, 43, 44]	Neural networks, Polynomial projection, Hilbert transform, lorentzian modelling, Radon transform, interleaving.
[24]	R-R lossless compression	[35]	Max-Lloyd quantizer

17.3.3.3 EMG Compression

Based on international publications, compression of EMG has not been considered to the same degree as the ECG or the EEG; nevertheless one has to mention some of the works done in this field which are basically wavelet based-techniques. For instance, the reader can refer to [20, 21, 22, 49] .

It is obvious that the number of publications can at any time increase depending of course on: (1) the future applications that will be developed, (2) clinical requirements.

17.4 Multimodal Compression

As one can notice from the techniques presented in the previous sections, most of the published approaches have been developed to compress only one signal at the same time and sometimes, they are dedicated to a particular biosignal. This means that one can face the following situations:

- If *N* signals are recorded, this implies that encoding/decoding process is executed *N* times (i.e. time consuming),
- If *N* different signals including images are acquired, then *N* different codes should be implemented. This might decrease the performances of some of the imbedded systems.

Nowadays, in various applications such as for the one we have seen in the example presented in Sect. 17.2; physicians require more than one signal to identify clinical anomalies. Therefore, in the context of telemedicine where one has to share, store or transmit the medical data; an appropriate compression system should be employed. Technically, a flow of data can be handled and optimized for the application. For this purpose, we will try in this chapter to highlight a new idea dedicated to jointly encoding a set of signals/images. Consequently, various configuration and schemes can be drawn up. For instance, one can use one of the following schemes (Fig. 17.5):

- Joint compression of multichannel biosignals (e.g., multichannel ECG, or EEG),
- Joint compression of a set of various biosignals,
- Joint image-biosignals compression,
- Joint video-biosignals compression.

Since the aim of this chapter is to present an introduction to multimodal compression, we will be evoking only the principles of the last two schemes, namely the joint image-biosignals compression and the joint video-biosignals compression.

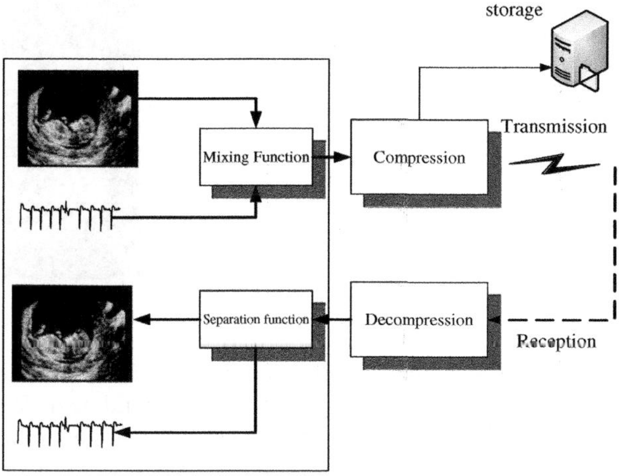

Fig. 17.5 Block diagram showing the principle of multimodal compression

17.4.1 Joint Image-Biosignal Compression

In this subsection, we show the basis of compressing jointly a medical image and a set of biosignals. For illustration purposes, we will consider here two examples. The first one is based on an insertion, in the wavelet domain. The obtained data mixture is compressed using the JPEG 2000 standard.

In the second example, the principle of inserting signal samples, directly into the spatial domain, is presented. In both cases, biosignals to be inserted into the medical image are gathered in a single *global-signal* as shown in Fig. 17.6. For instance, this *global-signal* might contain: ECG channels, EMG, acoustic signals (e.g. breathing, Mechanomyogram, etc.).

We point out here the fact that the proposed approaches shouldn't be considered as a watermarking method.The reader can also refer to [63], where the application concerns a joint image- multichannel ECG compression. We also underline the fact that all the data used here for the purpose of illustration can be downloaded from MeDEISA, "medeisa.net".

Example 1: Transform Domain

The insertion process is achieved according to the scheme shown in Fig. 17.7. We consider here the first level of decomposition of an ultrasound image. This decomposition produces four blocks (i.e. approximation (LL), horizontal details (HL), vertical details (LH) and diagonal details (HH)). The insertion region can be selected before the decomposition phase. It can be performed, either manually or automatically. For a given insertion region, four other regions are obtained after

Fig. 17.6 The principle of gathering biosignals in a single *global-signal*, before its insertion in an image or a video

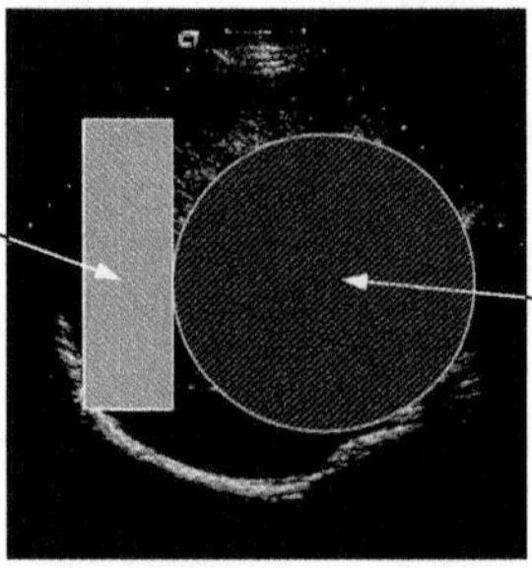

Fig. 17.7 Region of insertion. It should be different than the ROI

decomposition. They are denoted here by BLL, BHL, BLH and BHH (Fig. 17.8a). Of course, the selected insertion region should be different from the region of interest (ROI). This prevents critical regions from becoming distorted by the compression process.

Since the block HH, contains generally high frequencies having low amplitude values, one can neglect some of them so that they can be replaced by useful samples (i.e. *global-signal* samples). As we can see from the scheme shown in Fig. 17.8b, a simple decimation is performed on the block BHH, and each removed value is replaced by a *global-signal* sample.

The insertion process can be explained using various mathematical equations. For a selected rectangular region x_0, y_0, w, h, in the spatial domain, the detail coefficients to be replaced with signal samples are selected as follows;

$$C_i = HH(x' + 2k, y' + 2l) \tag{17.1}$$

where $x' = x_0/2$, $y' = y_0/2$ and $K = \dfrac{w}{4}, L = \dfrac{h}{4}$.

with $i = 0, \ldots, M$ $k = 0, \ldots, K$ and $l = 0, \ldots, L$.

Note that x_0, y_0, w and h should be even. If this is not the case, they should be rounded up to the nearest even integer.

Before achieving the insertion, signal samples should be scaled by a factor, denoted α. This will prevent signal samples from being truncated by the quantization step in the JPEG 2000 encoder. For a given α value, the block BHH is decimated in order to insert signal samples as follows:

$$HH(x' + 2k, y' + 2l) = \alpha.s_i \tag{17.2}$$

where s_i denotes *ith* signal sample.

When inserting the signal samples, an Inverse Discrete Wavelet Transform (IDWT) is performed. Hence, a new mixture image, denoted I' is obtained, the values of which are shifted outside the interval [0, 65535], (i.e. for a 16-bit image). Therefore, an offset, denoted $\beta = \min(I')$ should be subtracted from I' so that the input is properly conditioned to the standard JPEG 2000.

During the decoding phase, an inverted process is achieved according to the scheme shown in Fig. 17.9. The image I' is decompressed by JPEG 2000 then the value $\beta = \min(I')$ is added as follows:

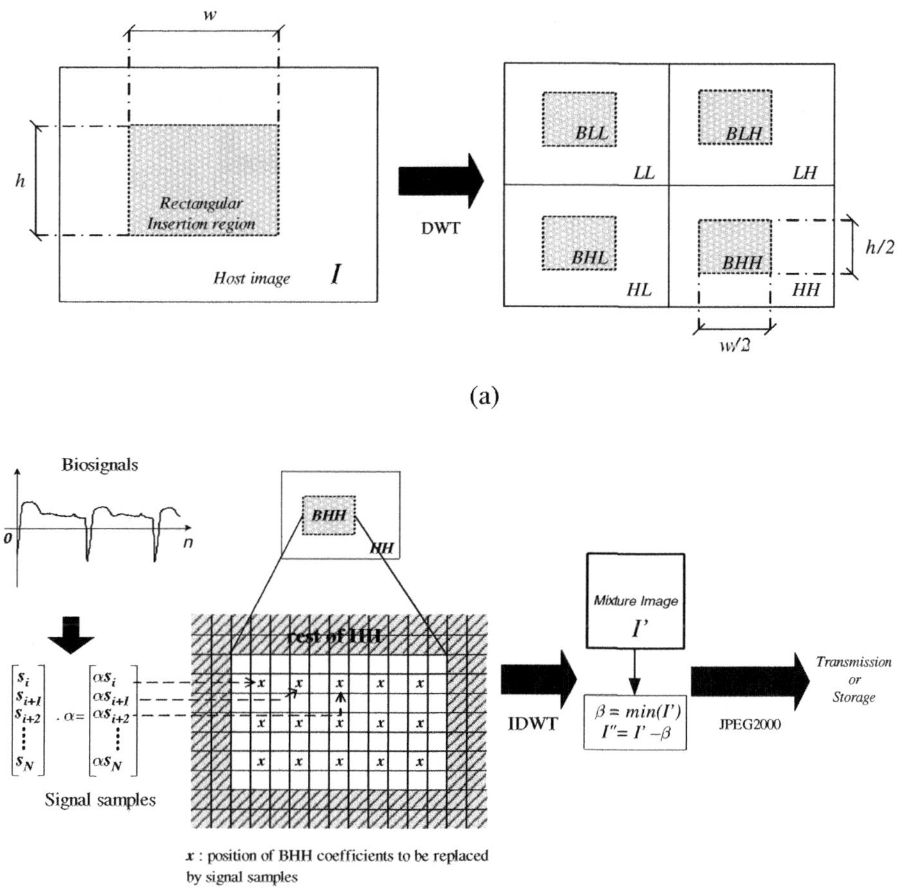

Fig. 17.8 Transform domain samples insertion. (**a**) Selection of the insertion region. (**b**) Insertion of interleaved signal samples in the BHH block

$$\overline{I}' = \overline{I}'' + \beta \tag{17.3}$$

Here $\overline{I}'$ is the decompressed mixture image, shifted by β.

DWT decomposition of the mixture image $\overline{I}'$ is calculated and detail coefficients (HH) are isolated to extract the signals as well as the reconstructed image as follows:

- Extraction the global-signal samples

From the Eq. (17.2), signal samples corresponding to the *global-signal* are easily extracted from BHH as follows:

$$\bar{s}_i = \frac{HH(x' + 2k,\, y' + 2l)}{\alpha} \tag{17.4}$$

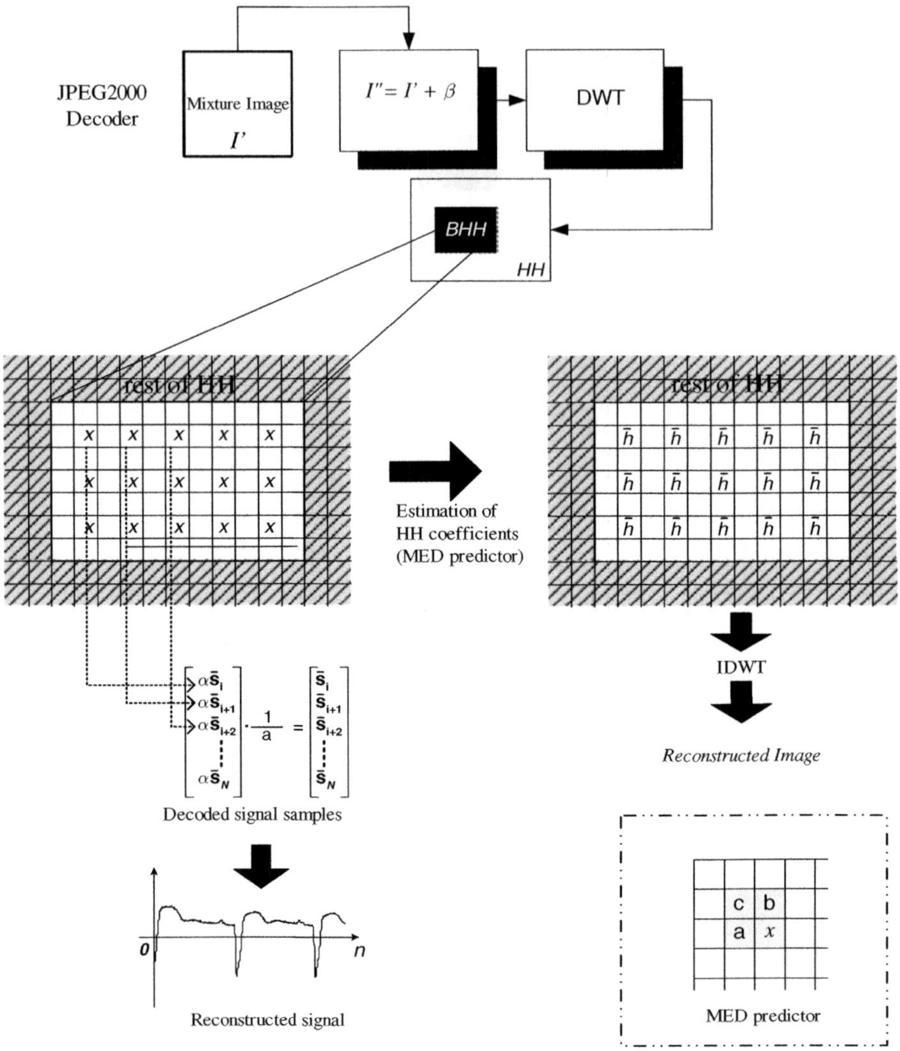

Fig. 17.9 Decoding process: signal samples are extracted from the BHH block. MED is then employed to reconstruct the image

- Reconstruction of the ultrasound image

As explained previously, some values from the block BHH have been replaced by signal samples. Consequently, the values removed during the compression phase should be estimated after decoding the mixture image. For this purpose, one can use the well known Median Edge Detector (MED) predictor. Although, MED is usually used in the spatial domain, on pixel values, it is employed here to estimate the suppressed values.

Example 2: Spatial Domain
In this example, we try to emphasize the idea of inserting biomedical signals, directly in the spatial domain of a medical image. In this case, no prior transform is required. In fact, when a clinician analyzes any medical image, usually the exploration is achieved in the ROI, which is generally located around the center of the image. Therefore, the idea consists in avoiding the ROI by inserting signal samples starting from the border of the image and according to a spiral way (see Fig. 17.10). The spiral path chosen should be decimated to allow signal samples to be inserted by a simple interleaving process. Therefore, one has to point out the following specificities:

1. The length of the spiral should be a multiple of the signal length to be inserted,
2. If *N* signals have to be inserted into the spiral path, they should be concatenated in order to form a single insertion signal,
3. Each signal should be scaled so that its values lie among the range of the image pixels. For instance, if an 8-bit image is used, signal sample values should belong to the range [0, 255],
4. The mixture image-signals is then compressed using any given specific encoder (e.g. JPEG 2000 standard).

After the decoding process, signal samples are extracted from the spiral path. Afterwards, an interpolation is required in order to estimate the missing image value pixels.

17.4.1.1 Extension to Video-Biosignal Compression

The basic idea evoked in the previous example can easily be extended further to compress jointly a video and a set of biosignals. It is also based on the fact that the most important information is mainly located in the central region of a video sequence; in other words, the ROI. For this purpose, a similar spatial insertion scheme can be adapted to a video encoder such as the H.264 (see Fig. 17.11). This video standard has been introduced as a recommendation by the International Telecommunication Union – Telecommunications Section (ITU) and it has been developed jointly between the Motion Picture Experts Group (MPEG) Community

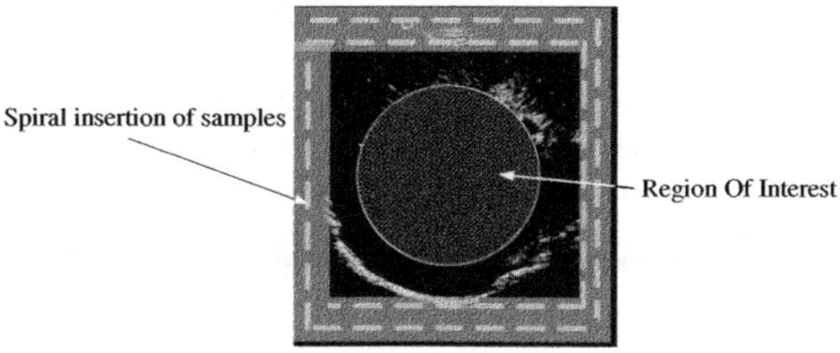

Fig. 17.10 Spatial insertion of the global-signal samples. ROI located at the center of the image should be avoided

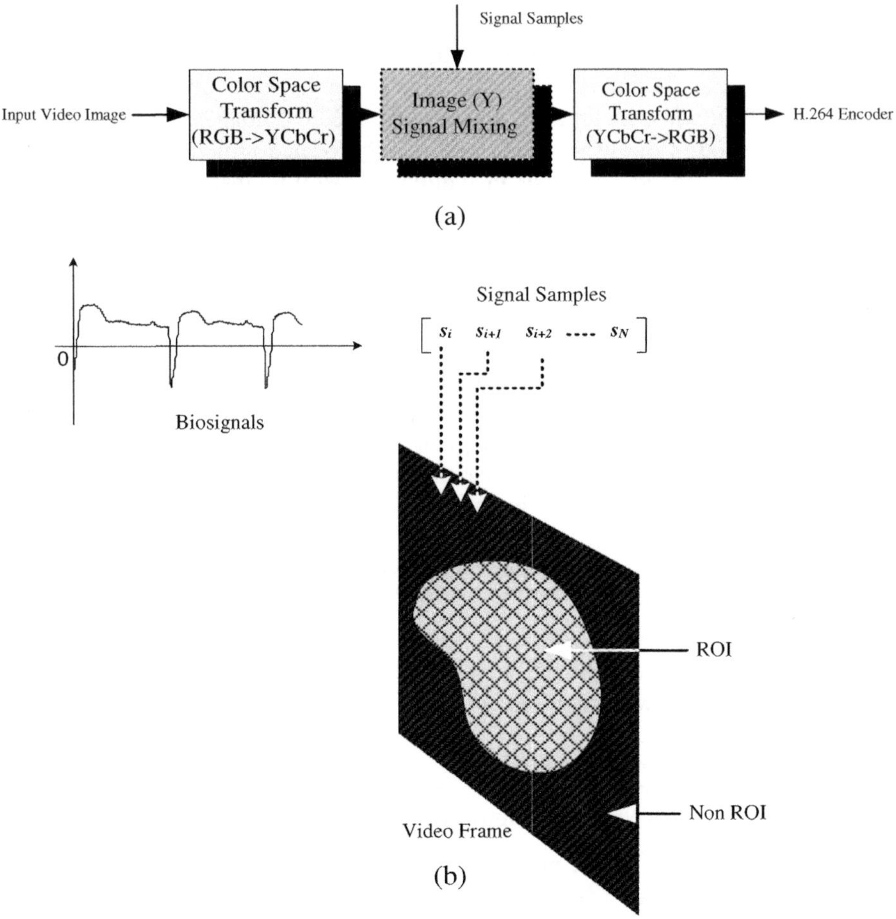

Fig. 17.11 Joint video-biosignals insertion. (**a**) Block diagram showing the global scheme of insertion after a RGB-YCbCr transform. (**b**) Principle of signal samples insertion in each Y frame

and ITU. This codec specification has been referred to as a separate layer in MPEG-4 (which is described in Part 10 of the standard). MPEG-4 also refers to this codec as MPEG-4 AVC (Advanced Video Coding).

H.264 codec scheme is a DCT based codec. It allows for many improvements over its predecessors in terms of image quality at a given Compression Ratio. In addition, this standard also allows an improvement in terms of functionalities. For illustration purposes, a H. 264 compression scheme is shown in Fig. 17.12. The first module consists in performing, for each frame, the RGB-to-YCbCr color space transform. This module is then followed by a prediction step where the frames are coded either using a spatial prediction or a temporal prediction. Resulting error images are then: (1) DCT transformed; (2) quantized; (3) encoded (i.e., entropy encoded).

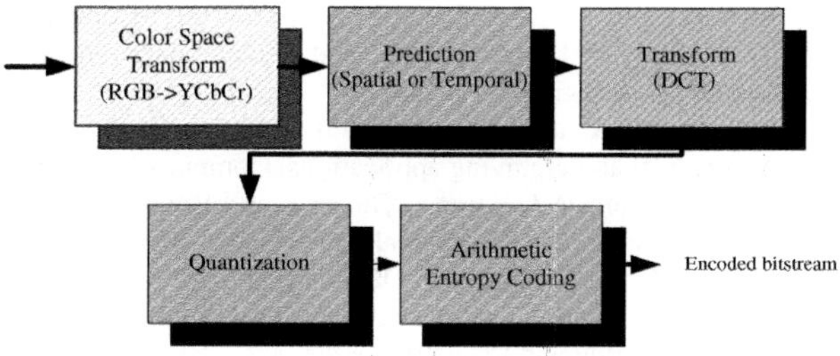

Fig. 17.12 High level block diagram of the H.264 encoder

17.5 Conclusion

As we have seen throughout this chapter, the principle of multimodal compression can be particularly beneficial for some applications which require storage or simultaneous transmission of various medical data, namely, biosignals, medical images and medical videos. Therefore, when biosignals are inserted in an image, standards such as JPEG 2000 can be used, whereas when they are inserted into a video, standards as H.264 are more appropriate. This will obviously avoid the use of one codec for each signal which might complicate the implementation of the codes. In addition, if the insertion function is well defined, higher performances, compared to the classical techniques can be achieved. As has been pointed out previously, this approach should not be regarded as a form of watermarking. In fact, the constraints and the context are absolutely different!

References

1. Agarwal A, Rowberg A, and Kim Y (2003) Fast JPEG 2000 decoder and its use in medical imaging. IEEE Trans Inf Technol Biomed 7:184–190
2. Ahmeda S, and Abo-Zahhad M (2001) A new hybrid algorithm for ECG signal compression based on the wavelet transformation of the linearly predicted error. Med Eng Phys 23:117–126
3. Al-Fahoum A, and Reza A (2004) Perceptually tuned JPEG coder for echocardiac image compression. IEEE Trans Inf Technol Biomed. 8:313–320
4. Alshamalı A, and Al-Smadi A (2001) Combined coding and wavelet transform for ECG compression. J Med Eng Technol 25:212–216
5. Antoniol G, and Tonnela P (1997) EEG data compression techniques. IEEE Eng in Med and Biol 44:105–114
6. Batista L, Melcher E, and Carvalho L (2001) Compression of ECG signals by optimized quantization of discrete cosine transform coefficients. Med Eng Phys 23:127–134
7. Beall D, Shelton P, Kinsey T et al. (2000) Image compression and chest radiograph interpretation: image perception comparison between uncompressed chest radiographs and chest radiographs stored using 10:1 JPEG compression. J Digital Imaging 13:33–38
8. Bilgin A, Marcellin M, and Albatch M (2003) Compression of ECG signals using JPEG2000. IEEE Trans on Cons Electr 49:833-840
9. Borsali R, Naït-Ali A and Lemoine J (2005) ECG compression using an ensemble polynomial modelling: comparison with the wavelet based technique. Biomed Eng 39:138–142

10. Brennecke R, U. Burgel, Rippin G et al. (2001) Comparison of image compression viability for lossy and lossless JPEG and Wavelet data reduction in coronary angiography. Int J Cardiovasc Imaging 17:1–12
11. Bruckmann A, and Uhl A (2000) Selective medical image compression techniques for telemedical and archiving applications. Comput Biol Med 30:153–169
12. Cardenas-Barrera J, Lorenzo-Ginori J, and Rodriguez-Valdivia E (2004) A wavelet-packets based algorithm for EEG signal compression. Med Inform Intern Med 29:15–27
13. Cavaro-Ménard C, Goupil F, Denizot B et al. (2001) Wavelet compression of numerical chest radiograph: quantitative et qualitative evaluation of degradations. Proceed of Inter Conf on Visual, Imaging and Image Processing (VIIP'01), 406–410, Spain
14. Chatterjee A, Naïtt-Ali A, and Siarry P (2005) An input-delay neural network based approach for piecewise ECG signal compression. IEEE Trans Biom Eng 52:945–947
15. Chen W, Hsieh L, and Yuan S (2004) High performance data compression method with pattern matching for biomedical ECG and arterial pulse waveforms. Comp Method Prog Biomed 74:11–27
16. Chen Y, and TAI S (2005) Enhancing ultrasound by morphology filter and eliminating ringing effect. Eur J Radiol 53:293–305
17. Chen Z, Chang R, and Kuo W (1999) Adaptive predictive multiplicative autoregressive model for medical image compression. IEEE Trans Med Imaging 18:181–184
18. Chiu E V J, and Atkins MS. (2001) Wavelet-based space-frequency compression of ultrasound images. IEEE Trans Inf Technol Biomed 5:300–310
19. Cho H, Kim J, and Ra J (1999) Interslice coding for medical three-dimensional images using an adaptive mode selection technique in wavelet transform domain. J Digit Imaging 12:173–184
20. de A Berger P, de O Nascimento FA, da Rocha AF et al. (2007) A new wavelet-based algorithm for compression of EMG signals. Conf Proc IEEE EMBS 1554–1557
21. de A Berger P, de O Nascimento F A, da Rocha A F et al. (2006) Compression of EMG signals with wavelet transform and artificial neural networks. Physiol Meas 457–465
22. Filho E, Silva Ed, and Carvalho Md (2008) On EMG signal compression with recurrent patterns. IEEE Trans Biomed Eng 55:1920–1923
23. Forchhammer S, Wu X, and Andersen J (2004) Optimal context quantization in lossless compression of image data sequences. IEEE Trans Image Process 13:509–517
24. Giurcaneanu C D, Tabus I, and Mereuta S (2001) Using contexts and R-R interval estimation in lossless ECG compression. Comput Meth Prog Biomed 67:177–186
25. Gruter R, Egger O, Vesin J et al. (2000) Rank-order polynomial subband decomposition for medical image compression. IEEE Trans Med Imaging 19:1044–1052
26. Hang X, Greenberg N, Qin J, et al. (2001) Compression of echocardiographic scan line data using wavelet packet transform. Comput Cardiol 28:425–427
27. Iyriboz T, Zukoski M, Hopper K et al. (1999) A comparison of wavelet and Joint Photographic Experts Group lossy compression methods applied to medical images. J Digit Imaging 12:14–17
28. Kalyanpur A, Neklesa V, Taylor C et al. (2000) Evaluation of JPEG and wavelet compression of body CT images for direct digital teleradiologic transmission. Radiology 217:772–779
29. Kaur L, Chauhan R, and Saxena S (2005) Space-frequency quantiser design for ultrasound image compression based on minimum description length criterion. Med Biol Eng Comput 43:33–39
30. Ko J, Rusinek H, Naidich D, McGuinness G, Rubinowitz A, Leitman B, and Martino J (2003) Wavelet compression of low-dose chest CT data: effect on lung nodule detection. Radiology 228:70–75
31. Ko JP, Chang J, Bomsztyk E et al. (2005) Effect of CT image compression on computer-assited lung nodule volume measurement. Radiology 237:83–88
32. Koski, Tossavainenn T, and Juhola M (2004) On lossy transform compression of ECG signals with reference to deformation of their parameter values. J Med Eng Technol 28:61–66

33. Liu Z, Xiong Z, Wu Q, Wang Y, et al. (2002) Cascaded differential and wavelet compression of chromosome images. IEEE Trans Biomed Eng 49:372–383
34. Lu M., and Zhou W. (2004) An EEG compression algorithm based on embedded zerotree wavelet (EZW). Space Med Eng 17:232–234
35. M. Rodriguez, Ayala A, Rodriguez S, et al. (2004) Application of the Max-Lloyd quantizer for ECG compression in diving mammals. Comp Meth Prog Biomed 2004:13–21
36. Miaou S, and Chao S (2005) Wavelet-based lossy-to-lossless ECG compression in a unified vector quantization framework. IEEE Trans Biomed Eng 52:539–543
37. Miaou S, and Chen S (2004) Automatic quality control for wavelet-based compression of volumetric medical images using distortion-constrained adaptive vector quantization. IEEE Trans Med Imaging 23:1417–1429
38. Miaou S, and Lin C (2002) A quality-on-demand algorithm for wavelet-based compression of electrocardiogram signals. IEEE Trans Biomed Eng 49:233–239
39. Miaou S, and Yen H (2001) Multichannel ECG compression using multichannel adaptive vector quantization. IEEE Trans Biomed Eng 48:1203–1209
40. Mitra S, Yang S, and Kustov V (1998) Wavelet-based vector quantization for high-fidelity compression and fast transmission of medical images. J Digit Imaging 11:24–30
41. Munteanu A, Cornelis J, Auwera G Vd, et al. (1999a) Wavelet image compression–the quadtree coding approach. IEEE Trans Inf Technol Biomed 3:176–185
42. Munteanu A, Cornelis J, and Cristea P (1999b) Wavelet-based lossless compression of coronary angiographic images. IEEE Trans Med Imaging 18:272–281
43. Naït-Ali A (2007) A New Technique for Progressive ECG Transmission using Discrete Radon Transform. Int J Biomed Sci 2:27–32
44. Naït-Ali A, Borsali R, Khaled W et al. (2007) Time division multiplexing based-method for compressing ECG signals: application for normal and abnormal cases. J MedEng Tech 31:324–331
45. Naït-Ali A, and Cavaro-Menard C (2008) Compression of biomedical images and signals. ISTE-WILEY
46. Nunes J, and Naït-Ali A (2005) ECG compression by modelling the instantaneous module/phase of its DCT. J Clin Monit Comput 19:207–214
47. Nygaard R, Melnikov G, and Katsaggelos A (2001) A rate distortion optimal ECG coding algorithm. IEEE Trans Biomed Eng 48:28–40
48. Ouamri A, and Naït-Ali A (2007) ECG compression method using Lorentzian functions Model. Digital Signal Processing 17:319–326
49. Paiva JP, Kelencz CA, Paiva HM, Gavão RK et al. (2008) Adaptive wavelet EMG compression based on local optimization of filter banks. Physiol Meas 29:843–856
50. Penedo M, Pearlman W, Tahoces P et al. (2003) Region-based wavelet coding methods for digital mammography. IEEE Trans Med Imaging 22:1288–1296
51. Peng K, and Kieffer J (2004) Embedded image compression based on wavelet pixel classification and sorting. IEEE Trans Image Process 13:1011–1017
52. Persons K, Palisson P, Manduca A, Erickson B, and Savcenko V (1999) An analytical look at the effects of compression on medical images. J Digit Imaging 10:60–66
53. Rajoub B (2002) An efficient coding algorithm for the compression of ECG signals using the wavelet transform. IEEE Trans Biomed Eng 49:355–362
54. Ricke J, Maass P, Hanninen EL et al. (1998) Wavelet versus JPEG (Joint Photographic Expert Group) and fractal compression. Impact on the detection of low-contrast details in computed radiographs. Invest Radiol 33:456–463
55. Sasikala M, and Kumaravel N (2000) Optimal autoregressive model based medical image compression using genetic algorithm. Biomed Sci Instrum 36:177–182
56. Schelkens P, Munteanu A, Barbarien J et al. (2003) Wavelet coding of volumetric medical datasets. IEEE Trans Med Imaging 22:441–458
57. Sung M, Kim H, Yoo S et al. (2002) Clinical evaluation of compression ratios using JPEG2000 on computed radiography chest images. J Digit Imaging 15:78–83

58. Suryanarayanan S, Karellas A, Vedantham S et al. (2004) A perceptual evaluation of JPEG2000 image compression for digital mammography : contrast-detail characteristics. J Digit Imaging 16:64–70
59. Ta S, Wu Y, and Lin C (2000) An adaptive 3-D discrete cosine transform coder for medical image compression. IEEE Trans Inf Technol Biomed 4:259–263
60. Wei J, Chang C, Chou N et al. (2001) ECG data compression using truncated singular value decomposition. IEEE Trans Inf Tech Biomed 5:290–299
61. Wu Y (2002) Medical image compression by sampling DCT coefficients. IEEE Trans Inf Technol Biomed 6:86–94
62. YG Y, and Tai SC (2001) Medical image compression by discrete cosine transform spectral similarity strategy. IEEE Trans Inf Technol Biomed 5:236–243
63. Zeybek E, Naït-Ali A, Olivier C et al. (2007) A Novel Scheme for joint Multi-channel ECG-ultrasound image compression. IEEE Proceedings of the 29th Annual Int. Conf. of the IEEE EMBS 713–716
64. Zhang Y, Pham B, and Eckstein M (2004) Automated optimization of JPEG2000 Encoder options based on model observer Performance for detecting variable signals in X-Ray coronary angiograms. IEEE Trans on Med Imag 23:459–474

Index

GPSR Compliance

The European Union's (EU) General Product Safety Regulation (GPSR) is a set of rules that requires consumer products to be safe and our obligations to ensure this.

If you have any concerns about our products, you can contact us on ProductSafety@springernature.com

In case Publisher is established outside the EU, the EU authorized representative is:

Springer Nature Customer Service Center GmbH
Europaplatz 3
69115 Heidelberg, Germany

Batch number: 10755284

Printed by Printforce, the Netherlands